Sustainable Agriculture: Principles and Practices

Sustainable Agriculture: Principles and Practices

Edited by John Williams Barrow

SYRAWOOD
PUBLISHING HOUSE

New York

Published by Syrawood Publishing House,
750 Third Avenue, 9th Floor,
New York, NY 10017, USA
www.syrawoodpublishinghouse.com

Sustainable Agriculture: Principles and Practices
Edited by John Williams Barrow

International Standard Book Number: 978-1-68286-573-6 (Hardback)

Cataloging-in-Publication Data

Sustainable agriculture : principles and practices / edited by John Williams Barrow.
    p. cm.
Includes bibliographical references and index.
ISBN 978-1-68286-573-6
1. Sustainable agriculture. 2. Agriculture. I. Barrow, John Williams.
S494.5.S86 S87 2018
630--dc23

TABLE OF CONTENTS

PREFACE

Sustainable agriculture aims to meet the growing demands of our society in a sustainable manner. Some practices of sustainable agriculture are minimizing usage of water, recycling crop waste, restoring soil health, etc. It focuses on utilizing the available natural resources in the most efficient and cost effective manner. While understanding the long-term perspectives of the topics, the book makes an effort in highlighting their impact as a modern tool for the growth of the discipline. This book includes contributions of experts and scientists which will provide innovative insights into this field.

After months of intensive research and writing, this book is the end result of all who devoted their time and efforts in the initiation and progress of this book. It will surely be a source of reference in enhancing the required knowledge of the new developments in the area. During the course of developing this book, certain measures such as accuracy, authenticity and research focused analytical studies were given preference in order to produce a comprehensive book in the area of study.

This book would not have been possible without the efforts of the authors and the publisher. I extend my sincere thanks to them. Secondly, I express my gratitude to my family and well-wishers. And most importantly, I thank my students for constantly expressing their willingness and curiosity in enhancing their knowledge in the field, which encourages me to take up further research projects for the advancement of the area.

Editor

Comparative Growth Performance of *Oreochromis* Hybrids and Selectively-Bred Strain (F$_8$) in Malawi

Daud Kassam[1] & Marcus Sangazi[1]

[1] Aquaculture and Fisheries Science Department, Bunda Campus, Lilongwe University of Agriculture and Natural Resources, P.O. Box 219, Lilongwe, Malawi

Correspondence: Daud Kassam, Aquaculture and Fisheries Science Department, Bunda Campus, Lilongwe University of Agriculture and Natural Resources, P.O. Box 219, Lilongwe, Malawi.
E-mail: dkassam@bunda.luanar.mw

Abstract

Most fish farmers in Malawi culture unimproved fish strains whose growth is slow and mature while still small. Four strains of *Oreochromis*, namely; selectively-bred/improved *O. shiranus* (F$_8$), two reciprocal F$_1$ *Oreochromis* hybrids, and *O. karongae* as a control (mean weight 2.5 ± 0.7 g) were stocked at a density of 5fish/m^2 in 9m^2 hapas replicated three times, and cultured for 90 days at Bunda Fish Farm. Fish were fed twice a day with feed formulated using maize bran and soybean containing 30% crude protein throughout the experimental period. The final mean weights were significantly different (p<0.05) across the treatments whereby; hybrid *O. shiranus* (male) X *O. karongae* (female) was 12.09g, hybrid *O. shiranus* (female) X *O. karongae* (male) was 9.72g, improved *O. shiranus* (F$_8$) registered 9.23g, and *O. karongae* was the least with 9.00g. Apparent food conversion ratio was also statistically different (p<0.05) across the treatments whereby; *O. karongae* was 3.63, hybrid *O. shiranus* (female) X *O. karongae* (male) was 3.25, improved *O. shiranus* (F$_8$) was 3.16 and hybrid *O. shiranus* (male) X *O. karongae* (female) was lowest with 2.26. There were no significant differences on the water quality parameters across the treatments throughout the experimental period and were within the required ranges for growth and survival of tilapias fish species. The results suggest that *Oreochromis* hybrids may be suitable candidates for aquaculture in terms of production as they performed better than the improved *O. shiranus* and the control *O. karongae*.

Keywords: food conversion ratio, hybridization, improved strain, *Oreochromis*

1. Introduction

Fish provides about 72% of animal protein and is a very important source of food and income for Malawians (Ecker & Qaim, 2011). High population growth and alarming levels of poverty in Malawi have caused an unprecedented overexploitation of freshwater ecosystem fisheries (Jamu, Banda, Njaya, & Hecky, 2011). Moreover, the combined effect of anthropogenic activities and urban extension have substantially reduced the natural fish habitat through siltation and water bodies' removal, resulting in a severe decrease in fisheries yield. Therefore, capture fisheries do not seem to be sustainable and aquaculture is the best alternative in sustaining fish supply to the population due to its potential to grow because land and water resources are still abundant. In order to enhance fish production, the Government of Malawi developed National Aquaculture Strategic Plan (NASP) in 2005 to spearhead the development of aquaculture. While there have been several initiatives to promote aquaculture development, the Presidential Initiative for Aquaculture Development (PIAD) stands out as the major initiative which targeted to raise by 10 folds (within 5 years) the national fish stock production, sustaining the local consumption and exports.

In Malawi, *Tilapia rendalli, Oreochromis shiranus, Oreochromis karongae* and *Clarias gariepinus* are some of the major fish species cultured by farmers in earthen ponds. Growth and survival rates in ponds are the most common complaints from farmers for these indigenous tilapia species (Maluwa & Gjerde, 2007). The national regulatory authority which define the legislation and laws in the fishery sector, promote a sustainable aquaculture through a close partnership between the public and private actors and make illegal some practices like exotic species introduction. Thus, the improvement of the indigenous species' growth performance and survival rates using the available technology has become the main target (Ross, Martinez Palacios, & Morales,

2008). Researchers mainly focused on the genetic improvement of fish species life-history traits, in particular, *O. shiranus,* the predominant farmed species, naturally characterized by a slow growth rate and small maturity-size. Fast growing species have a lower production cost and a higher commercial value than small species (Andrew, Weyl, & Andrew, 2003).

Growth performance tests done on *O. shiranus* (M'balaka, Kassam, & Rusuwa, 2012) and *O. niloticus* (Olesen, Gjedrem, Bentsen, Gjerde, & Rye, 2003) showed an improvement of 25.2% and 101% after 6 and 5 generations, respectively, compared to unimproved strains. The relative lower performance noted in the *O. shiranus* growth rate, issued from the selective breeding process, may potentially be corrected by hybridization. This argument is supported by the observations of Munthali (2011, unpublished report), suggesting that *Oreochromis* hybrids grow better than its parental lines. However, recommendations to farmers cannot be issued without resolving the scientific gap relative to the comparative growth performance between the selectively bred (F_8) *O. shiranus* strain and the hybridized one. Thus, the objective of the present paper is to accurately determine and compare the growth performance and survival rates of these two strains.

2. Materials and Methods

2.1 Breeding Phase

The experiment was conducted at Bunda fish farm. Four treatments were set using hapas of 9 m^2 fixed in a pond of 700 m^2 and each treatment was replicated twice. *O. shiranus* and *O. karongae* were reciprocally hybridized, a male of one species was crossed with a female of another species and vice versa. A single crossing was made between the improved strains issued from the seventh generation of *O. shiranus*. Finally a male and a female of *O. karongae* were crossed and used as controls. Broodstock of average weight 120g were stocked at sex ratio of one male for three females. The broodstock were fed with a diet formulated from a mixture of maize bran and soybean with 29% crude protein, using the Pearson's square method. As a supplement, the natural food were boosted by application of chicken manure to the pond at the rate of 500 kg/ha/week (Kang'ombe & Brown, 2008). The fish were hand fed twice a day at 5% body weight in line with recommendations of Lovell (1989). Fry were removed from the breeding hapas to fry rearing tanks every two weeks.

2.2 Grow out Phase

Fingerlings of mixed sex, with body weight ranging from 2.4 to 2.6 ± 0.7 g, from the rearing tanks were stocked in 9m^2 hapas at density of 5fish/m^2. A total of 135 fingerlings from each treatment was selected and 45 fish/hapa was randomly stocked. Feeding was twice every day at 5% body weight. A completely randomized design (CRD) was used to allocate four treatments, replicated three times each in 700 m^2 pond with the following treatments; hybrid *O. shiranus* (male) X *O. karongae* (female), hybrid *O. shiranus* (female) X *O. karongae* (male), improved *O. shiranus* (F_8) and pure *O. karongae*.

2.3 Water Quality Parameters

Water quality parameters were rigorously monitored before and during the experiment. The physical and chemical properties of the water were surveilled using appropriate devices and the titration method was used to determine the ammonia concentration. Measurements were conducted twice a day during the morning and in the afternoon.

2.4 Data Collection and Analysis

Data on growth were collected every fortnight whereby fish were weighed using an electrical analytical balance calibrated to 0.01g and total length (mm) measured using a stainless steel topped measuring board. Specific growth rate (SGR%/day) was calculated using the following formula:

Specific growth rate (SGR);

$$\%SGR = \frac{ln\,W1 - ln\,W0}{t1 - to} X100 \tag{1}$$

Where:
W_1= final mean weight (g) of fish
W_0 = initial mean weight (g) of fish at stocking time
t = time in days
ln = natural log

Survival rate

Survival rate was determined using the following formula:

$$\text{Survival rate (\%)} = \frac{Number\,of\,fish\,at\,the\,end\,of\,experiment}{Number\,of\,fish\,at\,the\,start\,of\,the\,experiment} \text{X}100 \qquad (2)$$

Food utilization efficiency

Apparent food conversion ratio (AFCR) was determined using the formula below

$$\textbf{AFCR} = \frac{\text{weight of dry feed given to the fish(kg)}}{\text{weight gained by the fish(kg)}} \qquad (3)$$

Data was analyzed using Genstat statistical package (15th edition); using One-way Analysis of Variance (ANOVA) at 95% level of confidence. When ANOVA revealed overall significant difference on the growth parameters, planned contrasts were used to separate the means among the treatments.

3. Results

The initial mean weights of the four treatments ranged from 2.5 ±0.07 g to 2.64 ± 0.07 g, with no significant differences whilst the final mean weights differed significantly among the treatments (p < 0.05, Table 1). The final mean weight of the hybrid *O. shiranus* (male) X *O. karongae* (female) which was 12.09 g, was significantly higher than the other three strains (p<0.001), whereby the final mean weight of the hybrid *O. shiranus* (female) X *O. karongae* (male) was 9.72 g, improved *O. shiranus* (F_8) was 9.23 g and *O. karongae* was 9.00 g and were not statistically different (Table 1 & 2).

Specific growth rates (SGR %/day) were significantly different for fish treatments (p < 0.05, Table 1), hybrid *O. shiranus* (male) X *O. karongae* (female) had higher specific growth rate (1.72%) while hybrid *O. shiranus* (female) X *O. karongae* (male) (1.52%), improved *O. shiranus* (F_8) (1.50%) and *O. karongae* (1.33%) were not statistically different (p>0.05).

Planned contrast showed that apparent food conversion ratio (AFCR) for the hybrid *O. shiranus* (male) X *O. karongae* (female) which registered 2.26, differed significantly from the other treatments (p <0.001, Table 1). The improved *O. shiranus* (F_8) had 3.25, hybrid *O. shiranus* (female) X *O. karongae* (male) had 3.16 and *O. karongae* had 3.63; there was no statistical difference among the later three treatments. Fish survival rates were not significantly different across the four treatments (p>0.05, Table 1). The hybrid *O. shiranus* (male) X *O. karongae* (female) was 98.5%, improved *O. shiranus* (F_8) was 91.9%, hybrid *O. shiranus* (female) X *O. karongae* (male) was 95.6%, whilst pure *O. karongae* was 86.7% (Table 2).

Table 1. Values of contrast and p-value of initial and final mean weights, specific growth rate (SGR), apparent food conversion ratio (AFCR), and survival rate of fish

Contrast	Parameter (mean difference $\mu_i - \mu_j$)				
	Initial weight (g)	Final weight (g)	SGR%/day	AFCR	Survival rate (%)
OS(♂) X OK(♀) - OS(F_8)	0.00	2.86	0.22	0.90	6.6
P-value	0.175	<0.001	0.018	<0.001	0.590
OS(♂) X OK(♀) - OS(♀) X OK(♂)	0.10	2.37	0.20	0.99	2.6
P-value	0.121	<0.001	0.028	<0.001	0.475
OS(♂) X OK(♀) - Ok	0.10	3.09	0.39	1.37	11.8
p-value	0.697	<0.001	0.004	<0.001	0.175
OS(F_8) - OS(♀) X OK(♂)	0.10	0.49	0.02	0.09	3.7
p-value	0.846	0.377	1.213	0.776	0.226
OS(F_8) – OK	0.10	0.23	0.17	0.47	5.2
p-value	0.082	0.744	0.213	0.089	0.060
OS(♀) X OK(♂) - OK	0.2	0.72	0.19	0.38	8.9
p-value	0.053	0.227	0.013	0.155	0.408

The mean difference is significant at the 0.05 level.

Table 2. Initial and final mean weights, specific growth rate (SGR), apparent feed conversion ratio (AFCR), and survival rate of fish

	Treatment			
Parameter	*OS*(♂) X *OK*(♀)	*OS*(F$_8$)	*OS*(♀) X *OK* (♂)	*Ok*
Initial mean weight (g)	2.5	2.5	2.4	2.6
Final mean weight (g)	12.09	9.23	9.72	9.00
SGR (%/day)	1.72	1.50	1.52	1.33
AFCR	2.26	3.16	3.25	3.63
Survival rate (%)	98.5	91.9	95.6	86.7

Water quality parameters in the pond were within the required ranges for the growth of tilapia fish species during the experiment. There were no significant differences for measured water quality parameters among treatments throughout the experimental period ($P>0.05$). Mean water temperature ranged from 24.74 ± 0.2^0C to $28.23 \pm 0.2°C$ across the treatments. Dissolved oxygen ranged from 4.64 to 6.82 mg/L, the pH between 8.10 and 8.41, ammonia ranged from 0.061 to 0.087 mg/L throughout the experimental period (Table 3).

Table 3. Water quality parameters (mean ± SE)

	TREATMENT				DESIRABLE LIMITS
PARAMETER	**OS**(♂) **X OK**(♀)	**OS (F$_8$)**	**OS**(♀) **X OK**(♂)	**OK**	
Temperature(^{o}C)	Am 24.7 ± 0.5a	25.1 ± 0.5a	25.2 ± 0.5a	25.2 ± 0.5a	20-30
	Pm 28.1 ± 0.7a	28.3 ± 0.7a	27.9± 0.6a	28.2 ± 0.7a	
DO (mg/L)	Am 4.7 ± 0.4a	4.9 ± 0.4a	4.6 ± 0.4a	4.8 ± 0.4a	5
	Pm 6.8 ± 0.2a	6.7 ± 0.2a	6.0 ± 0.2a	6.8 ± 0.2a	
pH	Am 8.1 ± 0.2a	8.2 ± 0.2a	8.1 ± 0.2a	8.1 ± 0.3a	6.5-9
	Pm 8.3 ± 0.2a	8.4± 0.2a	8.4 ± 0.2a	8.3 ± 0.2a	
Ammonia (mg/L)	0.087 ± 0.03a	0.061± 0.02a	0.073 ± 0.03a	0.062±0.02a	<0.1

Figures in rows with same letters are not significantly different.

4. Discussion

This study revealed that the growth of hybrid *O. shiranus* (male) X *O. karongae* (female), in terms of the weight gained and specific growth rate (SGR), were significantly higher than that of the improved *O. shiranus, O. karongae* and hybrid *O. shiranus* (female) X *O. karongae* (male). This higher growth performance of the hybrid strain may be attributed to the genetic superiority of this strain and, if so, lends support to the fact that hybrid tilapia are preferred because of their significant high yield and survival (Mair, 2007). The results can also be supported by the fact that hybrid individuals acquire traits from both parents; therefore higher growth rate and other growth parameters in the hybrids may be as a result of combined traits from both parents.

The improved *O. shiranus* strain has shown to start spawning at less than 10 g (in this study at 8.7g), yet the demand for the local tilapias is high for large sizes like 150-200g (Worldfish Centre, 2010). This early sexual maturation in production ponds can significantly reduce fish yield, as maturing fish can become aggressive, stop growing, and become more susceptible to disease. In this study, there was no fish observed to start spawning throughout the experimental period from both the reciprocal hybrids and *O. karongae*. The late maturity of the *Oreochromis* hybrids, resonate well with other findings that have reported that tilapia hybrids attain larger sizes at harvest (Mair, 2007). In terms of genetic gain over the control, this study has shown a 34% genetic gain, which is similar to what Munthali (2011) found, though in that study it ranged from 34-49% depending on the pair crossed. The results of this study also showed that there was no significant difference in fish survival among the strains, above 90% for reciprocal F$_1$ hybrids and improved *O. shiranus* and 86.7% for *O. karongae*. This may be attributed to the fact that tilapias are hard fish species, tolerate harsh conditions of water such as low dissolved oxygen of about 0.6 ml/l and wide range of temperatures 16 to 42 °C, as such, tilapias thrive in most fish ponds of Malawian farmers.

One of the most significant growth parameters in aquaculture is the efficiency of the fish to convert food into flesh; this was determined as apparent food conversion ratio (AFCR).The hybrid *O. shiranus* (male) X *O. karongae* (female) strain had lowest apparent food conversion ratio (AFCR); thus more efficient at utilizing feed and growing better than the improved *O. shiranus*, hybrid *O. shiranus* (female) X *O. karongae* (male) and *O. karongae*.

5. Conclusion

This study revealed that the growth performance of *Oreochromis* hybrid *O. shiranus* (male) X *O. karongae* (female), in terms of the weight gained and specific growth rate (SGR), were highest coupled with survival rate which was above 85%. The hybrids which performed better had lowest AFCR, thus more efficient in food utilization. Therefore, this study reveals that hybridization is more productive than the selective breeding program as improvement of the indigenous fish species is concerned.

References

Andrew, T. G., Weyl, O. L. F., & Andrew, M. (2003). *Aquaculture master plan development in Malawi: Socioeconomic survey report.*

Ecker, O., & Qaim, M. (2011). Analyzing nutritional impacts of policies: an empirical study for Malawi. *World Dev., 39,* 412-428. http://dx.doi.org/10.1016/j.worlddev.2010.08.002

Jamu, D., Banda, M., Njaya, F., & Hecky, R. E. (2011). Challenges to sustainable management of the lakes of Malawi. *J. Great Lakes Res., 37,* 3-14. http://dx.doi.org/10.1016/j.jglr.2010.11.017

Kang'ombe, J., & Brown, J. A. (2008). Effect of using low-protein diets in semi-intensive pond cage culture of tilapia, *Tilapia rendalli* (Boulenger). *J. Appl Aquac., 20,* 243-255. http://dx.doi.org/10.1080/10454430802498203

Lovell, T. (1989). *Nutrition and feeding of fish.* Van Nostrand Reinhold, New York, USA. http://dx.doi.org/10.1007/978-1-4757-1174-5

Mair, G. C. (2007). Genetics and breeding in seed supply for inland aquaculture. In M. G. Bondad-Reantaso (ed.), Assessment of freshwater fish seed resources for sustainable aquaculture (pp. 519-547). *FAO Fisheries Technical Paper.* No. 501. Rome.

Malawi Government. (2001a). National Fisheries and Aquaculture Policy, Department of Fisheries, Ministry of Natural Resources and Environmental Affairs, Lilongwe, Malawi.

Malawi Government. (2011b). Guidelines for Tilapia hatchery operators in Malawi, Ministry of Agriculture and food Security, Lilongwe, Malawi.

Maluwa, A. O., & Gjerde, B. (2007). Genetic evaluation of four strains of *Oreochromis shiranus* for harvest body weight in a diallel cross. *Aquac., 259,* 28-37. http://dx.doi.org/10.1016/j.aquaculture.2006.06.003

M'balaka, M., Kassam, D., & Rusuwa, B. (2012). The effect of stocking density on the growth and survival of improved and unimproved strains of *Oreochromis shiranus*. *Egypt. J. Aqua. Res., 38,* 205-211. http://dx.doi.org/10.1016/j.ejar.2012.12.013

Munthali K. (2011). Growth performance and survival of hybrids between *Oreochromis karongae* and *Oreochromis shiranus*. Aquaculture and Fisheries Science Department, Bunda College, Lilongwe, Malawi.

NASP. (2005). Ministry of Mine, Natural Resources and environment, Department of Fisheries, Zomba, Malawi No: 28.

Olesen, I., Gjedrem, T., Bentsen, H. B., Gjerde, B., & Rye, M. (2003). Breeding programs for sustainable aquaculture. *J. Appl. Aquac., 13,* 179-204. http://dx.doi.org/10.1300/J028v13n03_01

Ross, L. G., Martinez Palacios, C. A., & Morales, E. J. (2008). Developing native fish species for aquaculture: the interacting demands of biodiversity, sustainable aquaculture and livelihoods. *Aquac. Res., 39,* 675-683. http://dx.doi.org/10.1111/j.1365-2109.2008.01920.x

World Fish Center. (2010). Cage Aquaculture in Malawi, Briefing note 2119 World Fish Center.

An Integration of GIS and Simulation Models for a Cost Benefit Analysis of Irrigation Development

Monika Ghimire[1], Art Stoecker[1], Tracy A. Boyer[1], Hiren Bhavsar[2], & Jeffrey Vitale[1]

[1]Department of Agricultural Economics, Oklahoma State University, Stillwater, OK 74078, USA

[2]Hiren Bhavsar is an assistant professor at Tennessee State University, Nashville, TN 37209, USA

Correspondence: Monika Ghimire, Department of Agricultural Economics, Oklahoma State University, 557 AG Hall, Oklahoma State University, Stillwater, OK 74078, USA. E-mail:monika.ghimire@okstate.edu

Abstract

This study incorporates spatially explicit geographic information system and simulation models to develop an optimal irrigation system. The purpose of the optimized irrigation system was to save depleted ground water supplies. ArcGIS was used to calculate the area of potential irrigable soils, and EPANET (a hydrological simulation program) was used to calculate energy costs. Crop yield response functions were used to estimate the yield of cotton to the amount of irrigation and the accumulation of soil salinity over a 50-year period. Four irrigation designs (A, B, C, and D) were analyzed with different irrigation schedules.

Design A allowed all producers to irrigate simultaneously at 600 gallons per minute (gpm) or 2,271 liters per minute (lpm) while designs B and C divided the irrigable areas into two parts. Design D divided the areas into four parts to allow producers to irrigate one part at a time at 800 gpm (3,028 lpm). Irrigation scheduling not only lessened the water use and cost, but also amplified the profitability of the irrigation system. In design A, if all producers adopted 600 gpm (2,271 lpm) pivots and operated simultaneously, the cost of the 360,000 gpm (1363,000 lpm) pipeline would be prohibitive. In contrast, designs B, C, and D increased net benefits and lowered the breakeven price of cotton. The 50-year net present value for designs A, B, C, and D was profitable over 75, 70, 70, and 65 cents of cotton price per pound (454 g), respectively. Thus, this study endorses irrigation scheduling as a tool for efficient irrigation development and management, and increases water conservation.

Keywords: geographic information system, simulation model, net present value, irrigation system

1. Introduction

In the context of climate change reduced water supplies and increasing energy prices, it is crucial to identify methodologies, tools, and actions that optimize the use of water, energy, and other durable resources for environmental and economic benefits. The increasing dependence on irrigated agriculture requires efficient irrigation planning to conserve water and maximize net revenue. Several studies have used geographic information systems (GIS) for irrigation planning, management (Belmonte, González, Mayorga, & Fernández, 1999; Todorovic & Steduto, 2003; Satti & Jacobs, 2004; and Singh, Jhorar, Van Dam, & Freddes, 2006), and for irrigation scheduling (George, Raghuwanshi, & Singh, 2004; Fortes, Platonov, & Pereira, 2005). Geographically mapped spatially explicit soil data and simulation models can be combined to optimize cost efficient and dynamic management planning for scarce water supplies. However, the combined GIS and simulation models have not been used to select, design, and manage the pressurized irrigation system. This project design is methodologically unique as it integrates spatially specific data mapped in GIS software, hydrological simulation models, and economic optimization models to develop a feasible irrigation system. In addition, this paper also discusses the cost minimization benefits of irrigation system by irrigation scheduling which is an addition to the literature. Irrigation scheduling here refers to the application of irrigation water to different sections of land in different times rather than simultaneously. Thus, the general objective of this study was to develop the irrigation design using GIS, and the hydrological and mathematical simulation models to analyze the profitability of an irrigation system under different scheduling. Further, this study also conducts sensitivity of net returns of irrigation systems under different cotton prices and electrical conductivity (EC) or salinity levels. The specific objectives of this study were: (1) to identify irrigable areas and the length and route of pipelines using GIS, (2) to determine the cost of the pipeline and the net economic returns of irrigation from yield increment using an

EPANET hydrological simulation model and mathematical optimization model, respectively, and (3) to determine the net economic returns of an irrigation system under different irrigation scheduling, cotton prices, and salinity levels within the irrigation district. This study hypothesizes that irrigation scheduling increases the net benefit of the irrigation system or lowers the breakeven price for cotton.

1.1 Background and Study Area

The North Fork of the Red River of southwestern Oklahoma, USA is overburdened with salt from three canyons including Kaiser, Robinson, and Salton. This study area is a part of a chloride control and salinity management project in North Fork of Red River under which there is a potential for construction of the Cable Mountain Reservoir (CMR). The CMR was proposed to augment the water supply of Altus dam which is in the Lugert-Altus Irrigation District, about 29 km north of the city of Altus, OK. The Altus Reservoir has been losing its water storage capacity due to sediment accumulation (Bureau of Reclamation, 2005). Displacement of available reservoir capacity by sediment will diminish the project's capacity to supply water within 30 to 50 years. The CMR is one of the proposed alternatives to increase and augment the water supply of Lake Altus (Bureau of Reclamation, 2005). The estimated holding capacity of the proposed CMR is 12,335 hectare-meter, more volume than is required to replace water lost from Lake Altus. This study proposes the use of additional water to irrigate arable lands at lower elevations of the reservoir in Tillman Terrace Area (TTA), located to the southwest of the state of Oklahoma. Ground water has been extensively used since 1950 in Tillman County for irrigation and municipal uses resulting in depletion of ground water and encroaching salt water from the North Fork River (Osborn, 2002). Thus, irrigation development to TTA reduces the depletion of groundwater and salinity problem in North Fork River.

2. Materials and Methods

2.1 Potential Irrigable Areas

Irrigable areas were identified for allocation of irrigation water and for determination of the optimal quantity of irrigation water for the area. A GIS was used to determine irrigable areas from the Natural Resources Conservation Services (NRCS) -SSURGO database (NRCS, 2010) map of soil types and their classifications in the Tillman and Kiowa counties. The land classification capability categorizes land based on the suitability of soil quality for potential agricultural output. These land categories class are types I-VIII which represent progressively greater limitations and narrower choices for agriculture. There are land capability subclasses, denoted by codes e, w, s, and c, which are respectively related to problems with erosion, wetness, root zones, and climate. Irrigable soil types were defined as I, IIe, and IIw (National Soil Survey Handbook, USDA). The section shape files of Tillman and Kiowa counties were intersected with the SSURGO database to determine soil type in each township and sections of the counties.

Irrigable areas were further identified by slope, elevation, and plot size to form a set of potential sites for irrigation. The fields with 10 m slopes less than three percent were selected. The land area was filtered to the areas with elevation less than 436 meter above sea level (masl) since elevation of CMR is 436 masl. The fields with less than 4 ha of irrigable field soils were removed on an assumption that it would be uneconomical to irrigate those small areas.

2.2 Pipeline Network Design of the Irrigation System

A feasible irrigation pipeline network was designed given the area and shape of field sites using ArcMap 9.3. The elevation of irrigable areas was used to calculate the head pressure and the required power to deliver the water into the field. Global Mapper was used to create elevation of every pivot node of the pipeline (Global Mapper, 2011).

Settlement areas, railway tracks, and gullies that exemplified physical obstacles for an irrigation system were excluded. The editor in ArcGIS was used to draw the pipeline network to provide irrigation for each pivot circle. The pipeline route was designated to follow the maximum elevation level from the CMR to TTA to minimize the pumping cost of the irrigation system.

2.3 Cost of Earthwork

As the cost of earthwork depends on pipe size, the linear cost per foot of earthwork for different sizes of pipes, small and large was determined. Trenches of five feet or deeper have to be excavated with certain slopes for the safety of the workers and the durability of the trench (Occupational Safety and Health Administration (OSHA), 2011). The costs of trenching for larger and smaller pipelines were different because larger wall slopes are required for larger trenches (> 1.5 m deep) and smaller trenches require smaller wall slopes (<1.5 m deep) (OSHA, 2011). The cost of trenching was estimated with a regression model where the dependent variable was cost of trenching and the independent variables were trench width, trench depth, and square of the trench depth. Total earthwork cost was calculated as a sum of trenching, backfilling, and packing costs. Finally, pipe cost and total earthwork were summed as total piping cost. The total costs were then annualized for equal payments over a 50-year period at a discount rate of four, five, six, and eight percent. A spreadsheet was used to develop the cost for purchase and installation cost of alternative sizes of pipe from 6 to 120 inches (0.15 to 3 m), using data on the cost of pipe, excavation, and backfilling using estimates from RS Means (Mossman and Plotner, 2009).

2.4 Pipeline Design and Cost Estimation

The most economically beneficial design of a pipeline system to serve an irrigation district from a reservoir depends not only on the quantity and quality of the water, but also on the distance and elevation changes between the reservoir and irrigation district. As the diameter of pipe increases, the total cost of the pipe increases, but the energy required for pumping water (pumping cost or energy cost) through the pipe decreases. Pipe size can be optimized to contribute to decrease in the cost of pipeline to some extent. The optimal size of pipeline was determined using EPANET software (EPA, 2011).

The EPANET models water flow and pressure loss in distribution systems to help size the pipeline and determine the pump location, to minimize energy cost (EPA, 2011). The GIS provided the estimate of pipe lengths. The pipeline size was varied to minimize the annual cost of pipeline construction and pumping cost. The minimum annual cost involves a tradeoff between pipe size and energy cost. In EPANET software, the pipeline diameters were iteratively increased or decreased until the minimum size of pipeline was determined that gave the pressure required at different points. The pumps were added to low pressure points to meet the minimum pressure of 35 PSI (0.01 kg m^{-2}) for each pivot system operation. A standard capital recovery factor was used to annualize the pipe cost. The annual capital cost for the pipeline and the annual pumping costs were added to get the size of pipeline for minimal annual cost.

2.5 Water Demand and Energy Cost

The water demand at each pivot and number of pivots determines the diameter and cost of pipelines. Energy cost in this study is the cost of energy for pumping water to the fields. The pressure of at least 35 PSI (0.01 kg m^{-2}) in an individual pivot is obtained by adjusting the pumps in the required areas of the irrigation system. The energy cost for pumping was estimated with the use of the volume of water and pressure head required as indicated by the brake horse power methods as described in Keller and Bliesner (1990).

$$Bhp = \frac{GPM * Head\ (ft)}{3960 * Peff * Meff} \tag{1}$$

where gpm is gallons per minute, *Peff* is pumping efficiency, *Meff* is motor efficiency, and Head (ft) is the pressure. The head loss was calculated using Hazen William's formula (Jensen, 1983) as shown below:

$$Head\ loss = \frac{10.46(\frac{GPM}{C})^{1.85} * length}{D^{4.87}} + ELC + DelH \tag{2}$$

where *C* is the retardation constant, which is 120 for steel or aluminum, 140 for Cement Asbestos, and 150 for plastic, *D* is diameter of pipe (inches), *ELC* is elevation change, and *DelH* is delivery head .

Energy costs (EGC) were calculated using the following formula (Keller and Bliesner, 1990):

$$EGC = \frac{(GPM * Hd) * kwbhp * hpy * pelec}{3690 * Peff * Meff} \tag{3}$$

where *Hd* is head loss in feet, *kwBhp* is kilowatt per brake horse power, *hpy* is hours per year, and *pelec* is electricity cost per kwh. Thus, the pumps were chosen according to the pressure requirement to deliver water in each pivot according to output of EPANET. The pumping cost for the irrigation system was calculated using equation (3).

2.6 Crop Yield Response Function

The crop yield response function of a crop to irrigation supply and soil salinity is a crucial factor in an optimization model for irrigation systems with water salinity (Feinerman, 1993). Several studies have estimated response functions (Dinar & Knapp, 1986; Dinar, V. Sharma, & D. Sharma, 1991; Datta, 1998; Kiani & Abbasi, 2009). Because cotton is the major irrigated crop in the study region and also very resistant to salinity (up to 7.7 mmhos cm^{-1} EC), the responses of cotton yields for 50 years to different levels of irrigation and water quality for major irrigable soil types were simulated using Environmental Policy Integrated Climate (EPIC) model (Choi, 2011). The quadratic yield function for each individual soil types is:

$$Y_{st} = a_{s0} + a_{s1}W_{st} + a_{s2}S_{st} + a_{s3}NR_{st} + a_{s4}W_t^2 + a_{s5}S_{st}^2 + a_{s6}\frac{S_{st}}{W_{st}} \quad (4)$$

where W_{st} is the total water (i.e. sum of irrigation and rainfall) applied (ac-feet), S_{st} is the quantity of salt in irrigation water (tons/ac-ft), plus the salt in the soil profile, $\frac{S_{st}}{W_{st}}$ is the amount of total salt (soil +irrigation salt) divided by the total amount of water (irrigation plus rain fall) per acre, and NR_t is the precipitation in the non-growing season (ft).

2.7 Net Present Value Estimation

The net present value for a 50-year period was calculated for each individual pivot circle which contains one or more soil types. It was not possible to develop a single integrated optimization program. The net present value (NPV) is calculated using the following formula:

$$\max_I \text{NPV} = \sum_{t=1}^{T} \frac{1}{(1+r)^t} A_s \{P \cdot Y_t - C_{irr} \cdot I - C_o\} \quad (5)$$

Subject to,

$$Y_{st} = a_{s0} + a_{s1}W_{st} + a_{s2}Sh_{st} + a_{s3}NR_{st} + a_{s4}W_{st}^2 + a_{s5}Sh_{st}^2 + a_{s6}\frac{Sh_{st}}{W_{st}} \quad (6)$$

$$Sh_t = b_{s0} + b_{s1}I_{st} + b_{s2}Is_t + b_{s3}Sp_t + b_{s4}Rg_t * W_t \quad (7)$$

$$W_{st} = (Rg_t + Is_t) \quad (8)$$

$$Sh_{st} = (S_{st} + SI_{st}) \quad (9)$$

$$Sp_{st} = c_{s0} + c_{s1}Sh_{st-1} + c_{s2}R_{wt-1} \quad (10)$$

where Y_{st} is yield (lbs/acre) in soil *s*, year *t*, A_s is the acreage of a soil type *s* in the individual irrigation circles (number of soils types differ for each pivot circle), *P* is the price of cotton lint ($/lb) (price was assumed same over the year but a sensitivity of change in price ($0.50, $0.65, $0.70, and $0.9) was calculated), W_{st} is the total water applied (i.e., sum of the growing season rainfall (Rg_t) and irrigation water (Is_{st})), Sh_{st} is the total salt for soil *s* and year *t* (i.e., sum of salt in soil (S_{st}) and salt in irrigation water (SI_{st})), Sp_{st} is the soil salt at planting, Sh_{st-1} is soil salt at previous harvest, R_{wt-1} is non-season (winter) rainfall, C_{irr} is the irrigation cost ($/acre-feet), C_o is the operation cost, and *r* is discount rate.

3. Results and Discussions

3.1 Irrigable Soil Types and Areas

Potential irrigable areas in Tillman and Kiowa counties are shown in Figure 1.

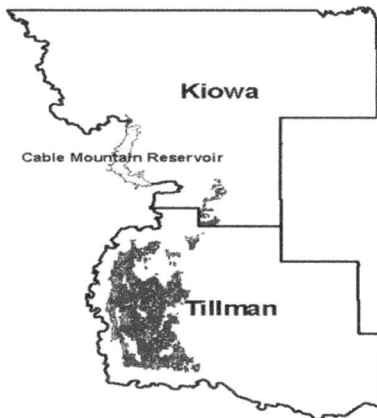

Figure 1. Potential irrigable areas in Tillman and Kiowa counties.

The area of potential irrigable soils totaled 67,868 acres (27,465 ha). Tipton Sandy Loam and Tipton Loam are the dominant soil types within the area. There were 5,196 acres (2,102 ha) which were not designated as irrigable but which producers were currently irrigating. These contributed approximately 45 more pivot circles. As people would likely continue irrigation in those areas, these areas were included in potential irrigable areas for the region. Most of the irrigable areas were covered by 543 pivot circles. There can be up to four pivot circles, each with an area of 125.6 acres (50.8 ha), in each section of land. Total irrigable areas, including identified irrigable soils, and non-irrigable soils that are currently under irrigation, were 73,064 acres (29,568 ha). Of this area, a total of 64,433 acres (26,075 ha) were covered by 513 full pivot circles and 2,764 acres (1,119 ha) were covered by 30 partial pivot circles.

3.2 Irrigation Network

Irrigation water flows through the main pipeline (North to South) from the reservoir to the lateral pipelines (East to West). Each lateral pipeline was connected with final pipelines which deliver water to the individual pivot in the fields. The entire pipeline outline is provided in Figure 2.

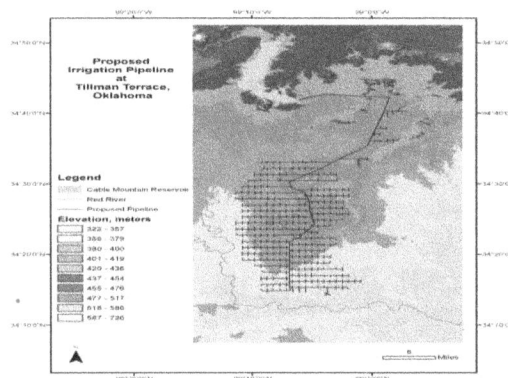

Figure 2. Outline of pipeline from the reservoir to Tillman Terrace with main, lateral, and final pipelines. North-South line is main pipeline and East-West lines are lateral pipelines overlaid on the elevation file.

The length of main, lateral, and final pipelines were 66 miles (106 km), 214 miles (344 km), and 243 miles (391 km), respectively. The size of main pipeline ranged from 48 to 120 inches (1.22 to 3 m), lateral pipelines ranged from 12 to 36 inches (0.3 to 0.9 m), and final pipes were 8 to 10 inches (0.2 to 0.25 m).

3.3 Trenching and Pipeline Cost

Total pipe costs, trenching costs, and total annualized pipeline costs for 50 years at four percent discount rate increased as the size of pipeline increased. Total pipe costs per linear foot (0.3 m) ranged from $151 for 0.6 m diameter pipe to $1,925 for 3 m diameter pipe. Total earthwork cost increased with increasing pipe size, ranging from $70 per linear foot (0.3 m) for 0.6 m diameter pipe to $271 for 3 m diameter pipe. Total cost was calculated as a sum of total pipe cost and total earthwork cost, which ranged from $221 to $2,196 per linear foot (0.3 m). The 50-year annualized cost at four percent discount rate was $10 per linear foot (0.3 m) for 0.6 m diameter pipe and reached up to $102 for 3 m diameter pipe. Diameter of smaller pipes ranged from 6 to 18 inches (0.15 to 0.45 m) while total pipe costs ranged from $8 (0.15 m) to $55 (0.45 m). Total earthwork cost increased with increasing pipe size ranged from $8 to $16. Total cost (pipe cost + earthwork cost) ranged from $16 to $71. The 50-year annualized cost at four percent discount rate ranged from $0.7 (0.15 m pipe) to $3.3 (0.45 m pipe).

3.4 Irrigation Systems

Four different irrigation system (A-D) designs that would deliver the water to every pivot were identified. Outline of all four designs are provided in Figure 3. Design A (Figure 3a) allowed all producers to irrigate simultaneously at 600 gpm (2,271 lpm) while designs B (Figure 3b) and C (Figure 3c) divided the irrigable areas into two parts. Design D (Figure 3d) divided the area into four parts to allow producers to irrigate one part at a time with 800 gpm (3,028 lpm) of individual pivot demand. The four designs were evaluated in terms of the annual fixed and variable costs.

3.5 Variable Costs

The variable costs of the irrigation systems included the costs of cotton production, cost of the pivot irrigation system, irrigation labor cost, and other related costs. Total non-irrigation variable cost was estimated to be approximately $500 per acre ($1,235 per ha) (Enterprise Budgets (Oklahoma State University), 2011). The annual pumping cost per acre foot was approximately $50 per acre ($124 per ha) for all four designs.

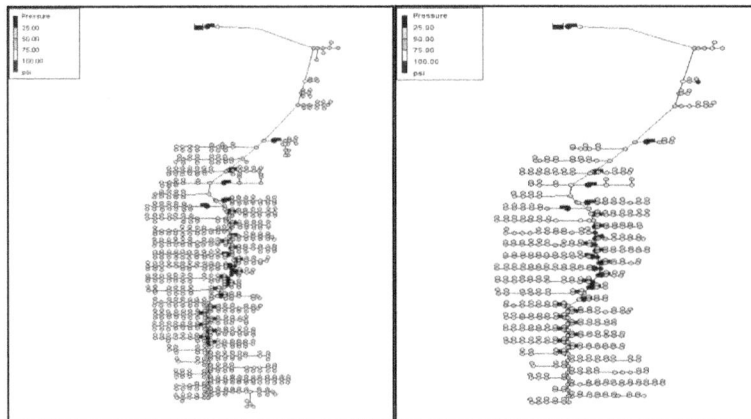

a. Design A: capacity 100% area at a time   b.   Design B: capacity 50% area at a time

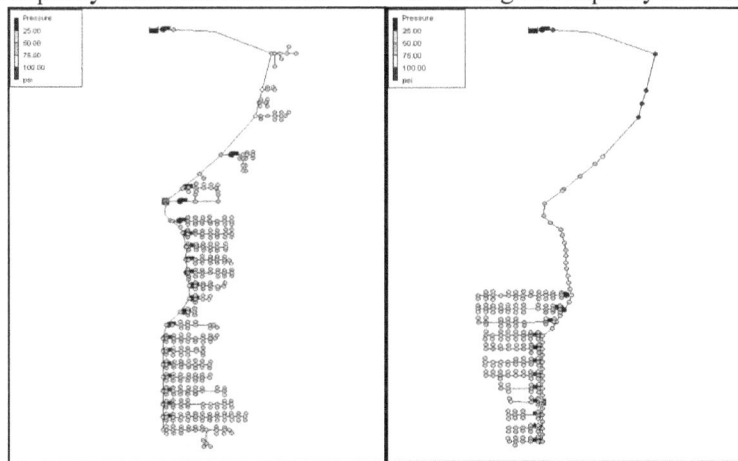

c. Design C: capacity 50% area at a time   d.   Design D: capacity 25% area at a time

Figure 3. Illustration of Irrigation Designs; a. Design A: Capacity 100% area at a time, b. Design B: Capacity 50% area at a time, c. Design C: Capacity 50% area at a time, and d. Design D: Capacity 25% area at a time.

3.6 Fixed Costs

3.6.1 Design A: Capacity 100% Area at a Time

This design was for the unrestricted irrigation system with demand of 600 gpm (2,271 lpm) for each pivot circle (Figure 3a). This design used main pipelines of 48 to 120 inches (1.22 to 3 m), lateral pipelines of 12 to 36 inch (0.3 to 0.9 m), and final pipelines of 8 and 10 inches (0.2 to 0.25 m). The total water demand was 326,800 gpm (1237,000 lpm) and annual fixed cost was $986 per ha at a four percent discount rate and a 1.5 mmhos cm^{-1} EC level (Table 1). Table 1 shows that the total annualized fixed cost per ha increased by 17%, from $986 to $1,153, when the discount rate was increased from four to five percent. It increased by 35% when the discount rate was increased from four to six percent, and increased by 74% when the discount rate was raised from 4% to 8%.

Table 1. Annual Fixed Costs of Design A Irrigation System at Different Discount Rates.

Cost	Cost/ha (4%)	Cost/ha (5%)	Cost/ha (6%)	Cost/ha (8%)
Pipe costs	$847	$990	$1,146	$1,472
Cost of centrifugal pumps and motors	$15	$17	$20	$25
Cost of pivot irrigation system	$124	$146	$168	$217
Total	$986	$1,153	$1,334	$1,714

With an annualized fixed cost of $986 per ha and total annual variable cost of $1,359 per ha of the irrigation system, the cotton lint price must be 75 cents or more per pound (454 g) to make this irrigation design economically feasible. A cotton price less than 75 cents per pound (454 g) resulted in a negative NPV for the project. The aggregate NPV for the non-scheduled irrigation system for cotton lint prices of 75 and 90 cents was $1,102 and $8,128 per ha, respectively for EC value of 1.5 mmhos cm^{-1}(Table 2).   At this level of EC, the total NPV for 67,197 acres (27,193 ha) of land at cotton price of 75 and 90 cents was approximately $31 million and $225 million (Table 2), respectively. The average NPV for a 50.8 ha system for EC value of 1.5 was approximately $413,367 for 90 cents and $55,970 for 75 cents per pound (454 g) cotton price. The sensitivity of the fluctuation in the NPV of the system to EC was also analyzed. The result showed that an increase in EC would decrease the NPV and decrease the optimal amount of irrigation water to maximize NPV. For design one, a decrease in EC from 1.5 to 0.9 mmhos cm^{-1} increased NPV by 12 and 30% for cotton lint prices of 90 and 75 cents per pound (454 g), respectively. The average optimum quantity of irrigation water increased by 35 and 60% for 90 and 75 cents of cotton prices, respectively.

Table 2. Aggregate Net Present Value (NPV) of Irrigated Cotton for Design A at Cotton Price of $0.75 and $0.90 Per Pound (454 g) at Different Electrical Conductivity (EC) Levels.

Cotton price $0.75				
EC level (mmhos cm^{-1})	0.9	1.5	2.2	3
Total NPV	$38,905,339	$30,391,486	$24,553,297	$21,585,240
Average NPV/50.8 ha	$71,649	$55,970	$45,218	$39,752
Average irrigation water (m/ha)	0.05	0.03	0.02	0.01
NPV/ha	$1,428	$1,102	$899	$790
Cotton price $0.90				
EC Level (mmhos cm^{-1})	0.9	1.5	2.2	3
Total NPV	$252,422,563	$224,458,253	$199,760,168	$183,232,769
Average NPV/50.8 ha	$464,867	$413,367	$367,882	$337,445
Average irrigation water (m/ha)	0.10	0.08	0.05	0.03
NPV/ha	$9,141	$8,128	$7,336	$6,728

An increase in EC value to 3.0 decreased the NPV by 17 and 28% for 90 and 75 cents of cotton prices, respectively. It also decreased the average optimum irrigation water by 65% for cotton prices of 90 cents and by 78% for cotton prices of 75 cents per pound (454 g).

3.6.2 Design B: Capacity 50% Area at a Time

This design was for the restricted irrigation system for instantaneous irrigation water supply (Figure 3b). The design A was modified for scheduling irrigation in design B to the north of each lateral at one time and the south the other time. The total water demand at a time was 217,600 gpm (823,700 lpm). Main pipelines from 36 to 108 inches (0.91 to 2.74 m), lateral pipelines from 12 to 30 inches (0.3 to 0.76 m), and final pipelines from 6 to 10 inches (0.15 to 0.25 m) were used in this design. The annual fixed cost for 50 years for design B decreased to $691 (Table 3) from $986 per ha for design A at four percent discount rate. The sensitivity of cost to discount rates to the cost of this system showed that increasing discount rate from four to five, six, and eight percent increased the cost by 18%, 36%, and 76%, respectively.

Table 3. Annual Cost per Acre of Design B Irrigation System at Different Discount Rates.

Costs	Cost/ha 4%	Cost/ha 5%	Cost/ha 6%	Cost/ha 8%
Cost of pipe and earthwork	$553	$650	$753	$971
Cost of pivots	$124	$143	$168	$217
Cost of centrifugal pumps and motors	$7	$7	$10	$12
Cost of valves	$7	$10	$10	$12
Total	$691	$810	$941	$1,212

With an annualized fixed cost of $691 per ha and total annual variable cost of $1,359 per ha, design B irrigation system was economically feasible for cotton prices above 70 cents per pound (454 g). At 70 cents per pound (454 g) cotton price this design was feasible for EC level less than and equal to 2.2 mmhos cm^{-1}. The NPV increased with increasing cotton price and decreased with increasing EC level linearly.

The aggregate NPVs per ha for the design B irrigation system for cotton lint prices of 70, 75, and 90 cents per pound (454 g) were $1,665, $4,434, and $13,254 (Table 4), respectively for an EC value of 1.5 mmhos cm^{-1}. At this level of EC, the total NPVs for 67,197 acres (27,194 ha) of land at cotton price of 70, 75 and 90 cents were approximately $47 million, $124 million, and $368 million (Table 4), respectively. The sensitivity of the NPV to the EC in the irrigation water was also analyzed. The result showed that an increase in EC would decrease the NPV per acreage and decrease the optimal average quantity of irrigation water to maximize NPV.

For design B, an increase in EC from 0.9 to 1.5 mmhos cm^{-1} decreased NPV by 40, 24, and 12% for cotton lint prices of 70, 75 and 90 cents per pound (454 g), respectively. An increase in cotton price also increased the total water, and increase in EC levels decreased the total water for the system. With an increase in EC level from 0.9 to 1.5 mmhos cm^{-1}, the average optimum quantity of irrigation water decreased by 23%, 22%, and 17% for cotton prices of 70, 75, and 90 cents per pound (454 g), respectively. An increase in EC value from 1.5 to 2.2 decreased the NPV by 64%, 32%, and 19% for 70, 75, and 90 cents of cotton prices, respectively.

Table 4. Aggregate NPV of Irrigated Cotton for Design B at Cotton Price Of $0.70, $ 0.75, and $0.90 per Pound (454 G) a Different Electrical Conductivity (EC) Levels.

Cotton price $0.70			
EC level (mmhos cm^{-1})	0.9	1.5	2.2	3
Total NPV	$77,467,321	$46,978,340	$18,121,976	-$2,502,726
Average NPV/50.8 ha	$142,665	$86,516	$33,374	-$4,609
Average irrigation water (m/ha)	0.14	0.10	0.06	0.04
NPV/ha	$2,786	$1,665	$603	-$153
Cotton price $0.75			
EC level (mmhos cm^{-1})	0.9	1.5	2.2	3
Total NPV	$162,560,744	$123,708,374	$85,685,646	$57,696,768
Average NPV/50.8 ha	$299,375	$227,824	$157,800	$106,256
Average irrigation water (m/ha)	0.15	0.11	0.07	0.05
NPV/ha	$5,861	$4,434	$3,036	$2,008
Cotton price $0.90			
EC level (mmhos cm^{-1})	0.9	1.5	2.2	3.0
Total NPV	$414,591,644	$367,712,681	$299,118,775	$245,904,800
Average NPV/50.8 ha	$763,521	$677,187	$550,863	$452,863
Average irrigation water (m/ha)	0.17	0.15	0.10	0.06
NPV/ha	$14,978	$13,254	$10,734	$8,778

3.6.3 Design C: Capacity 50% Area at a Time

This design was also for the restricted instantaneous irrigation supply by scheduling irrigation (Figure 3c). The irrigation system was divided into two sections, and irrigation was scheduled for one section at a time. The diameter of the largest pipe for this system was 108 inches (2.74 m). Main pipelines of 36 to 108 inches (0.9 to 2.74 m), lateral pipelines of 12 to 30 inches (0.3 to 0.76 m), and final pipelines of 6 to 10 inches (0.15 to 0.25 m) were used in this design. Water demand for each pivot was increased from 600 to 800 gpm (2271 to 3028 lpm) so that it would take less time to irrigate each section and the other section can be irrigated sooner. This design required 189,000 gpm (715,000 lpm) for a section. This scheduling has an annual fixed cost of $672 per ha at four percent discount rate (Table 5), which is approximately 32% less than that of design A. A one percent increase in discount rate increased the annual cost of the irrigation system per ha approximately by 17%. Increasing the discount rate from four to six percent and from four to eight percent increased the cost by 36% and 75%, respectively.

Table 5. Annual Fixed Costs of Design C Irrigation System at Different Discount Rates.

Cost	Cost/ha (4%)	Cost/ha (5%)	Cost/ha (6%)	Cost/ha (8%)
Pipe cost	$531	$625	$726	$934
Cost of centrifugal pumps and motors	$10	$12	$15	$17
Cost of pivot	$124	$146	$168	$217
Cost of valves	$7	$10	$10	$12
Total	$672	$793	$919	$1,180

At an annualized fixed cost of $672 per ha and total variable cost of $1,359 per ha, the irrigation system was only feasible for the cotton lint price of 70 cents or more per pound (454 g). Aggregate NPV per ha at different EC levels for design C at cotton prices of 70, 75, and 90 cents are presented in Table 6. The average NPVs per ha at EC level of 1.5 mmhos cm^{-1} for this scheduled irrigation system were $1,924, $4,770, and $13,797 at cotton prices of 70, 75, and 90 cents, respectively. The average optimal annual quantity of irrigation water used at EC level of 1.5 mmhos cm^{-1} was 0.10, 0.12, and 0.15 m ha^{-1} at cotton prices of 70, 75, and 90 cents, respectively. The NPV and optimal average irrigation water use were reduced when EC level was increased. A decrease in the EC level from 1.5 to 0.9 mmhos cm^{-1} increased NPV by 20%, and the average optimum irrigation water increased by 27% at cotton price of 90 cents. It increased the NPV by 24% and average optimum irrigation water increased by 15% at cotton price of 75 cents. Similarly, the NPV increased by 61% and the average optimum irrigation water increased by 38% at cotton prices of 70 cents. An increase in EC level from 1.5 to 2.2 and 3 decreased both the NPV and the optimum quantity of irrigation water. Increasing the EC level to 3 from 1.5 mmhos cm^{-1} decreased NPV by 35% (for 90-cent cotton), 53% (for 75-cent cotton), and 101% (for 70-cent cotton). In the same way, the average optimum quantity of irrigation water decreased by approximately 55% for both 90 and 75 cents of cotton prices.

Table 6. Aggregate NPV of Irrigated Cotton for Design C at Cotton Price Of $0.7, $ 0.75, and $ 0.90 per Pound (454 g) at Different Electrical Conductivity (EC) Levels.

Cotton price $0.70				
EC level (mmhos cm^{-1})	0.9	1.5	2.2	3
Total	84,426,767	$52,429,273	$21,858,011	($178,039)
Average NPV/50.8 ha	$155,482	$96,555	$40,254	($328)
Average irrigation water (m/ha)	0.14	0.10	0.07	0.04
NPV/ha	$3,100	$1,924	$800	$10
Cotton price $0.75				
EC level (mmhos cm^{-1})	0.9	1.5	2.2	3
Total	$163,589,347	$129,892,665	$89,989,453	$60,422,953
Average NPV/50.8 ha	$301,270	$239,213	$165,726	$111,276
Average irrigation water (m/ha)	0.13	0.12	0.08	0.05
NPV/ha	$5,926	$4,770	$3,302	$2,218
Cotton price $0.90				
EC level (mmhos cm^{-1})	0.9	1.5	2.2	3
Total	$443,274,771	$375,608,518	$304,758,898	$249,579,827
Average NPV/50.8 ha	$816,344	$691,728	$561,250	$485,426
Average irrigation water (m/ha)	0.19	0.15	0.10	0.07
NPV/ha	$16,285	$13,797	$11,194	$9,166

3.6.4 Design D: Capacity 25% Area at a Time

This design was also for the restricted instantaneous irrigation supply. In this design, the irrigation system was divided into four areas (Figure 3d). At one time, only one area would be irrigated. This reduced the water demand leading to a further reduction in pipeline size as compared to other designs. Water demand of 800 gpm (3028 lpm) per pivot for this design requires a total 108,600 gallons (411,096 liters) of water per minute to irrigate a section. Main pipelines of 36 to 84 inches (0.9 to 2.1 m), lateral pipelines of 12 to 24 inches (0.3 to 0.6 m), and final pipelines of 6 to 10 inches (0.15 to 0.25 m) were used in this design. The annualized fixed cost at a four percent discount rate for this design was $550 per ha (Table 7), which is approximately 44% less than that of design A and 19% less than that of designs B and C.

Table 7. Annual Fixed Costs of Design D Irrigation System at Different Discount Rates.

Cost	Cost/ha (4%)	Cost/ha (5%)	Cost/ha (6%)	Cost/ha (8%)
Pipe cost	$412	$487	$563	$726
Cost of centrifugal pumps and motors	$7	$7	$7	$10
Cost of valves	$7	$10	$10	$12
Cost of pivots	$124	$146	$168	$217
Total	$550	$650	$748	$965

With an annualized fixed cost of $550 per ha and total variable cost of $1,359 per ha, the irrigation system was feasible for the cotton price of 65 cents per pound (454 g) or more. The aggregate NPVs per ha for the non-scheduled irrigation system at 65, 75, and 90 cents were $1,314, $7,415, and $17,008, respectively at 1.5 mmhos cm^{-1} EC level (Table 8). When the EC level of the irrigation water was decreased from 1.5 to 0.9 mmhos cm^{-1} at 60 cents, the NPV increased by more than 100%, and the average optimum quantity of irrigation water increased by 22%. The NPV increased by 28% and average optimum quantity of irrigation water increased by 25% at 75 cents. At the same EC level and cotton price of 90 cents, both the NPV and the average optimal quantity of irrigation water increased by 21%. Increasing the EC level to 3 from 1.5 mmhos cm^{-1} decreased the NPV by 39% (for 90-cent cotton), 53% (for 75-cent cotton), and 193% (for 65-cent cotton). Similarly, the average optimum quantity of irrigation water decreased by approximately 54% for cotton prices of 75 and 90 cents per pound (454 g). The NPV for the area was negative at an EC level of 3 mmhos cm^{-1} and a cotton price of 65 cents per pound (454 g), indicating that the design D irrigation system was unfeasible at higher EC (2.2 or more) level and lower cotton prices (less than 65 cents).

Scheduling may require less water at each point in time so that the smaller pipes would be enough to meet the demand (Pereira, Calejo, Lamaddalena, Douieb, & Bounoua., 2003). Efficient irrigation scheduling and modifying irrigation system not only help to conserve water but also avoid excessive energy use, reducing the irrigation costs (Morris & Grubinger, 2015). Improved delivery scheduling is a feasible solution for better crop irrigation management (Zaccaria et al., 2010). In this study, irrigation scheduling reduced the size of pipeline, lowered the cost of the irrigation system, and ultimately increased net returns of irrigation systems. As expected, higher salinity level lowered the net returns of irrigation systems, consistent with the findings of a previous study (Tayfur et al., 1996). The soil salinity response functions had negative relation with the amount of irrigation water (Choi, 2011).

Table 8. Aggregate NPV of Irrigated Cotton for Design D at Cotton Price Of $0.65, $0.75, and $0.90 per Pound (454 G) at Different Electrical Conductivity (EC) Levels.

Cotton price $0.65				
EC level (mmhos cm^{-1})	0.9	1.5	2.2	3
Total	$75,327,627	$35,827,977	$(3993223)	$(34052596)
Average NPV/50.8 ha	$138,725	$65,982	$(7354)	$(62712)
Average irrigation water (m/ha)	0.16	0.12	0	0
NPV/ha	$2,766	$1,314	$(146)	$(1,233)

Cotton price $ 0.75				
EC level (mmhos cm^{-1})	0.9	1.5	2.2	3
Total	$259,850,367	$201,911,965	$140,963,134	$93,314,077
Average NPV/50.8 ha	$478,546	$371,845	$259,601	$171,849
Average irrigation water (m/ha)	0.19	0.15	0.11	0.07
NPV/ha	$9,544	$7,415	$5,175	$3,426

Cotton price $ 0.90				
EC level (mmhos cm^{-1})	0.9	1.5	2.2	3
Total	$550,210,680	$463,049,407	$367,823,753	$291,094,555
Average NPV/50.8 ha	$1,013,279	$852,761	$677,392	$536,086
Average irrigation water (m/ha)	0.22	0.18	0.13	0.08
NPV/ha	$20,212	$17,008	$13,508	$10,690

4. Summary and Conclusions

A combination of GIS and simulation models was used to identify irrigable areas, to develop the irrigation system, and to determine the cost-benefit of the irrigation systems in this study. The sizing of the pipeline involved a tradeoff between annual energy cost and pipeline cost. Total pipeline cost increases with increasing diameter of pipeline, but the energy required for pumping water through the pipe decreased with increasing pipe diameter. Pipe sizes were iteratively increased or decreased to determine the optimum cost of the irrigation system. Four design with different irrigation scheduling were evaluated in this study.

The total cost (fixed cost plus variable cost) of the Design A irrigation system (irrigation without scheduling) was approximately \$950 per acre (\$2,345 per ha) at 1.5 mmhos cm^{-1} EC level (Appendix A). At this cost and EC level, NPV was feasible for the cotton lint price of 75 cents or more per pound (454 g). Both Design B (irrigate alternately to the north and south of the laterals of the Design A) and Design C (scheduled to irrigate two areas alternatively) were feasible above cotton price of 70 cents for 0.9, 1.5, 2.2, and 3 mmhos cm^{-1} EC levels. At cotton price of 70 cents, the full and partial pivots in Design B and C were feasible till the EC level of 2.2 mmhos cm^{-1}. The total cost of the Design D (scheduled to irrigate one area at a time of four areas) irrigation system was approximately \$773 per acre (\$1,909 per ha) (Appendix A). The NPV was feasible at a cotton lint price of 65 cents per pound (454 g) at this cost and EC level of 0.9 and 1.5 mmhos cm^{-1}. However, the NPV for 2.2 and 3 mmhos cm^{-1} EC levels was feasible at prices higher than 65 cents for Design D. The sensitivity analysis for different salinity levels and different cotton prices showed that the NPV, average and total optimum irrigation increased linearly with increasing cotton prices. Increase in salinity decreased the yield of cotton (Razzouk and Whittington, 1990), and that decreased the net returns of the irrigation system.

The cost of water flowing through a pipe increased as the diameter of pipe increased. Design A had larger pipelines but scheduling water supply as in Designs B, C, and D allowed the use of smaller more cost effective pipes, which ultimately reduced the irrigation cost, increased the net returns, and lowered the breakeven price of cotton. Although there are several methods to control salinity levels, both salinity and cotton prices are out of the control of the individual agricultural operator. Thus, this study endorses irrigation scheduling as a tool for water conservation with effective irrigation development and management.

References

Addink, J. W., Keller, J., Pair, C. H., Sneed, R. E., & Wolfe, J. W. (1983). *Design and operation of sprinkler systems*, M.E. Jensen. ed., Design and operation of farm irrigation systems. The American Society of Agricultural Engineers, Michigan, USA.

ArcGIS v.9.3. ESRI, Redlands, California, USA.

Belmonte, A. C., González, J. M., Mayorga, A. V., & Fernández, S. C. (1999). GIS tools applied to the sustainable management of water resources: Application to the aquifer system 08-29. *Agricultural Water Management, 40*(2), 207-220. http://dx.doi.org/10.1016/S0378-3774(98)00122-X

Bureau of Reclamation. (2005). *Water supply augmentation: Appraisal Report, W. C. Austin Project, Oklahoma.* Bureau of Reclamation, Oklahoma-Texas Area Office, Austin, Texas. https://www.usbr.gov/gp/otao/austinappraisal0505.pdf

Choi. J. S. (2011). *Economic Approach on Allocation Irrigation Water under Salinity based on Different Soils for Potential Irrigated Agriculture using EPIC Crop Model.* Oklahoma State University, Oklahoma, USA.

Datta, K. K., Sharma, V. P., & Sharma, D. P. (1998). Estimation of Production Function for Wheat under Saline Conditions. *Agricultural Water Management, 36*(1), 85-94. http://dx.doi.org/10.1016/S0378-3774(97)00015-2

Dinar, A., & Knapp, K.C. (1986). A Dynamic Analysis of Optimal Water Use under Saline Conditions. *Western Journal of Agricultural Economics,* 11(1), 58-66. http://www.jstor.org/stable/40987692

Enterprise Budgets. (2011). Department of Agricultural Economics, Oklahoma State University. (Available at http://agecon.okstate.edu/budgets).

Environmental Protection Agency (EPA). (2011). *Software that models the hydraulic and water quality behavior of water distribution piping systems.* U. S. Environmental Protection Agency. (Accessed August 2011, software and user manuals are available at http://www.epa.gov/nrmrl/wswrd/dw/epanet.html#downloads).

Feinerman, E. (1994). Value of information on crop response function to soil salinity in a farm-level optimization model. *Agricultural Economics*, 10(3), 233-243. http://www.sciencedirect.com/science/article/pii/0169515094900256

Ferrari, R. L. (2008). *Altus Reservoir 2007 Sedimentation Survey, Tech. Serv. Center, Water and Environmental Resources Division.* Sedimentation and River Hydraulics Group, Bureau of Reclamation, U.S. Dept. of the

Interior, Denver, Colorado, USA.
http://www.usbr.gov/tsc/techreferences/reservoir/Altus%20Reservoir%202007%20Sedimentation%20Survey.pdf

Fortes, P. S., Platonov, A., & Pereira, L. S. (2005). GISAREG-A GIS based irrigation scheduling simulation model to support improved water use. *Agricultural water management, 77*(1), 159-179. http://dx.doi.org/10.1016/j.agwat.2004.09.042

George, B., Raghuwanshi, N. S., & Singh R. (2004). Development and testing of a GIS integrated irrigation scheduling model. *Agricultural water management, 66*(3), 221-237. http://dx.doi.org/10.1016/j.agwat.2003.11.004

Geospatial Data Gateway. Soil Survey Geographic Database, NRCS- USDA. Retrieved December 20, 2010, from http://datagateway.nrcs.usda.gov

Global Mapper. (2011). GIS Mapping Software, Blue Marble Geographics, Maine, USA. Retrieved August 10,2011, from http://www.bluemarblegeo.com/global-mapper

Keller, J., & Bliesner, R. D. (1990). *Sprinkler and Trickle Irrigation.* Van Nostrand Reinhold, New York, USA.

Kiani, A. R., & Abbasi, F. (2009). Assessment of the Water-salinity crop production function of wheat using experimental data of the Golestan Province, Iran. *Irrigation and Drainage, 58*(4), 445-455. http://onlinelibrary.wiley.com/doi/10.1002/ird.438/epdf

Morris, M., & Grubinger, V. (2015). *Introduction to Energy Efficient Irrigation.* Retrieved August 11, 2015 from http://articles.extension.org/pages/27775/introduction-to-energy-efficient-irrigation

Mossman, M. J., & Plotner, S. C. eds. (2009). *RS Means Facilities Construction Cost Data.* 24th Annual Edition. RS Means Company, Inc. Construction Publishers and Consultants, Massachusetts, USA.

Osborn, N. I. (2002). *Update of the Hydrological Survey of the Tillman Terrace Groundwater Basin, Southwestern Oklahoma.* OWRB Technical Report GW 2001-1.Oklahoma, USA.

Pereira, L. S., Calejo, M. J., Lamaddalena, N., Douieb, A., & Bounoua, R. (2003). Design and performance analysis of low pressure irrigation distribution systems. *Irrigation and Drainage Systems, 17*(4), 305-324. http://dx.doi.org/10.1023/B:IRRI.0000004558.56077.d4

Razzouk, S., & Whittington, W. J. (1991). Effects of salinity on cotton yield and quality. *Field Crops Research, 26*(3), 305-314. http://www.sciencedirect.com/science/article/pii/037842909190007I

Satti, S. R., & Jacobs, J. M. (2004). A GIS-based model to estimate the regionally distributed drought water demand. *Agricultural Water Management, 66*(1), 1-13. http://dx.doi.org/10.1016/j.agwat.2003.10.003

Singh, R., Jhorar, R. K., Van Dam, J. C., & Feddes, R. A. (2006). Distributed ecohydrological modelling to evaluate irrigation system performance in Sirsa district, India II: Impact of viable water management scenarios. *Journal of Hydrology, 329*(3), 714-723. http://dx.doi.org/10.1016/j.jhydrol.2006.03.016

Tayfur, G., Tanji, K.K., House, B., Robinson, F., Teuber, L., & Kruse, G. (1995). Modeling deficit irrigation in Alfalfa production. *Journal of Irrigation and Drainage Engineering, 121*(6), 442-451. http://ascelibrary.org/doi/abs/10.1061/(ASCE)0733-9437(1995)121:6(442)

Todorovic, M., & Steduto P. (2003). A GIS for irrigation management. *Physics and Chemistry of Earth, PT A/B/C, 28*(4), 163-174. http://dx.doi.org/10.1016/S1474-7065(03)00023-8

U.S. Department of Agriculture, Natural Resources Conservation Service. *National Soil Survey Handbook, title 430-VI.* Retrieved October 1, 2010 from http://soils.usda.gov/technical/handbook/

U.S. Department of Labor, Occupational Safety and Health Administration (OSHA). Retrieved July 11, 2011, from https://www.osha.gov/dts/osta/otm_v/otm_v_2.html /

U.S. Geological Survey. Retrieved January 16, 2012 from http://waterdata.usgs.gov/nwis/nwisman/?site_no=07303400&agency_cd=USGS

U.S. Geological Survey. Retrieved November 2010, from http://www.usgs.gov/pubprod/

Zaccaria, D., Oueslati, I., Neale, C. M., Lamaddalena, N., Vurro, M., & Pereira, L. S. (2010). Flexible delivery schedules to improve farm irrigation and reduce pressure on groundwater: a case study in southern Italy. *Irrigation science, 28*(3), 257-270.

Appendix A. Total annual fixed and variable costs at 4% discount rate for four different designs of irrigation system.

	Design A	Design B	Design C	Design D
Annual fixed costs/ha				
Cost of pipe and earthwork	$847	$553	$531	$412
Cost of pumps and motors	$15	$7	$10	$7
Cost of pivot irrigation system	$124	$124	$124	$124
Cost of valves		$7	$7	$7
Annual variable cost/ha				
Non-irrigation variable cost	$1,235	$1,235	$1,235	$1,235
Pumping cost	$124	$124	$124	$124
Total cost	$2,345	$2,050	$2,031	$1,909

Timed Strategy for Control of Bollworm for Sustainable Sorghum Crop Yield under Varied Regimes of Rainfall, Temperature and Soil Fertility

Daniel L. Mutisya[1], Canute P. M. Khamala[2], Jacob J. J. O. Konyango[3], Clement K. Kamau[1] & Lawrence K. Matolo[3]

[1]Kalro-Katumani. P. O. Box 340. 90100. Machakos

[2]University of Nairobi. P. O. Box 30197. 00100. Nairobi

[3]Machakos University College. P. O. Box 136. 90100. Machakos

Correspondence: Daniel L. Mutisya, Kalro-Katumani. P. O. Box 340. 90100. Machakos.
E-mail: dlmutisya@gmail.com

Abstract

Various environmental factors influence yield of sorghum grain, *Sorghum bicolor* (L) in Sub-Sahara Africa. Various production conditions of rainfall amount, temperature regimes, soil fertility levels and bollworm *Helicoverpa armigera* density at specific sorghum grain stage were evaluated for effect to sorghum grain yield. High rainfall amount, high temperature and soil fertility levels were positively correlated to sorghum grain yield at three test sites at Ithookwe, Katumani and Kampi of eastern Kenya. The warmest Kampi site achieved the highest seed viability on germination test at 43, 87 and 99% for grain stage of light-green, cream-dough and hard dough, respectively. High *H. armigera* density was inversely correlated to grain yield. Comparatively, yield loss of < 10% was observed when grain was at early soft dough and > 35% as the grain ripened to early hard dough stage. Thus initial *H. armigera* damage occurred at late soft dough stage and increased exponentially as the grain ripened to early hard dough stage. The right time to spray against *H. armigera* was determined as at soft dough stage of sorghum grain to prevent economic damage of the crop. Thus fertility level, rainfall amount and time of bollworm pest attack were deemed worth considerations towards sustainable yield of sorghum.

Keywords: sorghum, environmental factors, soft dough, hard dough, yield

1. Introduction

The United States of America lead in sorghum (*Sorghum bicolor* L.) production with 25% of world total of 33,963,000 MT followed by Mexico (21%), Nigeria (18%), Sudan (16%) and India (2%) (FAO, 2016). While in the America and Asian countries the cereal has various uses in industrial and animal production sectors the same cannot be said in the African continent where the crop is mainly the second staple food after maize (Ratnavathi & Patil, 2013). Similarly farmers in Sub-Sahara (SSA) countries continue to realize low yields of staple cereals like maize, rice and wheat (Mastersa et al., 1998; Kilambya & Witwer, 2013). This is due to reduced and irregular rainfall patterns characterized by climate variability scenarios (Oehmke & Crawford, 1996; Ndjeunga & Bantilan, 2005). The remedy to this could lie on production of millets and sorghum which could diversify premium incomes (Van Wajk & Kwak, 2011). In SSA sorghum grain is a major staple in the low-income countries where in rare cases it is exported to some trade partners for foreign exchange (Ndjeunga & Bantilan, 2005; Orr et al., 2013). Largely much relief from crop yield loss could be attained through sorghum production due to its low moisture and low nutrient requirements as well as fast development as an annual crop (Muui et al., 2013; Orr et al., 2013).

Nevertheless various production constraints exist as both biotic and abiotic factors (Fitt, 1989; Guo, 1997). These can be listed majorly as low soil fertility in marginal areas, insect damage, low yield due to low genetic potential from landraces, minimal adoptions of improved varieties, bird menace causing close to 100% grain loss in some areas and poor government policies on incentives to productions (Muui et al., 2013; Gitz III et al., 2013).

A good number of African governments are slowly addressing policy issues and farmers doing all they can to address bird menace by physically guarding crop against birds' damage though with minimal success (Oehmke

& Crawford, 1996; Pretty et al., 2011; Hirron et al., 2014). Other constraints of insect damage do exist where stem borers like *Chilo partellus* Swinhoe and *Busseola fusca* Fuller cause some 15-23% crop loss in SSA where sorghum is relied upon as food security staple (van den Berg & van Runsburg, 1991). Otherwise another insect which occur in large density during some production seasons and cause yield loss especially on white varieties is the African bollworm *Helicoverpa armigera* Hübner (Pedgley, 1985; Pogue, 2004; Kumar et al., 2009). A close field observation has indicated that if the arrival of the larvae worms of *H. armigera* occur during the grain ripening stage of sorghum it leads to heavy yield loss than if the grain is at dry dough (Duraimurugan & Regupathy, 2005; Moral-Garcia, 2006). There was need to elucidate yield loss of sorghum due to *H. armigera* damage under specific rainfall, temperature and fertility regimes at three site plots of eastern Kenya to determine a safe time to implement control of *H. armigera* to prevent yield grain loss. Specific considerations were to be assessed: (1) at what time of grain stage a spray insecticide can be applied to prevent yield loss, (2) how much yield loss could occur if no control mechanisms are put in place. A further consideration was: (3) what other environmental factors influence and correlate to final yield like fertility, rainfall and temperature, and lead to sustainable yield achievement by farmers in marginal lands.

2. Materials and methods

2.1 Sites Plot Establishment

The field study was carried out during the short rains of October-December 2014, long rains of March-July 2015 and short rains of October 2015-January 2016 periods of production seasons. Plot sizes of 8m x 10m were established at the sites of Katumani, Kampi-Mawe and Ithookwe of eastern Kenya. The white sorghum spacing was 60cm inter-row by 25cm intra-row of three plants per hill. The plots consisted of three blocks, one of insecticide treatment, one where birds were expected fed on the worms and a control in a randomized complete block design (RCBD). Bird scaring was introduced at late soft dough stage when bollworm damage peaked towards hard grain. Each of blocks had three treatments of pure sorghum, sorghum-millet alternate row and sorghum-edge rows. The insecticide used was Duduthrin® (Lambdacyhalothrin 17.5g/L as active ingredient of synthetic pyrethroid insecticide applied once at the rate of 200ml /ha three weeks after crop emergence against shoot fly and stem borers. After vegetative growth one of the blocks was sprayed with same Duduthrin® at anthesis stage of sorghum crop and repeated after seven days to prevent bollworm *H. armigera*. A four metre path separated the three blocks. Fertilizer application of di-ammonium phosphate (DAP) was carried out at the rate of 40kg per hectare for uniform nutrition. Crop top dressing with calcium ammonium nitrate (CAN) was carried out once at the rate of 30kg/ha after plants attained eight leaves at the experimental sites. Weeding by hand hoe (jembe) was done three times before crop maturity during each season. The amount of rainfall (mm) during the production period was recorded for analysis and comparable production potential of sorghum at the sites. Sorghum growth development from flowering (anthesis) deep-green, light-green, light-cream, cream-white, full-cream and full grain maturity were observed for *H. armigera* density and damage on the grain among the treatments.

2.1.1 Katumani Site

The site at KALRO-Katumani Research Station (01°20'51" S, 037° 16' 56" E, Elev.1609m asl) was established on a former cassava field, 11 kilometers south of Machakos Town. A yearly bimodal rainfall occurs at the site of short and long seasons respectively. Crop developmental days to grain filling, grain colour and dry dough was taken for comparative analyses between the plots. After grain yield data collection at the end of each cropping season, the field was ploughed by tractor for repeat of experiment.

2.1.2 Kampi-Mawe Site

The Kampi-Mawe KALRO-Sub-Centre Station is located (01° 50'54" S, 037° 39'29" E, Elev. 1164m asl) 10 kilometers east of Wote Town. The field had previously been planted cowpea before start of experiment in October 2014. A bimodal rainfall occurs per year at the site of short and long seasons. Data collection was carried out similar to Katumani site.

Similarly crop field was prepared by tractor for the repeat of the experiment after each season data collection.

2.1.3 Ithookwe Site

The Ithookwe KALRO-Sub-Centre Station is located (01° 22'34" S, 037° 58'43"E, Elev. 1147m asl) three kilometers in the western side of Kitui Town. A similar bimodal type of rainfall occurs at the site like as in the above other sites. As with other sites crop developmental days to grain filling, grain colour and dry dough was taken for comparative analyses. Data collection of grain yield was taken in the same procedure as the other two fields. Likewise the crop field was prepared for next season by ploughing by tractor.

2.2 Soil Fertility at Sites

The sorghum variety used for the experimental test was one being scaled up to farmers in eastern Kenya region, simply referred as "Gadam". The variety is white in colour and known by farmers as drought resistant and thrives even in low fertility areas. Soil analyses was taken in April 2016 after the experiments were ended. The aim was to determine the residual fertility conditions of the plots. At each of the sites, 10 soil samples were collected randomly across the intercrop treatment plots with soil auger at depth of 0-20cm and sent to National Agricultural Laboratories of Kenya Agricultural & Livestock Research Organization (KALRO) for nutrient element analyses on soil fertility factor suitability for sorghum production at the end of experimental period. Analyses on macro-element and micro-element nutrient levels were also scored. Residual fertility element compositions levels at each site were compared to sorghum crop requirement. The rationale here was to find out if sorghum production had been produced throughout the experimental seasons with the right fertility conditions reflected in the final yield.

2.3 Rainfall-Temperature Effects

Rainfall amount was recorded at each site each production season. Incidentally the three sites had functioning meteorological stations where daily records of temperature, rainfall, humidity and wind speeds were recorded. Monthly rainfall distribution and temperature regimes were averagely analyzed for site comparison. Effects of site conditions to grain yield were analyzed for comparison on insecticide applied and control plots.

2.4 Grain Maturity and Viability

Sorghum growth development days from flowering (anthesis) to deep-green, light-green, light-cream, cream-white, full-cream and full grain maturity were recorded. Grain samples (100 seeds) for each grain stage harvest interval were subjected to viability test of percentage germination at each specific stage and for each site plot. Net harvest sample was undertaken from three rows at the middle of the plots inclusive of all three plants (plus tillers ears).

2.5 Grain Ripening and Bollworm Damage

At each site grain ripening stage *H. armigera* infestation level and the resultant yield loss were correlated to elucidate the relationship. It was important to determine when highest damage occurred and recommend the right time to spray against bollworms and prevent subsequent grain loss. Some 10 randomly damaged panicle grains were compared to similar undamaged (10) heads of sorghum plants. The cumulative yield loss was expected to be different from specific interval harvest but reflective of damage at specific grain stage.

2.6 Effect of Environmental Factors to Yield

The effect of various environmental factors of abiotic like rain amount, temperature fluctuations and soil fertility were elucidated to explore impact to sorghum yield. Likewise bollworm damage impact to yield was determined as well the right time to spray against the insect. This would separate major production constraint in sorghum production. A multiple factor correlation was carried out to determine various factors influencing final sorghum yield.

2.7 Data Analyses

Analyses of variance (ANOVA) was carried to test significant difference of rainfall amounts, temperature levels, grain weight and grain yield parameters among the sites using SAS General Linear Model Procedure (GLM PROC) where Student Neumann's Keuls Post Hoc Test was used in separating the variable means. Mean nutrient element test results from the sites were comparatively analyzed for fertility levels. Multiple Linear Regression Model (MLRM PROC) was carried out to determine effect to various variables to grain yield. Correction analyses was carried out to determine effect of varied environmental factors to sorghum yield using Multiple Linear Regression Model (MLRM PROC) at 5% level. Percentage grain maturity and viability was analyzed from 100 grains per site pooled from the 10 ears sample size per plot on each treatment.

3. Results

3.1 Soil Fertility Levels

The soil sample analyses from the three sites at the end of cropping period showed that pH level was within sorghum crop requirements (Table 1). As for the macronutrients Nitrogen (N) levels were inadequately low at 0.11-0.15g/kg at the three sites. Ithookwe site led with the lowest levels of N fertility. Similarly, Phosphorous (P) was inadequately low for the crop production. Potassium, Magnesium and Calcium nutrient elements were adequate in the three site. The levels of the micronutrient elements of Cu, Fe, and Mn (mg/kg) were assessed as adequate at Katumani and Kampi sites. At Ithookwe Cu was inadequately low for sorghum production. The soil texture was closely similar as loam sandy at the three sites. Overall the soil samples were found deficient of organic matter from the three sites.

Table 1. Soil fertility residual level evaluation results on sorghum plots at three sites in eastern Kenya after two year period of crop production in 2016

Site plot	Macronutrients (g/kg)						Micronutrients (mg/kg)			Soil texture (%)		
	pH	N	P	K	Mg	Ca	Cu	Fe	Mn	Sand	Clay	Silt
Katumani	5.8	0.15	0.06	1.0	2.9	5.0	42.9	22.0	0.3	58	30	12
*Level	medium	low	low	ad	ad	ad	ad	ad	ad			
Kampi	5.5	0.12	0.02	0.8	3.4	4.2	31.3	182.0	0.4	61	23	16
*Level	medium	low	low	ad	hh	ad	ad	ad	ad			
Ithookwe	5.9	0.11	0.02	0.5	2.1	3.4	23.0	161.0	0.3	56	30	14
*Level	medium	low	low	ad	ad	ad	low	ad	ad			

Key: ad = adequate, hh = high: *Soil nutrients diagnosis given by National Agricultural Laboratories-Kenya Agricultural & Livestock Research Organization (KALRO)-April 2016.

3.2 Rainfall-Temperature Effects to Yield

Rainfall distribution within the production period indicated that Ithookwe received significantly ($P < 0.0001$) highest amount (247mm) followed by Kampi (134mm) (Table 2). The warmest environment was Kampi (25 °C) and the second being Ithookwe (22 °C) within a range of 21-25 °C at the three sites. The resultant yield level was significantly ($P < 0.05$) highest at Kampi at 25.6 t ha^{-1} while at Katumani and Ithookwe were 20.3 and 25.4 t ha^{-1}, respectively.

Table 2. Level of monthly rainfall and temperature fluctuations at three sites in 2015-2016 and resultant grain yield during 3-month development period of sorghum

Site plot	Monthly RF (mm) fluctuation	Monthly Temp (°C) fluctuation	Grain wt. (mg)	Yield Tons /Ha
Katumani	123 (14)^C	21 (6)^A	2.9 (0.4)^A	20.3 (3.1)^{AB}
Kampi	134 (16)^B	25 (5)^A	3.1 (0.2)^A	25.6 (2.6)^A
Ithookwe	247 (21)^A	22 (4)^A	3.6 (0.2)^A	25.4 (2.8)^B
F	576.9	0.2	1.6	3.74
P	< 0.0001	0.9164	0.3248	0.0453

Similar upper case superscript letters within columns of the four variables indicate no significant ($P > 0.05$) difference among the sites at 5% level (df=2, 8)-GLM PROC of SNK Test.

3.3 Grain Maturity and Viability

The results from the three sites showed that Kampi site had significantly ($P < 0.05$) highest level of seed viability (43%) at the light-green grain stage with Katumani and Ithookwe at 22 and 9% levels respectively (Table 3). Cream-dough stage at Kampi had significantly ($P < 0.05$) higher viability than the other sites. At hard dough grain stage, all the site fields indicated over 90% seed viability score.

Table 3. Sorghum grain viability (%) index measure (and standard deviation) from milk to mature ripening stage of crop at three production sites

Site plot	Light-green	Cream-dough	Hard dough
Katumani	22 (9)^B	70 (6)^B	96 (3)^A
Kampi	43 (7)^A	87 (6)^A	99 (1)^A
Ithookwe	9 (3)^C	48 (12)^C	95 (6)^A
F	6.2	55.8	1.4
P	0.0200	0.0004	0.4289

Different upper case superscript letters within columns at the three sorghum grain ripening stages indicate significant ($P < 0.05$) difference at the sites at 5% level (df=2, 8-GLM PROC of SNK Test) among the three treatments.

3.4 Grain Ripening and Bollworm Infestation

The vegetative stage to bloom took about 80 days where averagely at 83rd day the deep-green grain stage was attained (Fig. 1). Later by 87th day the grain turned light-green. Light-cream, full-cream, cream-white and full white seed colours were attained on 92, 97, 102 and 107 days, respectively at the sites. The dry seed maturity was attained averagely at 112 days since crop establishment at the three sites for 50% of the sorghum seed.

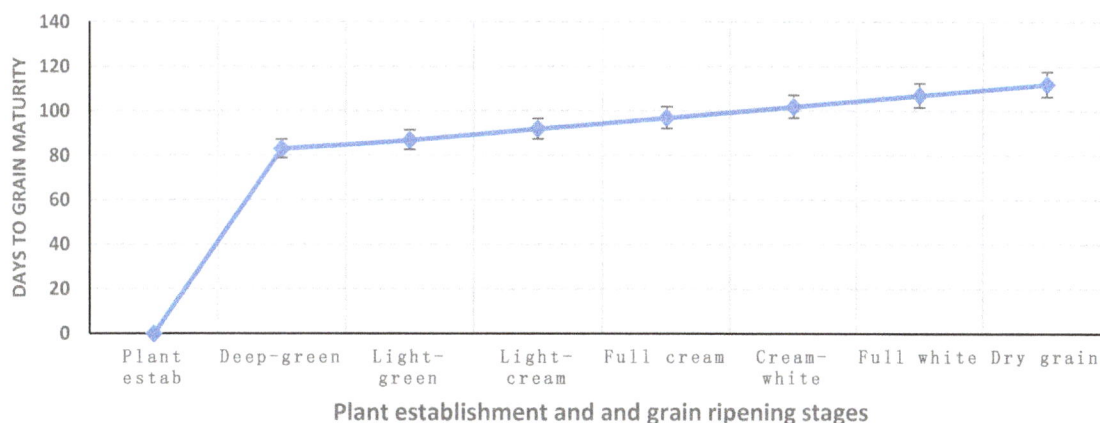

Figure 1. Mean number of days to full grain maturity stage of white sorghum crop

As the plant panicles progressed towards ripening, the insect *H. armigera* infestation levels peaked (R^2 = 0.5814) to 14 larvae per panicle at cream-white stage of grain with the insect feeding on the grain (Fig.2). At full-white and with decreased soft dough grain stage *H. armigera* infestations dropped to 8 larvae per plant panicle. By that time the larvae appeared to be on the fourth to sixth instar stage feeding on the grain increased. These larvae were found to be feeding on the grain and rendering the seed unviable on germination test. Our observation was that even a single larva could damage half of grains in one panicle by the time grain turned hard dough during a period of two to three weeks. The intercrop systems of millets did not show bollworm infestation difference on the plant ears. Bollworm larvae numbers did not differ between bird-fed plots and control plots.

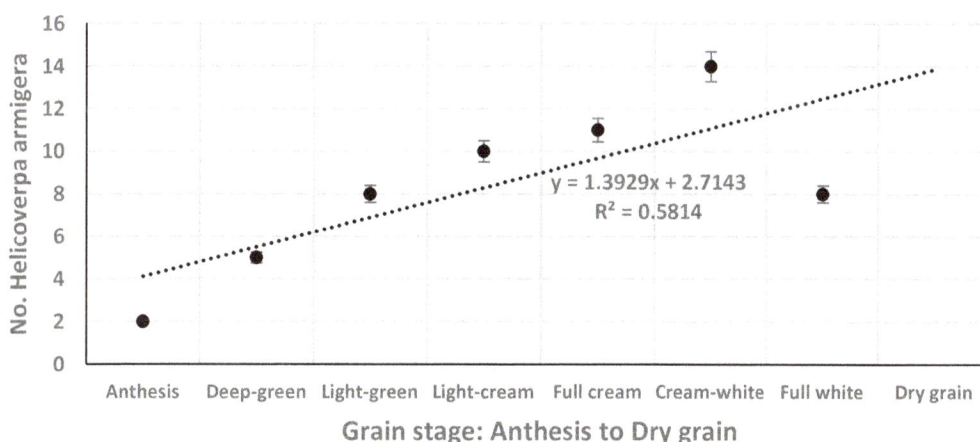

Figure 2. *Helicoverpa armigera* infestation on sorghum panicle towards full grain ripening stage

3.5 Grain Damage Accumulation

Sorghum grain damage was not visible from anthesis to deep-green with *H. armigera* density at 2-5 larvae. Grain damage visibility was from light-cream and full-cream grain stages at 5 and 10%, respectively (Fig. 3). The start of dry dough stage of grain at cream-white and full-cream indicated *H. armigera* density exponentially reaching 12-13 larvae per panicle and damage level being 35-42% (R^2 = 0.8474). Comparatively the 10 panicles randomly sampled from the control (no insecticide spray) indicated yield loss level of < 10% when grain was at early soft dough and > 35% as the grain ripened to early dry dough stage. This was observed to be the stage when much of the grain milk sap had turned into starch (dry sap).

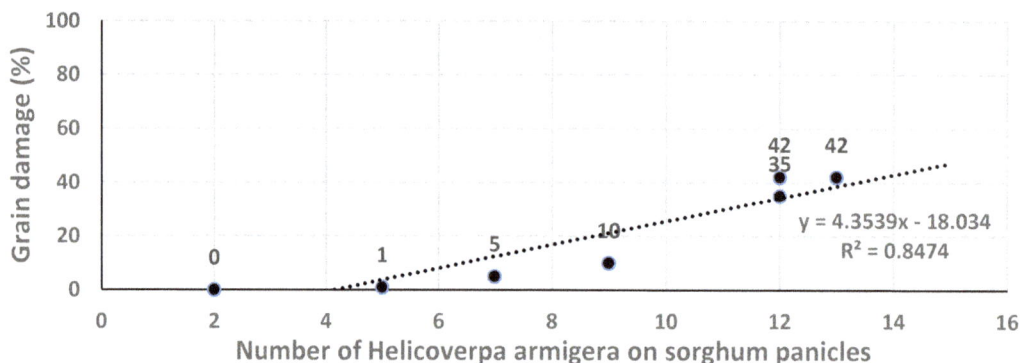

Figure 3. Damage accumulation towards grain ripening and bollworm increase on panicles of sorghum

3.6 Effect of Environmental Factors to Yield

Altitude was not significantly ($P > 0.05$; $R^2 = 0.0578$) correlated to yield as well as the macro-nutrient elements of Nitrogen, Phosphorous and Potassium (Table 4). Other factors like soil pH and temperature were strongly ($P < 0.05$; $R^2 > 0.6$) corrected to yield. The combined variables of rainfall (mm) and temperature demonstrated strong ($P < 0.05$; $R^2 = 0.9817$) correlation to yield. Likewise temperature-pH-rainfall combined variables indicated yet another strong correlation ($P < 0.05$; $R^2 > 0.90$) between yield and environmental factors. Nevertheless bollworm density was inversely correlated to yield of sorghum grain ($P < 0.05$; $R^2 = 0.9086$).

Table 4. Effects of various environmental variables to sorghum yield in Kenya eastern sites in 2015-2016

Variable	t	F	R^2	P
Altitude (m)	-0.50	0.25	0.0578	0.6464
N element	0.30	0.09	0.0215	0.7815
P element	-0.48	0.23	0.0552	0.6539
K element	0.49	0.24	0.0577	0.6466
pH	-2.98	8.88	0.6895	0.0407
Temp (°C)	2.23	4.97	0.5541	0.0897
RF (mm)	1.26	1.58	0.2832	0.2771
Temp (°C) x RF (mm)	9.55	80.68	0.9817	0.0024
Temp (°C) x RF (mm) x pH	0.29	69.31	0.9905	0.0143
Bollworm density	-6.31	39.70	0.9086	0.0032

Correlation results of above indicated independent single variable to yield at the sites (df = 1, 17) and two and three variables (df = 1, 23; df= 1, 29) Multiple Linear Regression Model (MLRM PROC) at 5% level.

4. Discussion

Results of soil analyses showed that the three sites had low inadequate macronutrients elements of nitrogen (N) and phosphorous (P) required for sorghum production. On the other hand potassium (K) was reported as adequate in the three site plots. The micronutrient element compositions showed that the levels of Cu, Fe, and Mn were fairy adequate for sorghum production in the test sites. The three fields were low in organic matter and hence the low fertility— consequently needing manure fertilization to improve organic nutrient levels for sorghum production (Balay et al., 2002; Fornara et al., 2008). Much soil fertility requirement in crop production has been disseminated in Kenya though various assessment studies have reported low impact adoptability of such technologies due to lack of capital and low-income among farming communities (Marenya & Barrett, 2007; Adolwa et al., 2013). Most countries in Africa suffer similar constraints of technology adoptability mainly due to little government policy on resource mobilization in the farming communities (Ajayi et al., 2007; Camara et al., 2013).

Rainfall amount was highest (247mm) at Ithookwe and the site had close similar yield (25.4 t /ha) to Kampi site (25.6 t/ ha). Gichangi et al. (2015) have demonstrated that both long and short rainfall regimes have remained average at 255 and 300mm respectively between years of 1961-2011 for eastern Kenya. The recorded levels during the present study slightly reflect above average rainfall regimes when considered cumulatively for three months production periods. Considering the low moisture requirement for sorghum the crop has low risk level for production in case of insufficient rainfall in the region. Bryan et al. (2013) have demonstrated that farmers need to weigh their options of crop enterprise with prevailing predicted rainfall patterns and plant the least risk ones like sorghum. As the climate variability continue occurring in most parts of Africa, evidently rainfall amounts are diminishing with each subsequent year and much hope for the farming communities lies with cereals like sorghum and millets and less with maize production (Tabo et al., 2007; Gichangi et al., 2015). The reason for the close similar yield levels among the two sites could be attributed to the closely related temperature regimes at Ithookwe and Kampi at 22 ± 4 °C and 25 ± 5 °C respectively.

Analyzed temperature regimes since 1978 to 2011 indicate increased temperatures of 1 °C and 0.3 °C for maximum and minimum levels in eastern Kenya (Gichangi et al., 2015). These temperature increase levels are not expected to affect sorghum in most areas in Africa as the crop requires moderate similar high temperature regimes. So long as rainfall is sufficient the predicted global temperature levels rise of 4 °C will likely lead to higher sorghum yields in most areas as long as moisture levels are within optimum (Christensen et al., 2007). Various predictions of what climate variability and finally the change in short and long term effects in different regions have mostly presented a bleak future to most country economies especially in SSA (Calzadilla et al., 2013). Probably for sorghum it might not lead to such despair as with other crops like maize and wheat whose yields have been reportedly reduced by such climate variability leading to low rainfall amounts (Yegbemey et al., 2013). Pretty et al. (2011) have indicated that African agriculture growth on most crop commodities will increase with increased intensification of production factors by maximum utilization of natural resources and technical knowledge on inputs and their applications.

The final physiological maturity of the harvested grain was found related to the environmental conditions of the sites. The warmest Kampi site achieved the highest seed viability on germination test at 43, 87 and 99% for grain stages of light-green, cream-dough and hard dough, respectively. Presence of bird-feeding plots did not alter infestation levels of the bollworm insect on the sorghum plants. The highest *H. arimigera* grain damage peaked exponentially at late grain soft dough (cream-white) and indicated this as the right time to spray against *H. armigera* on sorghum as pooled data from the sites showed. As for the economic injury level (EIL) we observed that even one larva of *H. armigera* could lead to serious grain damage before the grain reached hard dough stage at which no more feeding occurs. Martin et al. (2010) has recommended insecticide spray against *H. zea* (Boddie) in North America when an egg butch or several larvae are observed on cotton leaves. Our recommendation would be to spray immediately a single larvae is noticed on the panicles since such larva can live on the ear for over 10 days causing substantial damage as starch accumulation builds at late soft dough stage of grain (Czepak & Albernaz, 2013). The polyphagous pest is reported to be migratory aided by heavy winds at adult stage where moths fly searching for fresh fields and has a vast diversity of host plants in east Africa (Pedgley, 1985; Mutegi et al., 2010). Species of *Trichogramma* genus have been cited as some of the natural enemies of *H. armigera* but their presence in some areas is in doubt due to other advance environmental conditions not favouring survival of the wasp (Duraimurugan & Regupathy 2005). When *H. armigera* density levels reach economic injury levels other control mechanisms include use of botanical derivatives like neem or suitable chemical spray (Karim, 2000; Duraimurugan & Regupathy, 2005). In USA and Latin America countries various chemicals have been used and exchanged after either resistant development or for environmental reasons in addition to sexual pheromone trap use (Czepak & Albernaz, 2013). Depending on specific climatic factors like temperature and humidity, higher levels of these two could lead to higher density of *H. armigera* and hence need to monitor the pest presence frequently on sorghum ears. Sorghum fields in warm localities need insecticide spray control more frequent than cool areas as reported on cassava mite pest (Mutisya et al., 2015).

The present work has analyzed the existing factors which lead to attainment of yield levels of white sorghum, Gadam variety. The findings show that while low yield could be due to low fertility and or insufficient rainfall amounts at the production sites, other factors contribute significantly to sustainable yield of sorghum in the marginal areas. The correlation test showed that sufficient rainfall and temperature regimes led to highest yield of sorghum. To sustain the yield, control of *H. armigera* has to be carried out before dry dough grain stage, specifically at soft dough stage as the larvae start feeding on the ripening grains on the panicles. At farm level the farmer has to start with (i) adequate fertility on the production plot, (ii) some irrigation could be required where it is possible to sustain plant development and final yield in case of low rainfall amounts, and (iii) apply insecticide to suppress bollworm density growth early enough before economic damage occurs, otherwise temperature regimes are not expected to overshoot beyond the expected optimum levels.

Acknowledgement

We acknowledge the financial support from European Union (EU) presented by European Commission (EC) grant (CONTRACT NO.FED/2012/291-241) project name: Agricultural Productivity Research Project (APRP) and Government of Kenya (GoK) which enabled the timely carrying out of the evaluation activities at the three sites in the Counties of Machakos, Makueni and Kitui. Messrs.' Robert Mutweti, Daniel Kitheka and late Duncan Mutinda of KALRO Centers of Ithookwe, Kampi-Mawe and Katumani, respectively are acknowledged for their efforts to prepare the fields early enough for seasons' subsequent planting. Mr. Noah Mwangangi and Ms Isabel Wanza are acknowledge for yield processing and data collection. Meteorological persons serving at Kampi-Mawe (ICRISAT) Ithookwe- KALRO Sub-Centre and KALRO-Katumani in 2015-16 are acknowledged for providing climatic data of the sites.

References

Adolwa, I. S., Okoth, P. F., Mulwa, R. M., Esilaba, A. O., Mairura, F. S., & Nambiro, E. (2012). Analysis of communication and dissemination channels influencing the adoption of integrated soil fertility management in western Kenya. *The Journal of Agricultural Education and Extension, 18*(1), 71-86. http://dx.doi.org/10.1080/1389224X.2012.638782

Ajayi, O. C., Akinnifesi, F. K., Sileshi, G., & Chakeredza, S. (2007). Adoption of renewable soil fertility replenishment technologies in the southern African region: lessons learnt and the way forward. *Natural Resources Forum, 31*(4), 306-317. http://dx.doi.org/10.1111/j.1477-8947.2007.00163.x

Belay, A., Claassen, A. S., & Wehner, F. C. (2002). Soil nutrient content, microbial properties and maize yield under long-term legume-based crop rotation and fertilization: a comparison of residual effect of manure and NPK fertilizers. *South African of Plant Soil Journal, 19*, 104-110. http://dx.doi.org/10.1080/02571862.2002.10634447

Bryan, E., Ringler, E., Okoba, B., Roncoli, C., Silvestri, S. & Herrero, M. (2013). .Adapting agriculture to climate change in Kenya: household strategies and determinants. *Journal of Environmental Management, 114*, 26-35. http://dx.doi.org/10.1016/j.jenvman.2012.10.036

Calzadilla, A., Zhu, T., Rehdanz, K., Told, R. S. J., & Ringler, C. (2013). Economy-wide impacts of climate change on agriculture in Sub-Sahara Africa. *Ecological Economics, 93*, 150-165. http://dx.doi.org/10.1016/j.ecolecon.2013.05.006

Camara, B., Camara, F., Berthe, A., & Oswald, A. (2013). Micro-dosing of fertilizer-a technology for farmers' needs and resources. *International Journal of AgriScience, 3*(5), 387-399.

Christensen, J. H., Hewitson, B., Busuioc, A., Chen, X. G., Held, I., Jones, R., ... Busuoic, A. (2007). Regional Climate Projections, Chap. 11. In: Solomon S, Qin D, Manning M, Chen Z, Marquis M, Averyt KB, Tignor M and Miller HL (eds). Climate Change 2007-The Physical Science Basis: contribution of working Group I to the fourth assessment report of the intergovernmental panel on Climate Change, Cambridge University Press, Cambridge, UK, pp 849-940, http://www.ipcc.ch/publications and data/ar4/wgl/en/ch11.htl.

Czepak, C., & Albernaz, K. C. (2013). First reported occurrence of *Helicoverpa armigera* Hübner (Lepidoptera: Noctuidae) in Brazil. *Pesquisa Agropecuaria Tropical Goiânia, 43*(1), 110-113. http://dx.doi.org/10.1590/S1983-40632013000100015

Duraimurugan, P., & Regupathy, A. (2005). Mitigation of insecticide resistance in *Helicoverpa armigera* (Hübner) (Lepidoptera: Noctuidae) by conjunctive use of trap crops, neem and *Trichogramma chilonis ishii* in cotton. *International Journal of Zoological Research, 1*(1), 53-58. http://dx.doi.org/10.3923/ijzr.2005.53.58

Fitt, G. P. (1989). The ecology of *Heliothis* species in relation to agro ecosystems. *Annual Review of Entomology, 34*(1), 17-52. http://dx.doi.org/10.1146/annurev.en.34.010189.000313

Food and Agriculture Organization (FAO) (2016). Sorghum production ranking. FAO Corporate Repository. www.fao.org/docrep/t08189/T0818EO3.htl

Fornara, D. A., & Tilman, D. (2008). Plant functional composition influences rates of soil Carbon and Nitrogen accumulation. *Journal of Ecology, 96*, 314-322. http://dx.doi.org/10.1111/j.1365-2745.2007.01345.x

Gichangi, E. M., Gatheru, M., Njiru, E. N., Mungube, E. O., Wambua, J. M., &Wamuongo. J. W. (2015).

Assessment of climate variability and change in semi-arid eastern Kenya, *Climate Change, 130*, 287-297. http://dx.doi.org/10.1007/s10584-015-1341-2

Gitz, III D. C., Baker, J. T., Xin, Z., Lascano, R. J., Burke, J. J., & Duke, S. E. (2013). Bird resistant pollination bags for sorghum breeding and germplasm maintenance. *American Journal of Plant Sciences, 4*, 571-574. http://dx.doi.org/10.4236/ajps.2013.43074

Guo, Y. Y. (1997). Progress in the researches on migratory regularity of *Helicoverpa armigera* and relationships between the pests and its host plants. *Acta Entomologica sinica, 4*, 1-6.

Hirron, M., Reuben, P., Mweresa, C. K., Ajamma, Y. O. U., Owino, E. A., & Low, M. (2014). Crop damage despite protection efforts by human bird scares in a sorghum field in western Kenya. *Journal of African Ornithology, 85*, 153-159. http://dx.doi.org/10.2989/00306525.2014.937368

Karim, S. (2000). Management of *Helicoverpa armigera*: a review and prospectus for Pakistan. *Pakistan Journal of Biological Sciences, 3*(8), 1213-1222. http://dx.doi.org/10.3923/pjbs.2000.1213.1222

Kilambya, D., & Witwer, M. (2013). Analysis of incentives and disincentives for sorghum in Kenya. Technical Note No. 17. MAFAP, FAO, Rome.

Kumar, S., Sain, S. K., & Ram, P. (2009). Natural mortality of *Helicoverpa armigera* (Hubner) eggs in the cotton ecosystem. *Journal of Agricultural Science and Technology, 11*(1), 17-25.

Marenya, P. P., & Barrett, C. B. (2007). Household-level determinants of adoption of improved natural resources management practices among smallholder farmers in western Kenya. *Food Policy, 32*(4), 515-536. http://dx.doi.org/10.1016/j.foodpol.2006.10.002

Martin, D. E., Lopez Jr., J. D., Lan, Y., Bradley, K., Hoffmann, C., & Duke, S. E. (2010). Novaluron as an ovicide for bollworm on cotton; disposition and efficacy of field-scale aerial application. *Journal of Cotton Science, 14*, 99-106.

Mastersa, W. A., Bedigarb, T., & Oehmke, J. F. (1998). The impact of agricultural research in Africa- aggregate and case study evidence. *Agricultural Economics Journal, 19*, 81-86. http://dx.doi.org/10.1016/S0169-5150(98)00023-1

Moral-Garcia, F. J. (2006). Analysis of the spatiotemporal distribution of *Helicoverpa armigera* (Hubner) in a tomato field using a stochastic approach. *Biosystemic Engineering, 93*(3), 253-259. http://dx.doi.org/10.1016/j.biosystemseng.2005.12.011

Mutegi, E., Sagnard, F., Muraya, M., Kanyenji, B., Rono, B., ... Labuschagne, M. (2010). Eco-geographical distribution of wild, weedy and cultivated *Sorghum bicolor* (L.) Moench in Kenya: implications for conservation and crop-to-wild gene flow. *Genetic Resource Crop Evolution, 57*, 243-253. http://dx.doi.org/10.1007/s10722-009-9466-7

Mutisya, D. L., El Banhawy, E. M., Khamala, C. P. M., & Kariuki, C. W. (2015). Management of cassava green mite *Mononychellus progresivus* (Acari: Tetranychidae) in different agro-ecological zones of Kenya. *Systematic & Applied Acarology, 20*(1), 39-50. http://dx.doi.org/10.11158/saa.20.1.5

Muui, C. W., Muasya, R. M., & Kirubi, D. T. (2013). Baseline survey on factors affecting sorghum production and eastern Kenya. *Journal of Food Nutrition Development, 13*(1), 7339-7353.

Ndjeunga, J., & Bantilan, M. C. S. (2005). Uptake of improved technologies in semi and tropics of West Africa. Why are agricultural transformations lagging behind? *Electronic Journal of Agricultural Economics, 2*(1), 85-102.

Oehmke, J. F., & Crawford, E. W. (1996).The impact of agricultural technology in Sub-Saharan Africa. *Journal of African Economics, 5*, 271-92. http://dx.doi.org/10.1093/oxfordjournals.jae.a020905

Orr, A., Mwema, W., & Mulinge, W. (2013).The value chains for sorghum beer in Kenya. International Crops Research Institute for the Semi-Arid Tropics (ICRISAT)-Series Paper No.16.

Pedgley, D. E. (1985). Windborne migration of *Heliothis armigera* (Hubner) (Lepidoptera: Noctuidae) to the British Isles. *Entomologists Gazette, 36*(1), 15-20.

Pogue, M. G. (2004). A new synonym of *Helicoverpa zea* (Boddie) and differentiation of adult males of *H. zea* and *H. armigera* (Hubner) (Lepidoptera: Noctuidae, Heliothinae). *Annual Entomological Society of America, 97*(6), 1222-1226. http://dx.doi.org/10.1603/0013-8746(2004)097[1222:ANSOHZ]2.0.CO;2

Pretty, J., Toulmin, C., & Williams, S. (2011). Sustainable intensification in African agriculture. *International*

Journal of Agricultural Sustainability, 9(1), *5-24.* http://dx.doi.org/10.3763/ijas.2010

Ratnavathi, C. V., & Patil, J. V. (2013). Sorghum utilization as food. *Journal of Nutrition and Food Science, 4,* 247. http://dox.doi.org/10.4172./2155-9600.1000247

Tabo, R., Bationo, A., Gerard, B., Ndjeunga, J., Marchal, D., Amadou, B., & Koala, S. (2007). Improving cereal productivity and farmers' income using a strategic application of fertilizers in West Africa. In A. Bationo, B. Waswa, J. Kihara & J. Kimetu (Eds.), *Advances in Integrated Soil Fertility Management in Sub-Saharan Africa: Challenges and Opportunities* (pp. 201-208) Springer Netherlands. http://dx.doi.org/10.1007/978-1-4020-5760-1_18

Van den Berg, J., & van Rensburg, J. B. J. (1991). Infestation and injury levels of system borers in relation to yield potential of grain sorghum. *South African Journal of Plant and Soil, 8*(3), 127-131. http://dox.doi.org/10.1080/0257/1862.1991.10634819

Van Wajk, J., & Kwak, K. (2011). Beer multinationals supporting Africa's development? How partnership include smallholder into sorghum-beer supply chains. In: Van Dijk MP and Trienekens J (eds). Promoting sustainable value chains: the role of governance. Amsterdam, Amsterdam University Press, pp 71-88.

Yegbemey, R. N., Yabil, J. A., Tovignan, S. D., Gantoli, G., & Kokoye, S. H. E. (2013). Farmers' decisions to adapt to climate change under various property rights: a case study of maize farming in Northern Benin (West Africa). *Land Use Policy, 34,* 168-175. http://dx.doi.org/10.1016/j.landusepol.2013.03.001

4

Structure of Cocoa Based Vegetable Seed System for Selected Locale in Ghana

Jonas Osei-Adu[1], Offei Bonsu[1], Seth Obosu Ekyem[1], Victor Afari-Sefa[2] & Micheal Kwabena Osei

[1]CSIR-Crops Research Institute, Ghana

[2]World Vegetable Center Eastern and Southern Africa, P. O. Box 10 Duluti Arusha, Tanzania

Correspondence: Jonas Osei-Adu, CSIR-Crops Research Institute, Ghana. E-mail: aduseiz@yahoo.co.uk

Abstract

The vegetable seed industry in Ghana is still at its formative stages. Farmer access to quality improved seed is still a daunting challenge. As a response, very few improved vegetable lines have been evaluated and tested in the country for dissemination to farmers. Using multistage sampling, a total of 137 vegetable farmers in the Offinso South municipal of the Ashanti region of Ghana were interviewed using structured questionnaires to characterize vegetable seed supply and distribution system. Results from the study indicated 45.3% of respondents acquired seed from commercial seed growers. Farmer saved seed accounted for 37.2% of sampled respondents while 32.1% of respondents sourced seeds from other farmers. The role of the formal seed system through private seed companies was minimal (10.2%). Only 10.9% of respodents treated their seeds before storage with 38.7% of respondents doing so prior to planting. This led to 23% of seed loss in storage with some farmers losing as much as 100%. The development of a vibrant vegetable seed system will require strong actor linkages within the seed supply chain to identify solutions to critical bottlenecks. An enabling policy environment for establishing dynamic and operational private seed companies, is a critical determinant of success in targeted farming communities. Provision of cold room facilities will also be necessary to ensure seeds are well stored.

Keywords: cocoa-based farming systems, crop diversification, vegetable seed, seed policy, Offinso South

1. Introduction

Vegetables are increasingly becoming an important commodity for both domestic and export markets. They have great potential to improve nutrition and health of consumers as most of them, particularly traditional species are good sources of vitamins, minerals and proteins needed for the proper functioning and development of the human body (Wills *et al.,* 1998). Despite the nutritional importance of vegetables and their high farm gate values per unit area for generating employment and income (Weiberger and Lumpkin, 2007), the sub sector is faced with a myriad of challenges. The importance of seed to any crop-based production system cannot be overemphasized since it is the fundamental unit of any production system (Etwire et al; 2013). High-quality seed in particular is fundamental to enhancing agricultural productivity, increasing food security and improving rural livelihoods (Kuhlmann &Zhou, 2016).

Access to quality seed is one of the major production challenges in Ghana. The seed industry in Ghana is still at its formative stage with access to improved seed constituting a huge challenge for many smallholder producers. Very few improved vegetable lines have been evaluated and tested in the country for dissemination to farmers. Seed companies are also limited and spatially located far from most producers, especially in rural communities. Weak linkages exist between vegetable producers and actors of the supply chain preventing access to quality seed for production. Information on the seed industry is limited which affects investments to the industry. This paper therefore provides an insight to the local vegetable seed system in terms of its dynamics and how to strengthen it for improved productivity. Critical attention is given to seed source, source of information, use of certified seeds, practice of seed storage and testing.

2. Materials, Area Description and Techniques

This study was undertaken in the Offinso South municipal district in the Ashanti region of Ghana. The district lies between longitudes 10' 50 W and 10 '45 E and latitudes 70' 20 N and 60 '50 S. The total land area is about

741 kilometers square. The municipal shares boundaries with the Offinso North District Assembly in the North, Ahafo Ano South District Assembly in the West and Afigya Skyere District Assembly in the East, Atiwma Nwabiagya District Assembly and Kawbre District Assembly in the South. Using multistage sampling technique, 137 vegetable producers who cultivate various crops including cabbage, okra, tomatoes, garden eggs and leafy vegetables within a predominantly cocoa based system were randomly sampled for the study. The first stage was the purposive selection of Offinso South Municipal where the AVRDC cocoa based vegetable project was been implemented. The second stage was the purposive sampling of communities involve in the project with the last stage been the random selection of farmers from project communities. Sampled farmers were interviewed with the aid of a structured questionnaire. Data generated included source of seed, practice of seed treatment, seed storage, sources of information on seed and role of actors in the seed supply chain. Data was analysed using descriptive statistics and inferential graphs.

3. Results and Discussions

The Ghana seed system is a hybrid of the formal and informal system with scanty information (Etwire et al., 2013). Two main segments of the system are the formal and informal seed systems. The formal seed system is well structured and coordinated by governmental agencies such as the research institutions and the Ministry of Food and Agriculture (MoFA). The Plant Protection and Regulatory Directorate of MoFA is the responsible directorate for monitoring and certifying seed development and multiplication. It does this through its seed inspection and certification division.

Breeding and variety development in the country is carried out by research institutions under the Council for Scientific and Industrial Research (CSIR) in collaboration with other international partners. The Universities in the country also develop and test varieties leading to seed development. New variety development requires the institutions with the requisite mandate to go through a vigorous process of gene identification, crosses and or use of biotechnology, setting up of on-station and on farm trials and several years of multi location trials. Economic analysis is carried out prior to the release of a variety to ascertain the economic viability or otherwise to farmers in reducing either cost of production or enhanced yields for higher incomes.

Multiplication and distribution of seed is through the formal and informal seed system. With regards to the formal system, research institutions and the universities develop the breeder and foundation seeds. Breeder seeds are passed on to MoFA for multiplication and production of foundation seed. Foundation seed gets to seed companies and seed producers to produce certified seed. Farmers access certified seed from seed companies and seed producers.

Due to the weak linkages and poor financing of the formal seed system, the role of the informal seed system cannot be over emphasized. Farmers often recycle their own seeds from their harvest. Use of farmer saved seed typically leads to high incidence of pest and diseases thereby affecting the quality and quantity of yield.

The vegetable seed system is not different from the main seed system in the country. It is however, weak in terms of seed supply for crops such as tomatoes, garden eggs, okra and others. However, pepper seeds can be found relatively easier in some agricultural input shops. The informal seed system is the main routine for vegetable seed distribution in the country. Vegetable Seed marketing and quality assurance is relatively weak and developing.

The source of seed for vegetable production like any other crop is very critical to ensure good quality seedlings and optimal growth in the field after transplanting. Seed viability has direct effect on yield performance and thus ensuring use of quality seed is very essential. The result indicates that 45.3% of respondents acquired their vegetable seeds from private seed growers who were in the communities or elsewhere (Figure 1). This is in consonant with Etwire et al. 2013 who found that 88.2% of Ghanaian farmers generally obtained seeds through agro input dealers. What was difficult to determine was the quality of seed been offered to these farmers. Farmer saved seed was also another important source (37.2%) as well as sourcing of seed from other farmers (32.1%). This confirms the fact that majority (over 80%) of smallholder farmers in Africa mainly obtain their seeds from informal channels which include farmers' own saved seeds, seed exchanges among farmers and finally purchases from the local seed markets (Rajendran et. al., 2016; Louwaars and De Boef, 2012; Maredia et al., 1999; Crissman et al., 1993).

Development of a seed system requires diverse complementary supply channels of distribution so as to strengthen supply points and distribution outlets. The role of private seed companies was found to be minimal as depicted by 10.2% of respondents sourcing seed from them. This was due to the spatial, time, value and information gaps that exist between them and producers.

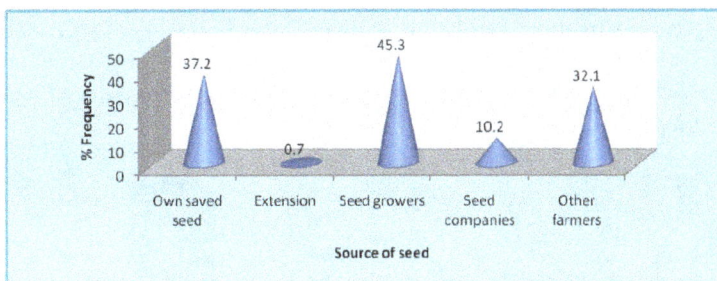

Figure 1. Source of seed

Source: Field Survey, 2015

Seed quality is an important determinant of seed germination. As indicated, most farmers sourced their seeds from sources for which it was difficult to ascertain the quality. About 49% (Figure 2) of respondents claimed they were able to determine seed quality. They could however, only assess seed quality in the field through what is often referred to as "neighbour certification" (Lyon andAfikorah-Danquah, 1998). Neighbour certification is the physical observation of the vegetative performance of a crop as a proxy for seed quality. This method is not scientifically robust and is done after planting reducing the probability of detecting poor quality seeds before planting. Building capacity of farmers in seed testing will be an important means to strengthen the current seed supply and distribution system.

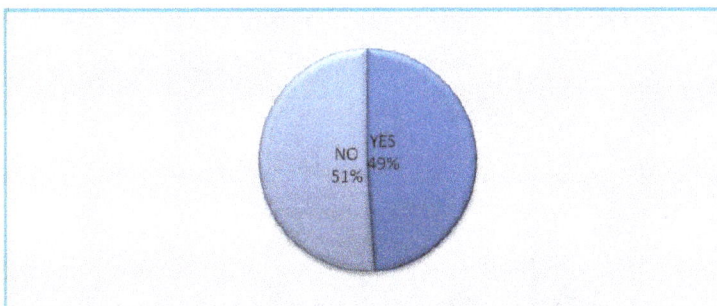

Figure 2. Ability to determine seed quality

Source: Field Survey, 2015

One of the biggest problems confronting vegetable farmers in Ghana is the high incidence of diseases and pests which ravage their crops (Asare & Micah, 2014). To prevent rodent attacks and other insects, seed treatment is very critical in ensuring the quality of seeds stored is not compromised during storage and before planting. However, respondents in the study area gave little attention to seed treatment. Only 10.9% treated their seeds before storage and 38.7% (Figure 3) before planting. This result is similar to findings by Tripp et al; (1998) who indicated the mean germination rate for cowpea farmers in Wenchi was 65.7% and 70.3% for Akatsi lower than the Ministry of Food and Agriculture's acceptable standard of 75%. The success of a viabrant seed system in the area will therefore require capacity building in seed treatment.

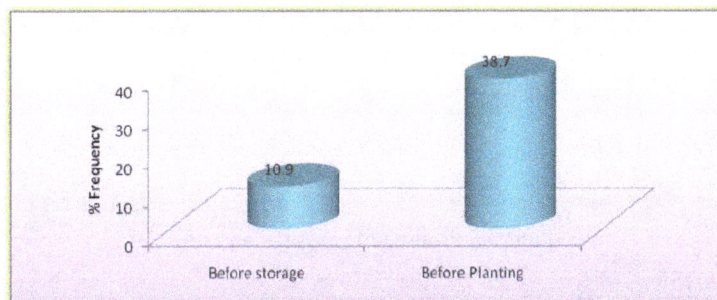

Figure 3. Seed treatment

Source: Field Survey, 2015

With only 10.9% of respondents treating seed before storage, 23% (Figure 4) of seed stored was lost in stoarge with some farmers lossing 100%. This affects the amount of seed available for planting and reduces the farmers ability to cultivate larger acreage. The end result is reduced output leading to reduced income and its effects on food and nutritional security.

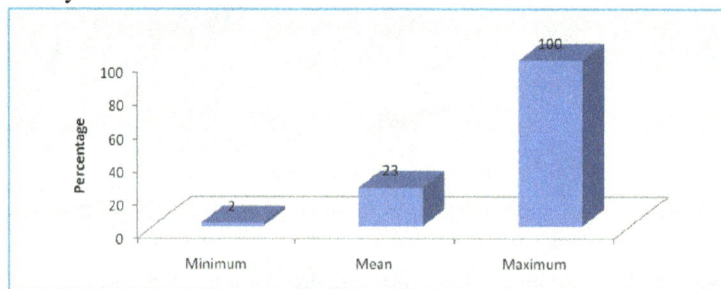

Figure 4. Percentage waste in storage

Source: Field Survey, 2015

To avoid rodent and other insect attack, the use of botanical extracts like neem is highly recommended. This is due to its availability and cost effectiveness. Its use in the study area was however, very limited such that only 2% (Figure 5) of respondnets indicated the use botentical extrct for seed treatment. Capacity building on the preparartion and use of these extracts will therefore be very important in stregthen the seed system in the area.

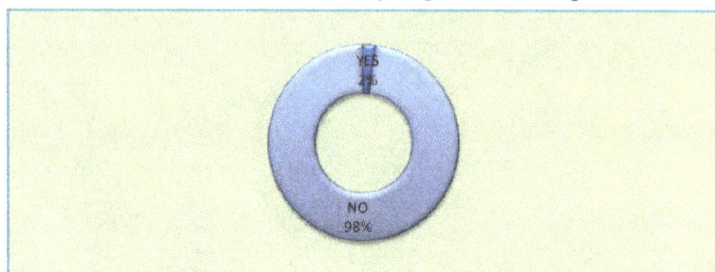

Figure 5. Use of Botanical extract for seed treatment

Source: Field Survey, 2015

Farmers within the study area receive technical information about seed from different sources . The role of farmer to farmer dissemination was confirmed in terms of serving as a major source of information on seed. About 68% (Figure 6) of respondents had information on seed from other farmers. This was followed with 16% from Agricultural extension and 6% from electronic media. Improved seed production technology dissemination should therefore focus on farmer to farmer approaches with an intergration of electronic media. Electronic media has the advantage of wide coverage and exploring its use will help reach a larger audience.

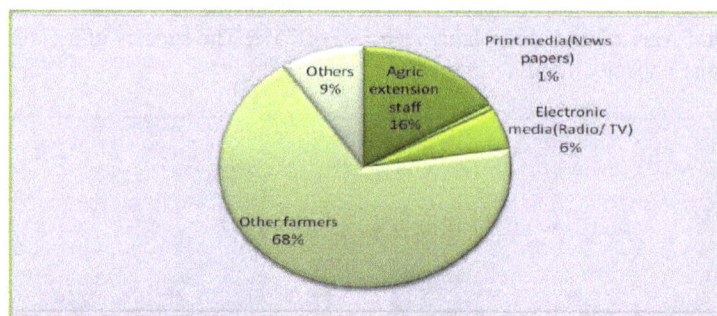

Figure 6. Source of message on seed

Source: Feild Survey, 2015

The role of farmer to farmer technology dissemination is still relevant in current production systems. Findings of this study show that 53.3% of respondents have a strong association with other farmers. About 40.1% and 38.7% (Table 1) also indicated their interection with extension services and land onwers respectively. Strengthening the seed system will therefore required close collabration with lead farmers in the communities, extension services and land owners for effective technology disseminationa and adoption.

Table 1. Association with supply chain actors

Actors of the seed chain	% Frequency
Seed grower/companies	26.3
Other farmers	53.3
Extension services	40.1
Financial institutions	21.2
Land owners	38.7
Water supplies	0.7
Spraying gangs	5.1

Source: Field Survey, 2015

The importance of any association is the benefits that accrue from that association. From the study, four major benefits accrue from association of respondents with supply chain actors namely; sale of seed, training, land for production and access to credit. Benefits in terms of association from other farmers was mainly on the sale of seed (32.8%). Extension supported respondents with training (29.2%) and financial institutions with credit (16.1%).

Table 2. Type of association with seed supply chain actors

	Sale of seed	Training	Credit	Land for production	Spraying services	Control of Chemicals
Seed grower/companies	24.8	5.1				
Other Farmers	32.8	6.6	5.8	2.2	2.2	
Extension services	1.5	29.2	0.7	0.7		2.9
Financial institutions		2.9	16.1			
Land owners				26.3		

Source: Field Survey, 2015

Seed certification is another important parameter in ensuring seed quality. With some respondents claiming they are able to determine seed quality, it was important to find out whether they have access to certified seed. From the study, 39% (Figure 7) had access to certified seed. This creates a high deficit for which the market potential exists for certified seed production and supply.

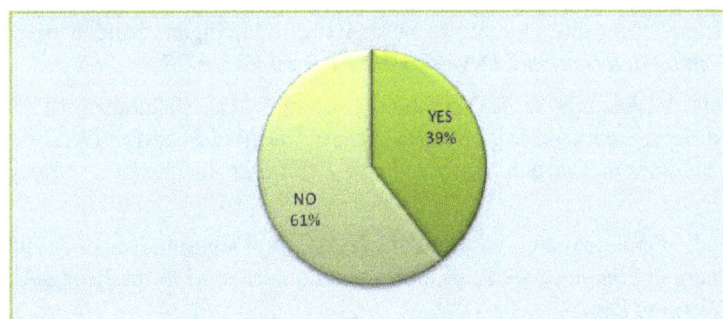

Figure 7. Use of Certified Seed

Source: Field Survey, 2015

4. Conclusion and Recommendations

The vegeble seed supply and distribution system in the study area is at the rudimentary stages of developmment and requires a lot of support to become vibrant to meet growing demands of smallholders. Farmers still relied on farmer saved seed (37.2%) and seed from other farmers (32.1). Acces to technical information was heavily dependent on collegue farmers (68%). Seed distribution channels were poorly linked with limited access to financial credit. Seed viability cannot be assured since very few farmers had access to certified seed (39%) and had no scientific way of deremining viability. Poor supply chain actor linkges was a constraint to developing a vibrant seed system. The low level of seed treatment before storage (10.9%) and before planting (38.7%) led to huge losses (23%) affecting quality and quantity for planting. The rate of gemination is also affected resulting in low yields. It is therefore important to find pertinent solutions to constraints affecting the vegetable seed systemto increase vegetable crop productivity. The following are therefore recommended;

- Building capacity of farmers in seed germination test and treatment.
- Development of strong supply chain actor linkages using innovation platform approaches that encourage private sector involvement
- Create enabling policy environment for the establishment of a vibrant formal seed system to enhance widespread access to quality seed.

Acknowledgments

We would like to acknowledge Humidtropics and the CGIAR Fund Donors for their provision of core and project-specific funding through the World Vegetable Center without which this research could not deliver results that eventually positively impact the lives of millions of smallholder farmers in tropical Americas, Asia and Africa.. We are also grateful to the MoFA extension agents, technical staff of the CSIR-Crops Research Institute (Horticulture and Socio economics sections), Ghana. Our drivers and others who directly or indirectly contributed to the survey are also acknowledged.

References

Asare, B. E., & John, A. M. (2014). *Vegetable Crop Protection Practices and Policy Related Issues in the Rural and Peri-Urban Areas of the Cocoa Belts of the Ashanti and Western Regions of Ghana.* Draft Consultant Report Submitted to the World Vegetable Centre

Crissman, C. C., Crissman, L. M., & Carli, C. (1993). *Seed Potato Systems in Africa: A Case Study.* Lima, International Potato Center.

Etwire, P. M., Ibrahim, D. K., Atokple, S. S. J. B., Alhassan, L. A., Afia, S. K., & Peter, A. (2013). Analysis of the seed system in Ghana. *International Journal of Advance Agricultural Research.* 7-13

Kuhlmann, K., & Yuan, Zh. (2016). Seed Policy Harmonization in ECOWAS: The Case of Ghana.Syngenta Foundation for Sustainable Agriculture working paper

Louwaars, N. P., & De Boef, W. S. (2012). Integrated seed sector development in Africa: A conceptual framework for creating coherence between practices, *Programs, and policies. J. Crop Imp. 26*, 39-59. http://dx.doi.org/10.1080/15427528.2011.611277

Lyon, F., & Seth, A. D. (1998). Small-scale seed provision in Ghana: social relations, contracts and institutions formicro-enterprise development. *Agricultural Research & Extension Network paper, 84.*

Maredia, M., Howard, J., Boughton, D., Naseen, A., Wanzala, M., & Kajisa, K. (1999). Increasing Seed System Efficiency in Africa: Concepts, Strategies and Issues. Michigan State University. *Department of Agricultural Economics- International Development Working Paper, 77.*

Rajendran, S., Afari-Sefa, V., Karanja, D. K, Musebe, R., Romney, D., Makaranga, M., Samali S, & Kessy, R. (2016). Farmer-led Seed Enterprise Initiatives to Access Certified Seed for Traditional African Vegetables and its Effect on Incomes in Tanzania. *International Food and Agribusiness Management Review, 19*(1), 1-24.

Tripp, R., Walker, D. J., Opoku-Apau, A., Dankyi, A. A., & Delimini, L. L. (1998). *Seed management bysmall-scale farmers in Ghana.* A study of maize and cowpea seed in the Brong-Ahafo and Voltaregions. NRI Bulletin 68, Working Paper

Weinberger, K., & Lumpkin, T. A. (2007). Diversification into horticulture and poverty reduction: A research agenda. *World Development, 35*(1), 1464-1480. http://dx.doi.org/10.1016/j.worlddev.2007.05.002

5

Improving French Bean (*Phaseolus Vulgaris* L.) Pod Yield and Quality Through the Use of Different Coloured Agronet Covers

Munywoki James Ngelenzi[1], Saidi Mwanarusi[1] & Ogweno Joshua Otieno[1]

[1]Department of Crops, Horticulture and Soils, Egerton University, P.O Box 536-20115 Egerton, Kenya

Correspondence:Saidi Mwanarusi, Department of Crops, Horticulture and Soils, Egerton University, P.O Box 536-20115 Egerton, Kenya. E-mail: msaidi@egerton.ac.ke

Abstract

French bean (*Phaseolus vulgaris* L.) is among important vegetables in supplying proteins, vitamins, minerals and dietary fiber to humans worldwide. Its successful production in the tropics is, however, constrained by abiotic and biotic stresses as the crop is predominantly grown in open fields. Netting technology has been proved successful in protecting crops against adverse weather and insect pests. Coloured net technology is an emerging technology, which introduces additional benefits on top of the various protective functions of nettings. Two trials were conducted at the Horticulture Research and Teaching Field, Egerton University, Kenya to evaluate the effects of different coloured agronet covers on growth, pod yield and quality of French bean. A randomized complete block design (RCBD) with six treatments and four replications was used. French bean plants were grown under a white, blue, yellow, tricolour or grey net cover with open field production as the control. Variables measured included days to emergence and emergence percentage (%), stem collar diameter, plant height, number of branches and internodes, internode length and crop yield. French bean grown under the different coloured net covers showed relatively better growth and crop performance marked by more pods and higher total yields and percentage of marketable yields compared to those grown in the open field. Growing French bean under net covers hastened the rate of pod maturation more-so under the light-coloured colour-nets. Findings of this study demonstrate the potential of coloured net covers in improving French bean pod yield and quality under tropical field conditions.

Keywords: Snap bean, Green bean, Modified environment, yield, pod quality

1. Introduction

French bean (*Phaseolus vulgaris* L.) is grown by both large and small holder farmers in East Africa and Africa in general (Center for International Agricultural Technology [CIAT], 2006). Total world production exceeds 17 million tonnes, with China, Indonesia, Turkey, India and Egypt among the largest producers and consumers of the crop (Food and Agriculture Organization Corporate Statistical Database [FAOSTAT], 2010). The crop is a major export vegetable in Kenya and ranks second to cut flowers in terms of foreign exchange earnings generated from the dynamic horticultural sub-sector (Horticultural Crops Development Authority [HCDA], 2011). Much of Kenyan French bean is intensively grown by small scale farmers (Ndegwa *et al.*, 2010) mainly for fresh export and as a source of family income (Monda, Ndegwa, & Munene, 2003). French bean is an important crop in the social economic systems and livelihoods of many Kenyans supporting people directly or indirectly (Odero, Mburu, Ogutu, & Nderitu, 2012). Moreover, the immature pods are an important food source in many parts of the world supplying protein (Arulbalachandran & Mullainathan, 2009), vitamins, minerals and dietary fibre (Kelly & Scott, 1992). The crop is also important for its nitrogen fixing capability (Amannuel, Kiihne, Tanner, & Vlek, 2000) and can be used in crop rotation systems to improve soil conditions.

Despite the economic importance and potential of French bean in addressing food insecurity, improving incomes and alleviating poverty, low yields remain a common scenario with most growers in many sub-saharan countries. As low as 6 to 8 tonnes per hectare of French bean have been realised in Kenya (Wahome, Kimani, Muthomi, & Narla, 2013), compared to 15 to 20 tonnes per hectare recorded in developing countries in South America and south East Asia (Wahome *et al.*, 2013). Throughout its production cycle, French bean is subject to many abiotic and biotic stresses including erratic rainfall, prolonged dry spells and heavy infestation by insect pest, which present major challenges to French bean growers particularly in the tropics as the crop is predominantly grown in open fields. French bean yield losses due to insect pests alone have been estimated to range from 35% to 100% annually (Singh & Schwartz, 2011) while drought stress can cause more than 50% yield loss (Zlatev & Lidon, 2012).

A number of simple technologies have been tested in different parts of the world and proved successful in protecting crops against adverse weather conditions and insect pests. Netting technology has been used in agriculture to protect crops against environmental hazards like excessive solar radiation, wind, hail and flying insects and improve plant microclimate through reduction in heat/chill, drought stresses, and moderation of rapid climatic stresses leading to improved crop yield and quality (Shahak, Gussakovsky, Cohen, & Lurie, 2004). The use of net covers in crop production offers a cheaper and less energy consuming technology than greenhouses (Shahak, 2008). Coloured net technology is on the other hand, an emerging technology, which introduces additional benefits on top of the various protective functions of nettings. Photoselective nets which include the coloured nets are unique in that they both spectrally-modify as well as scatter the transmitted light, absorbing spectral bands shorter or longer than the visible range. Spectral manipulation has a potential for promoting physiological responses in plants while the scattering of light improves penetration into the inner canopy, all of which contribute towards better crop performance (Rajapakse & Shahak, 2007). Different crops respond differently to the spectral manipulation induced by different colours of net covers. Evaluations to identify ideal net colour(s) that maximize French bean pod yield and quality are therefore imperative for better exploitation of the potential of the crop as a food and income source for the rural populations. This study therefore aimed at determining the effects of different coloured agronet covers on growth, pod yield and quality of French bean.

2. Materials and Methods

2.1 Experimental Site Description

The study was conducted at the Horticulture Research and Teaching Field of Egerton University, Njoro, Kenya. The field lies at latitude of 0°23′ S and longitude 35°35′ E in the Lower Highland III Agro Ecological Zone (LH3) at an altitude of ≈2238 m above sea level. The soils are predominantly vitric mollic andosols with a pH of 6.0 to 6.5 (Jaetzold & Schmidt, 2006). The site mean temperature and rainfall during the study period are presented in Table 1.

Table 1. Average monthly air temperature (°C) and precipitation (mm) during French bean production over the two trials (July to Oct. 2015 and Nov. 2015 to Feb. 2016)

	Trial 1				Trial 2			
	July	August	September	October	November	December	January	February
Air temperature	20.2	20.8	21.7	21.2	19.1	20.2	20.6	22.0
Precipitation	53.4	41.6	85.7	90.31	198.8	82.9	86.6	23.4

Source: Egerton University Engineering Department (2016)

2.2 Planting Material, Experimental Design, and Treatments

French bean seeds of cultivar Source [Amiran (K) Ltd., Nairobi, Kenya] were used. 'Source' is a determinate variety and one of the most popular among the French bean growers in Kenya. The experiment was laid out in a Randomized Complete Block Design (RCBD) with six treatments replicated four times. The treatments comprised of growing French bean under white net, blue net, yellow net, tricolour net, grey net, and open field (control). The tricolour net was predominantly white in colour with blue and yellow stripes. The agronets used are made from high-density polyethylene fully recyclable monofilament of 100 denier knitted into a mesh with 0.9 mm x 0.7 mm average pore size. They are ultraviolet protected for extended shelf life. They were obtained from A to Z Textile Mills Ltd., Arusha, Tanzania.

Each trial covered an area of 20.5 m by 11 m with individual blocks measuring 20.5 m by 2 m separated by a 1 meter buffer. Individual experimental units within a block measured 3 m by 2 m with an inter-plot spacing of 0.5 m. On net covered treatments, poles 75 cm long and ≈ 5 cm thick were installed before planting to provide support for the net covers. The poles were driven 25 cm into the ground at each corner and at the center of the plot to anchor the agronets. Agronets were then mounted completely covering the plots and pegged at each corner to minimize wind interference. Once covered, plots were maintained permanently covered and only opened during routine plant management and data collection periods.

2.3 Data Collection

Data collection commenced from the start of seedling emergence and continued to the last harvest.

2.3.1 Days to Emergence and Emergence Percentage

The number of days from sowing to first emergence of French bean in the different experimental units were monitored and recorded. Thereafter, seedling numbers were counted at a two-day interval for a period of one week and progressive emergence percentages computed for each treatment using the formula:

$$\frac{\text{Number of emerged seedlings}}{\text{Number of seeds sown}} \times 100\%$$

(1)

2.3.2 Plant Growth Variables

Four plants from the inner rows of each experimental unit were randomly selected and tagged for collection of plant growth variables data. In this study, plant growth variables measured were stem collar diameter, plant height, number of branches and internodes and internode length. Plant growth data were collected on a weekly basis beginning 21 Days after Planting (DAP) to first harvest. During each data collection date, stem collar diameter of the four tagged French bean plants was measured at ≈ 4 cm from the ground level using a digital vernier caliper (Model 599-577-1/USA) and data obtained used to compute the average stem collar diameter of plants for the different treatments in millimeters (mm). Height of each tagged plant was also measured in centimeters (cm) using a meter ruler from the ground level to the point of growth and the number of internodes and branches counted and recorded as number of branches or internodes per plant (no. /plant). Thereafter, the length of each internode on the main stem was measured using a ruler and data obtained used to compute the average internode length of the plants in cm.

2.3.3 Yield

Harvesting began at 62 DAP. The crop was harvested thrice per week for a period of four weeks removing pods that had attained horticultural maturity during each harvest. At each harvest, the pods harvested from each tagged plant were separately counted and the number of pods obtained recorded and later used to compute the average number of pods per plant (no./plant). The pods were then weighed using a weighing balance (Advanced Technocracy Inc. Ambala) and weight obtained recorded in grams (gms) and later used to compute the average pod weight per plant for the individual treatments. Thereafter, a composite sample was made from the harvest of tagged plants of each experimental unit, 100 gms of fresh pods drawn from the sample, oven dried at 70°C to a constant weight and the dry weight determined in gms which was later used in the computation of total plant biomass.

French bean pods harvested from each experimental unit were then separately sorted as marketable and non-marketable. Non-marketable pods included the overgrown pods, off type, those damaged by pest or disease and those with physical damage or physiological defects. Marketable French bean pods were then graded based on pod sizes as extra fine grade, fine grade and bobby beans according to the French bean grading system (HCDA, 2011). The weight of each grade was then measured and recorded as weight in grams per grade per treatment.

Total plant biomass was determined at three different plant growth stages; at the trifoliate leaf, flowering and podding stage using four plants from each experimental unit. The plants were dug out recovering most of the roots, cleaned and oven dried at 70°C to constant weights and their dry weights determined. The data obtained were used to compute the average total biomass per plant as;

$$\text{Total Biomass (g DW/plant)} = \frac{\text{Total Plant Biomass} + \text{Total Pod Dry Weight}}{4}$$

(2)

Where; Total pod dry weight was computed by multiplying pod dry weight of the 100 grams of fresh pods drawn from the composite sample for biomass determination at each harvest by the total fresh weight obtained for the individual treatments at that harvest.

2.4 Data Analysis

The Proc univariate procedure of SAS (Version 9.1; SAS Institute, Cary, NC) was used to check for normality of the data before analysis. Data were then subjected to analysis of variance (ANOVA) using the GLM procedure of SAS at $P \leq 0.05$. Data were analyzed using the statistical RCBD model:

$$Y_{ij} = \mu + \alpha_i + \beta_j + \varepsilon_{ij} \qquad (3)$$

Where; Y_{ij} is the French bean response, μ is the overall mean, α_i is the effect due to the ith treatment, β_j is the effect due to the jth block, ε_{ij} is the random error term. Means for significant treatments at the F test were separated using Tukey's honestly significant difference (THSD) test at $P \leq 0.05$.

3. Results

3.1 Effects of Colour of Agronet Cover on Days to First Emergence and Emergence Percentage

French bean seedling emergence was enhanced by use of the different coloured agronet covers (Table 2). Regardless of the colour of agronet cover used, seedling emergence occurred earlier under net covers compared to under open field production. The first seedling emerged 5 DAP under all net covers compared to after 7 DAP under the control treatment. Percent seedling emergence was also influenced by the use of agronet covers especially at the early days of emergence up to 7 DAP when percent emergence of seedlings was significantly higher under agronet covers compared to the control treatment. There was no significant difference in percent seedling emergence noted among the different coloured agronets during these sampling dates (Table 2). Beyond 7 DAP, percentage seedling emergence remained higher under the net covered treatments although the differences with the control treatment were not significant. On overall, the highest and lowest germination percentage of 81.29% and 76.21% was obtained under the blue and yellow net covers, respectively by the final day of data collection

Table 2. Effects of colour of agronet cover on days to first emergence and percent emergence of French bean during production

Treatment	Day after Planting			
	5	7	9	11
White	3.4a*	58.1a	75.4**	79.5
Yellow	2.6a	51.6a	72.6	76.2
Tricolour	4.3a	53.3a	74.0	78.7
Grey	2.2a	47.6a	73.1	78.5
Blue	3.4a	56.3a	77.0	81.3
Control	0b	22.2b	66.8	76.3

*Means followed by the same letter within a column are not significantly different according to Tukey's honestly significant difference test at p≤0.05.

**Means within a column not followed by a letter are not significantly different according to F-test at p≤0.05.

3.2 Effects of Colour of Agronet Cover on French bean Plant Growth

Regardless of the colour of agronet cover used, French bean plants showed relatively better growth under net covers than under the control treatment (Table 3). French bean plants grown under the white, yellow, grey and tricolor net covers had significantly larger collar diameter in all sampling dates compared to control plants. Collar diameter recorded for plants grown under the blue net was also slightly higher than that for control plants, although the difference among the two treatments was not significant in most sampling dates. Among the net covered treatments, plants under white cover had the thickest stems followed by those under tricolour cover, then those under grey cover, followed by those under the yellow cover, with the thinnest stems obtained under the blue net cover.

Growing French bean under net covers also improved plant height compared to control plants. In all sampling dates, plants were tallest under the yellow net cover and shortest under the control treatment. Among the other treatments, plants tended to be taller under the grey net cover followed by those under the blue net cover, then those under tricolour net and shortest under the white net cover in most sampling dates. The use of agronet covers also improved on the plant branching ability regardless of the colour of the net cover used. The number of branches per plant was significantly higher for plants grown under all net covers compared to control plants at all sampling dates except for plants grown under blue cover at the final sampling date (56 DAP). Up to 49 DAP, the number of branches per plant was not significantly different among the different net covers used. However, by56 DAP, the number of branches per plant was significantly higher for plants grown under the white cover compared to those under the blue cover. Among the other net covers, plants grown under the tricolour cover also tended to have slightly more branches at the later stages of plant growth (49 and 56 DAP) compared to those grown under yellow, grey and blue covers.

Table 3. Effects of colour of agronet cover on the growth of French bean during production

Treatment	Days after Planting					
	21	28	35	42	49	56
Collar diameter (mm)						
White	2.75a*	3.37a	4.09a	5.03a	5.74a	6.33a
Yellow	2.48bc	3.06bc	3.72bc	4.42bc	5.06bc	5.60bc
Tricolour	2.66ab	3.32a	4.00ab	4.86ab	5.48ab	6.06ab
Grey	2.54abc	3.15ab	3.82ab	4.53b	5.21b	5.80b
Blue	2.35cd	2.92cd	3.44cd	4.05cd	4.68cd	5.24c
Control	2.21d	2.80d	3.27d	3.84d	4.22d	4.69d
Plant height (cm)						
White	6.08c	9.01c	12.85c	21.70b	29.29b	35.17b
Yellow	7.91a	12.86a	21.12a	34.01a	40.69a	46.88a
Tricolour	6.48bc	9.74bc	13.88bc	22.86b	29.87b	36.14b
Grey	6.57bc	11.18ab	16.95b	25.90b	32.85b	38.74ab
Blue	7.09ab	10.16bc	14.81bc	23.80b	30.82b	36.88b
Control	4.89d	6.21d	7.77d	12.11c	16.18c	19.97c
Branches (no./plant)						
White	3.25a	5.22a	6.69a	7.69a	8.22a	8.69a
Yellow	2.97a	5.34a	6.34a	7.38a	7.63a	7.81ab
Tricolour	3.03a	4.97a	6.06a	7.28a	7.75a	8.34ab
Grey	3.00a	4.97a	6.06a	7.19a	7.66a	7.97ab
Blue	2.72a	4.53a	5.69a	6.75a	7.19a	7.50bc
Control	1.06b	2.88b	3.94b	5.25b	5.91b	6.63c
Number of internodes (no./plant)						
White	3.97a	5.44a	6.84a	7.84a	8.28a	8.47a
Yellow	4.03a	5.63a	6.75a	7.44a	7.69bc	7.81ab
Tricolour	3.91a	5.44a	6.75a	7.75a	8.03ab	8.19ab
Grey	3.91a	5.56a	6.66a	7.72a	7.94ab	8.00ab
Blue	3.84a	5.13a	6.31a	7.34ab	7.69bc	7.84ab
Control	3.25b	4.47b	5.75b	6.78b	7.34c	7.75b
Internode length (average/plant in cm)						
White	1.53c	1.65c	1.88d	2.76b	3.54b	4.16b
Yellow	1.96a	2.28a	3.11a	4.54a	5.27a	6.00a
Tricolour	1.66bc	1.79bc	2.06cd	2.95b	3.73b	4.42b
Grey	1.68bc	2.00b	2.52b	3.32b	4.12b	4.84b
Blue	1.85ab	1.98b	2.33bc	3.21b	4.00b	4.71b
Control	1.51c	1.39d	1.35e	1.78c	2.19c	2.57c

*Means followed by the same letter within a column and a parameter are not significantly different according to Tukey's honestly significant difference test at p≤0.05.

Plants grown under net covers also had significantly more internode numbers than control plants during the early stages of plant growth up to 35 DAP. At all sampling dates during this period, there were no significant differences in the number of internodes for plants grown under the different colours of net covers although plants under the white net cover tended to have slightly more internodes while those under the blue cover had the least number of internodes. Significantly more internodes were registered for plants grown under yellow, grey, tricolour and white net covers compared to internode numbers recorded for control plants at 42 DAP with no significant differences noted in internode numbers for plants grown under blue net cover and control plants. At 49 DAP, plants under the grey, tricolor and white net covers still registered significantly higher internode numbers than control plants with no statistical difference in internode numbers between plants grown under blue or yellow net covers and the control plants. By 56 DAP only plants under the white net were significantly different from control plants in internode numbers. Generally, plants grown without a net cover (control) had the lowest number of internodes throughout the study while among the net covered treatments, growing French bean under a blue net cover yielded plants with the lowest number of internodes during all sampling dates. Plants grown under net covers also had significantly longer internodes than control plants. During all sampling dates, internode length was longest under the yellow net and shortest under the control treatment. Among the net covers, plants grown under yellow net were significantly different from those under other net covers in internode length in all sampling dates except during the initial sampling date (21 DAP) when the internode length for plants grown under the blue net was not significantly different from that of plants grown under the yellow net cover. Internode length tended to be longer in plants under the grey net cover followed by those under the blue net cover, then those under tricolour net and shortest under the white net cover in most sampling dates.

3.3 Effects of Colour of Agronet Cover on the Yield of French bean

Growing French bean under the different coloured net covers improved the crop yield compared to open field production (Table 4). Regardless of the colour of the net cover used, plants grown under the net covers produced more pods per plant yielding higher total and marketable pod weights compared to control plants. On overall, the highest pod numbers per plant and total and marketable pod weight was obtained under the white net cover while pod numbers and total and marketable pod yields were lowest under open field production. Among the other treatments, plants under the yellow cover yielded better pod numbers and total and marketable pod weight while those under the blue net cover had the least pod numbers and total and marketable pod yields. There was no statistical difference in the number of pods per plant and total and marketable pod weight recorded for plants grown under the grey, tricolor and blue net covers. Growing the crop under the different coloured agronet covers substantially improved marketable French bean yields by between 42.0 – 103.3%. The highest increase in marketable pod yield of 103.3% was obtained under the white net cover while lowest increase of 42.0% was under the blue net covers. Growing French bean under the other net covers resulted in intermediate increase in French bean yield of 53.7% under the grey net cover, 56.3% under the tricolor net cover and 91.7% under the yellow net cover.

Table 4. Effects of colour of agronet cover on yield of French bean

Treatment	Pods/plant	Pod yield	Marketable	% increase in marketable yields
	no.	g/plant	g/plant	
White	87.47a*	132.55a	120.22a	103.3
Yellow	80.50ab	121.27a	113.23a	91.5
Tricolour	69.59ab	101.78ab	92.43ab	56.3
Grey	73.44ab	103.92ab	90.92ab	53.7
Blue	64.03ab	90.65ab	84.00ab	42.0
Control	53.38b	70.73b	59.14b	-

*Means followed by the same letter within a column and a parameter are not significantly different according to Tukey's honestly significant difference test at p≤0.05.

Growing the crop under the different coloured nets also showed some effect on the rate of pod maturation judged by the differences in the pod weight under the different pod grades (Figure 1). A higher percentage weight of marketable yields of French bean represented by extra fine pods was obtained under the blue net cover and control treatments compared to the other treatments while white and tricolour net covers registered lowest values of pods under the extra fine pod grade. On the other hand, higher fine grade pods were obtained under the white and tricolour net covers while the blue cover and control treatments registered lowest values of fine grade pod with the yellow and grey net covers registering intermediate values of pods under the extra fine and fine grades.

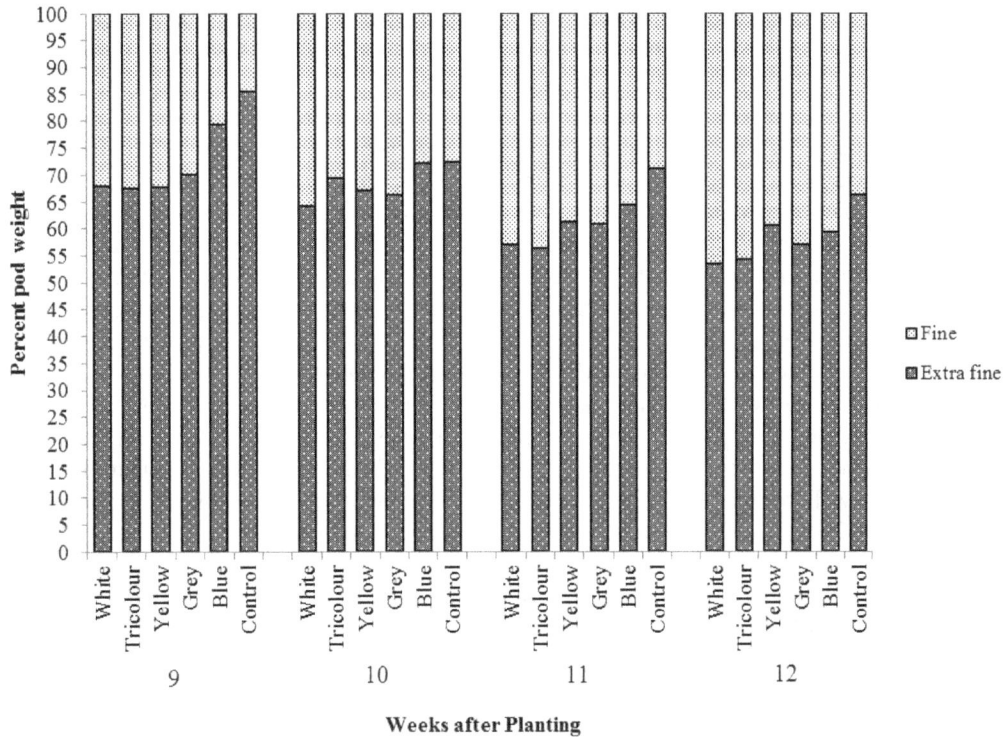

Figure 1. Effects of colour of agronet cover on weekly average percent pod weight per plant of the extra fine and fine French bean pods

Biomass production was also enhanced when French bean was grown under the different coloured net covers at all stages of plant growth compared to the control treatment (Table 5). During all the sampling dates, plants in the control treatment had the lowest total plant biomass while among the net covered treatments, biomass was lowest for plants grown under the blue net cover. At the early stage of French bean plant growth (trifoliate stage), plants under tricolour net cover had the highest plant biomass while at later stages of growth (flowering and podding stage) biomass was highest for plants grown under the white net cover. Total biomass for plants grown under the tricolour net cover was significantly higher than that of plants grown under the yellow and grey net covers at the trifoliate stage while at flowering and podding stage there was no statistical difference in plant biomass among the different net covers.

Table 5. Effects of colour of agronet cover on French bean total plant biomass (gms/plant)

Treatment	Trifoliate stage	Flowering stage	Podding stage
	g/plant		
White	1.46ab*	6.81a	30.22a
Yellow	1.24bc	6.44a	28.43ab
Tricolour	1.54a	6.67a	25.64ab
Grey	1.37abc	6.37a	23.97ab
Blue	1.12c	5.10a	20.46ab
Control	0.76d	2.77b	17.68b

*Means followed by the same letter within a column are not significantly different according to Tukey's honestly significant difference test at p≤0.05.

4. Discussion

Growing French bean under the different coloured agronet covers proved of potential benefit in French bean production. Regardless of the colour of agronet cover used, net covers advanced seedling emergence by two days and resulted in a higher percent emergence compared with the open field (control). Net covers have been reported to modify the immediate crop environment characterized by maintaining higher soil moisture content and air temperatures compared to open field conditions (Saidi, Gogo, Itulya, Martin, &Ngouajio, 2013; Gogo, Saidi, Ochieng, Martin, &Ngouajio, 2014). Adequate moisture and warmth are necessary conditions for activation of enzymes involved in seed germination (Raven, Ray, Evert, & Eichhorm, 2005). Moisture also ensures reduced resistance for the cotyledons of developing seedling as they move through the soil to reach the surface. These arguments lend support for the early and higher emergence registered under the different coloured net covers compared to open field production observed in the current study. Similar to the findings of the current study, Muleke, Saidi, Itulya, Martin, and Ngouajio (2013) observed early emergence and higher percent emergence of cabbage (*Brassica oleracea* var. *capitata*) seeds under net covered compared to open field nurseries. Among the different coloured net covers, final percent seedling emergence was highest under the blue net cover at 81.3% and lowest under the yellow net cover at 76.2% against a 76.3% emergence under the control treatment. Coloured shade nettings not only influence the microclimate to which the plant is exposed but also exhibit special optical properties to optimize desirable physiological responses of plants (Costa *et al.*, 2010). Depending on the pigmentation of the plastic threads, coloured nets provide varying mixtures of natural unmodified light together with spectrally modified scattered light which improves light penetration into the inner canopy of plants as well as promotes specific photomorphogenetic and physiological responses in plants (Rajapakse & Shahak, 2007). Coloured shade nets absorbs spectral bands shorter, or longer than the visible light (Shahak*et al.*, 2008). Differences in the final percent emergence of seedlings observed in the current study may be attributed to the differences in light intensity and quality under the different coloured nets marked by differences in the amount of filtered red, far-red or blue light by the different net covers. Yellow net covers have been documented to scatter more far-red light than red light thus decreasing the R/FR ratio (Goren, Alkalai-Tuvia, Perzelan, Fallik, & Aharon, 2011). On the other hand, R/FR ratio under blue net cover has been shown to be the same as under natural light (Oren-Shamir *et al.*, 2001). According to Yerima, Esther, Madugu, Muwa, and Timothy (2012) seed germination is inhibited by far-red light and stimulated by red light while the effects of blue light cannot be reversed by far-red light possibly explaining the higher and lower final emergence percentages recorded for the blue and yellow net covers, respectively under the current study.

French bean plant growth in the current study was also enhanced by the use of net covers compared to open field production. Plants under net covers had more branches and internodes, longer internodes length, thicker and taller stems compared to those grown in the open. Net covers have been reported to improve crop performance as a result of modified and stabilized crop microclimate under the covers marked by lower diurnal temperature ranges and higher volumetric water content (Gogo *et al.*, 2014) as well as reduced wind speed (Arthurs, Stamps, & Giglia, 2013). Besides the general advantages associated with net covers, colored nets selectively filter solar radiation to promote specific wavelengths of light (Arthurs *et al.*, 2013), and increase light scattering which influences plant branching and crop compactness (Abul-Soud, Emam, &Abdrabbo, 2014). In the current study, plant growth variables were differently influenced by the different coloured net covers. Growing plants under the yellow net cover stimulated stem and internode elongation resulting in taller but slender French bean plants. Growing plants under the white and tricolour nets, on the other hand enhanced stem collar diameter and internode numbers, resulting in stout and compact plants while those under the blue net cover exhibited reduced stem collar diameter, branching and number of internodes. Longer and thin plants observed under the yellow net elicit elongation of stems at the expense of their thickness which can be attributed to reduction of R/FR ratio (Kasperbauer, 1994). Yellow net covers scatter more far-red light that penetrates into the plant canopy which stimulates internode and stem elongation (Goren *et al.*, 2011). Far-red light promotes conversion of inactive gibberellins (GA) to active forms (Rajapakse, Young, McMahon, & Oi, 1999) which are potent promoters of stem elongation (Cummings, Foo, Weller, Reid, & Koutoulis, 2008) and regulate internode length in response to altered light condition (Maki, Rajapakse, S., Ballard, & Rajapakse, N., 2002). Despite having the highest internode numbers, plants under the white and tricolour (predominantly white) net covers tended to be shorter and bushy compared to plants under the other net covers which is attributable to the lower internode lengths recorded for plants under these treatments. According to Oren-Shamir *et al.* (2001), light transmitted through neutral-coloured nets is the same as the natural light but with consistently increased scattering over the natural light, making light reach a larger volume of the plant in a more homogenous way. Scattering of light may be as effective as manipulation of the light spectrum, in influencing the growth of plants (Nissim-Levi *et al.*, 2008). Inhibition effects of blue net on plant growth were expressed in form of reduced internode numbers, branching and collar diameter of plants. Blue net covers substantially reduce radiation reaching the plant underneath (Abul-Soud*et al.*, 2014) and the lack of RF or high R/FR ratio under blue net cover (Oren-Shamir *et al.*, 2001) have been implicated as major inducers of reduced plant growth, possibly explaining the lower growth observed under blue net in the current study.

Better growth of French bean plants observed under the different net covers reflected in to higher total plant biomass and pod yields per plant compared to open field production. The higher biomass and pod yield obtained under net covers compared to control plants in the current study can be associated with the better plant development recorded under net covers possibly favoured by the modified microclimate under these treatments. Plants under net covers tended to be taller with thicker collar diameters and more branches depicting better biomass accumulation and providing a greater bearing surface and more stored food reserves for translocation to developing pods compared to control plants. Proper light distribution favours photosynthesis and metabolites translocation for better plant growth (Setiawati, Hasyim, Hudayya, & Shepard, 2014). Colour shade nets prevent excess sunlight and retain soil moisture for proper plant growth and productivity (Ilic, Milenkovic, Durovka, & Kapoutas, 2011). Better performance observed on plants under net covers could also be attributed to improved light compensation under net covers as a result of favorable microclimatic conditions (Nangare, Singh, Meena, Bhushan, & Bhatnagar, 2015). Growing French bean under net covers also substantially improved marketable yields of the crop. Marketable pod yield was 103.3% higher under the white net cover, 91.7% higher under the yellow cover, 56.3% higher under the tricolour net, 53.7% higher under the grey net and 42.0% higher under the blue net compared control plants. Better marketable yields obtained under net covers in this study can be attributed to enhanced pod production rates and reduced number of non-marketable pods due to reduced pest damage and physiological defects. Similarly, Ilic *et al.* (2011) reported higher export quality pepper fruit yield under coloured shade nets with pepper grown under the net covers resulting in a 113% to 131% increase in total fruit yield compared to open field.

Apart from total and marketable yields, the different coloured nets differentially affected the rate of pod maturation depicted by the proportion of marketable yield represented by extra fine and fine grade pods. The higher percentage of the weight of marketable yields of French bean represented by extra fine pods obtained under the blue net cover and control treatments indicates delayed pod maturation under these treatments compared to other treatments. On the other hand, more fine grade pods obtained under the white and tricolour net covers indicate enhanced rate of pod maturation under these treatments. Based on these findings, growing French bean under net covers and especially light-coloured covers hastens the rate of French bean pod maturation which can potentially reduce the harvest interval. Similar to these observations, Shahak*et al.* (2008) recorded advanced maturation of a number of table grape cultivars under light-coloured nets (pearl, white) compared to blue netting and open field growing which have now been incorporated by the growers for earliness and improved quality.

5. Conclusions and Recommendations

Results of the present study reveal growing French bean under the different coloured agronet covers as a useful technology for enhancing growth and pod yield and quality of the crop. With higher yields and better quality being the ultimate goal of every grower, we recommend the use of a white net cover in French bean production in regions with similar climatic conditions to those of the site of the current study. Studies combining the use of blue net covers early in the growing period and a white net cover soon after emergence are also recommended to establish whether both seedling emergence and growth and yields can better be optimized to enhance crop performance. We also recommend additional studies on the subject using different cultivars of French beans, other colours and mesh size of the net covers and in different agroecological zones to further validate the results. An analysis of the effect of the different colours of net covers on the sensory attributes and nutritive value of French bean pods would also be beneficial.

Acknowledgements

This is part of Master of Science in Horticulture project by James N. Munywoki. The study was in part made possible by support of the National Science Council of Kenya under Award No. NACOSTI/RCD/ST&I/7TH CALL/MSc/107. Thanks to the Departments of Crops, Horticulture and Soils of Egerton University–Kenya for hosting the study.

Reference

Abul-Soud, M. A., Emam, M. S. A., & Abdrabbo, M. A. A. (2014). Intercropping of Some Brassica Crops with Mango Trees under Different Net House Color. *Research Journal of Agriculture and Biological Sciences, 10*(1), 70-79.

Amannuel, G. S., Kiihne, R. F., Tanner, D. G., & Vlek, P. L. G. (2000). Biological Nitrogen Fixation in Faba Bean (*Vicia faba* L.) in the Ethiopian Highlands as Affected by P Fertilization and Inoculation. *Biol. Fertil. Soils, 32,* 353-359. http://dx.doi.org/10.1007/s003740000258

Arthurs, S. P., Stamps, R. H., & Giglia, F. F. (2013). Environmental Modification Inside Photoselective Shadehouses. *Hort Science, 48*(8), 975-979.

Arulbalachandran, D., & Mullainathan, L. (2009). Changes on Protein and Methionine Content of Black Gram (*Vigna mungo* (L.)Hepper) Induced by Gamma Rays and EMS. *American- Eurasian Journal of Scientific Research, 4*(2), 68-72.

CIAT (Center for International Agricultural Technology). (2006). Beans (*Phaseolus vulgaris* L.) for Income Generation by Small Scale Farmers in East Africa. *Horticultural Insights, 2*(31), 69.

Costa, L. C. B., Pinto, J. E. B. P., Castro, E. M., Alves, E., Bertolucci, S. K. V., & Rosal, L. F. (2010). Effects of Coloured Shade Netting on the Vegetative Development and Leaf Structure of *Ocimum selloi.Bragantia Journal, 69,* 349-359. http://dx.doi.org/10.1590/S0006-87052010000200012

Cummings, I. G., Foo, E., Weller, J. L., Reid, J. B., & Koutoulis, A. (2008). Blue and Red Photoselective Shade cloths Modify Pea Height Through Altered Blue Irradiance Perceived by the Cry1 Photoreceptor. *The Journal of Horticultural Science and Biotechnology, 83*(5), 663-667. http://dx.doi.org/10.1080/14620316.2008.11512440

FAOSTAT. (2010). Food and Agricultural Commodities Production; Available online: http://faostat.fao.org (accessed 20th April 2016).

Gogo, E. O., Saidi, M., Ochieng, J. M., Martin, T., Baird, V., & Ngouajio, M. (2014). Microclimate Modification and Insect Pest Exclusion Using Agronet Improve Pod Yield and Quality of French Bean. *Hort Sci., 49*(10), 1298-1304.

Goren, A., Alkalai-Tuvia, S., Perzelan, Y., Fallik, E., & Aharon, Z. (2011). Photoselective Shade Nets Reduce Postharvest Decay Development in Pepper Fruits. *Advances in Horticultural Science, 25*(1), 26-31. http://www.jstor.org/stable/42882805

HCDA. (2011). Horticultural Validated Report. Agricultural Information Resource Centre, Nairobi, Kenya.

Ilic, Z., Milenkovic, L., Durovka, M., & Kapoutas, N. (2011). The effect of color shade nets on the greenhouse climate and pepper yield. *Proceedings 46th Croatian and 6th International Symposium on Agriculture* (pp. 529-532). Opatija, Croatia.

Jaetzold, R., & Schmidt, H. (2006). *Farm management handbook of Kenya. Natural conditions and farm information.*Vol.11/c. (2nd Ed.).Ministry of Agriculture, East Kenya, Kenya.

Kasperbauer, M. J. (1994). Light and plant development. In: R. E. Wilkinson (Ed.), *Plant-environment interactions* (pp. 83-123). Marcel Dekker Inc., NY, USA.

Kelly, J. F., & Scott, M. K. (1992). The nutritional value of snap beans versus other vegetables. In: G. Henry and W. Janssen (Tech. Eds.), *CIAT Proceedings of an International Conference on Snap Beans in the Developing World held from 16th to 20th October 1989* (pp. 23-46). Cali, Colombia.

Maki, S. L., Rajapakse, S., Ballard, R. E., & Rajapakse, N. C. (2002). Role of Gibberellins in Chrysanthemum Growth under Far Red Light-deficient Greenhouse Environments. *Journal of the American Society for Horticultural Science, 127*(4), 639-643.

Monda E. O., Ndegwa, A., &Munene, S. (2003). French beans production constraints in Kenya. *African Crop Science Conference Proceedings*, 6, 683-687.

Muleke, E. M., Saidi, M., Itulya, F. M., Martin, T., & Ngouajio, M. (2013).The Assessment of the Use of Eco-Friendly Nets to Ensure Sustainable Cabbage Seedling Production in Africa. *Agronomy, 3,* 1-12. http://dx.doi.org/10.3390/agronomy3010001.

Nangare, D. D., Singh, J., Meena, V. S., Bhushan, B., &Bhatnagar, P. R. (2015).Effect of Green Shade Nets on Yield and Quality of Tomato (*Lycopersicon esculentum* Mill) in Semi-arid Region of Punjab. *Asian Journal of Advances in Basic and Applied Science, 1*(1), 1-8.

Ndegwa, A. M., Muthoka, N. M., Gathambiri, C. W., Muchui, M. N., Kamau, M. W., & Waciuri, S. M. (2010). Snap bean production, post-harvest practices and constraints in Kirinyaga and Machakos Districts of Kenya. http://www.kari.org/biennialconference/conference12/docs/.pdf.

Nissim-Levi, A., Farkash, L., Hamburger, D., Ovadia, R., Forrer, I., Kagan, S., & Oren- Shamir, M. (2008). Light-scattering Shade Net Increases Branching and Flowering in Ornamental Pot Plants. *J. Hort. Sci. Biotechnol, 83*, 9-14. http://dx.doi.org/10.1080/14620316.2008.11512340

Odero, D. O., Mburu, J., Ogutu, C. A., & Nderitu, J. H. (2012). Competitiveness of smallholder snap bean production in Kirinyaga County. *Kenya International Review of Business and Social Science, 2*(1), 49-65.

Oren-Shamir, M., Gussakovsky, E., Eugene, E., Nissim-Levi, A., Ratner, K., Ovadia, R., ...Shahak, Y. (2001). Coloured Shade nets can Improve the Yield and Quality of Green Decorative Branches of *Pittosporum variegatum*. *The Journal of Horticultural Science and Biotechnology, 76*(3), 353-361. http://dx.doi.org/10.1080/14620316.2001.11511377

Rajapakse, N. C., & Shahak, Y. (2007). Light and plant development. In: G. Whitelam and K. Halliday (eds.), *Light quality manipulation by horticulture industry* (pp. 290-312). Blackwell Publishing, UK.

Rajapakse, N. C., Young, R. E., McMahon, M. J., & Oi, R. (1999). Plant height control by photoselective filter: Current status and future prospects. *Hortechnology, 9*(4), 618-624.

Raven, P., Ray, H., Evert, F., & Eichhorm, S. E. (2005).*Biology of plants* (7th ed.). W. H. Freeman and Company Publishers: New York, NY, USA, 504-508.

Saidi, M., Gogo, E. O., Itulya, F. M., Martin, T., & Ngaouajio, M. (2013).Microclimate Modification using Eco-friendly Nets and Floating Row Covers Improves Tomato [*Lycopersicon esculentum* (Mill)] Yield and Quality for Small Holder Farmers in East Africa. *Agricultural Sciences, 4*, 577-584. http://dx.doi.org/10.4236/as.2013.411078

Setiawati, W., Hasyim, A., Hudayya, A., & Shepard, B. M. (2014). Evaluation of Shade Nets and Nuclear Polyhedrosis Virus (SENPV) to Control *Spodoptera exigua* (Lepidoptera: Noctuidae) on Shallot in Indonesia. *AAB Bioflux, 6*(1), 88-97. http://www.aab.bioflux.com.ro/docs/2014.88-97.pdf

Shahak, Y. (2008). Photoselective Netting for Improved Performance of Horticultural Crops. A Review of Ornamental and Vegetable Studies Carried Out in Israel. *ActaHort, 770*, 161-168. http://dx.doi.org/10.17660/ActaHortic.2008.770.18

Shahak, Y., Gussakovsky E. E., Cohen, Y., & Lurie, S. (2004). ColorNets: A New Approach for Light Manipulation in Fruit Trees. *ActaHort, 636*, 609-616.

Shahak, Y., Ratner, K., Giller, Y. E., Zur, N., Or, E., Gussakovsky, E .E., ...Greenblat-Avron, Y. (2008). Improving Solar Energy Utilization, Productivity and Fruit Quality in Orchards and Vineyards by Photoselective Netting.*Acta Hort.*, 65-72. http://dx.doi.org/10.17660/ActaHortic.2008.772.7

Singh, S. P., & Schwartz, H. F. (2011). Review: Breeding Common Bean for Resistance to Insect Pests and Nematodes. *Can. J. Plant Sci., 91*, 239-250. http://dx.doi.org/10.4141/CJPS10002

Wahome, S. W., Kimani, P. M., Muthomi, J. W., & Narla, R. D. (2013). Quality and Yield of Snap Bean Lines Locally Developed in Kenya. *International Journal of Agronomy and Agricultural Research (IJAAR)*, 3(7), 1-10.

Yerima, J. B., Esther, M. A., Madugu, J. S., Muwa, N. S., & Timothy, S. A. (2012). The Effect of Light Color (wavelength) and Intensity on Vegetable Roselle (*Hibiscus Sabdariffa*) Growth. *Scholarly Journal of Scientific Research and Essay (SJSRE), 1*(2), 19-29.

Zlatev, Z., &Lidon, F. C. (2012). An Overview on Drought Induced Changes in Plant Growth, Water Relations and Photosynthesis. *Emir. J. Food Agric., 24*(1), 57-72.

6

The Adoption of Sustainable Management Practices by Mexican Coffee Producers

Simone Ubertino[1], Patrick Mundler[1] & Lota D. Tamini[1]

[1] Department of Agricultural Economics and Consumer Science, Laval University, Canada

Correspondence: Simone Ubertino, Department of Agricultural Economics and Consumer Science, Laval University, Canada. E-mail: simone.ubertino.1@ulaval.ca

Abstract

In order to maintain optimal growing conditions on coffee plots, producers in Mexico are encouraged to renovate their stock of coffee trees, use fertilizer, implement soil conservation measures and manage shade levels. The adoption of these sustainable management practices (SMPs) by smallholder coffee growers has become an important rural development objective, especially as a way to overcome low yields, poverty and land degradation. However, adoption rates for SMPs remain below expected levels, a situation that potentially threatens the long-term viability of the coffee sector in Mexico. To better understand the choices made by producers, a multivariate probit technique was used which modelled the adoption of possibly interrelated SMPs using data from a survey of 119 coffee producers. The analysis reveals that adoption of SMPs is related to the size of coffee holdings, the socio-economic characteristics of producers and the role of social capital, the latter being a key factor in the overall adoption process. Surprisingly, government subsidies to coffee growers were not tied to higher adoption rates, suggesting the need for policy reforms in order to better facilitate the uptake of new practices. The results indicate that efforts aimed at strengthening local institutions and organizing coffee growers into producer associations could increase the adoption of SMPs in smallholder coffee systems.

Keywords: adoption in agriculture, sustainable management practices, coffee

1. Introduction

Coffee production is central to the rural economy of Mexico and is one of the country's principal export crops. In light of its importance, recent studies have tried to explore the relationship between coffee growing and the livelihood strategies of Mexican rural households (Barham, Callenes, Gitter, Lewis & Weber, 2011; Padrón & Burger, 2015). From a development perspective, Mexico's coffee producing regions suffer from high levels of economic and social marginalization with most households continuing to live below the poverty line (CONAPO, 2010). As a result, the sustainability of coffee growing has a direct impact on poverty alleviation, food security and rural development in Mexico.

However, the performance of the coffee sector has failed to live up to expectations, characterized by several decades of stagnation and low yields. In response, Mexico's ministry of agriculture (SAGARPA) has recommended that producers adopt a series of sustainable management practices (SMPs) in order to improve physical conditions on coffee plots (AMECAFE, 2011). Practices considered to be a high priority include renovating coffee tree stocks, applying fertilizer, adopting soil conservation measures and managing shade levels. Depending on a plot's initial physical state, it is estimated that a full renovation cycle requires the implementation of SMPs for between seven and twelve years.

By considering the problem of low yields, the present study seeks to identify the social, economic and institutional factors that influence the adoption of SMPs by coffee growers in Mexico. For this purpose, an empirical application was tested using data from a survey of producers in the southern state of Oaxaca and a multivariate probit model. The study's importance derives from simultaneously modelling the effects of explanatory variables on the adoption of SMPs while taking into account possible correlations among practices.

This theoretical approach offers a new direction for research into smallholder coffee systems in Latin America. Most studies have tended to focus on the economic, social and environmental impact of certification schemes (Bacon, 2005; Valkila, 2009; Beuchelt & Zeller, 2011; Blackman & Naranjo, 2012). However, new research suggests that a greater effort should be made to understand the adoption choices of coffee producers themselves. As several impact studies conducted in Mexico and Peru have demonstrated, revenue differences between certified organic and non-certified coffee producers are largely attributable to differences in yields rather than to the premiums from certification (Barham et al., 2011; Barham & Weber, 2012; Ruben & Fort, 2012). A case can therefore be made that policymakers, NGOs and cooperatives looking to increase the revenues of coffee growing households should focus greater attention on promoting the adoption of better management practices with a view to increasing yields rather than pursuing smallholder certification (Barham et al., 2011). In this respect, a deeper understanding of the adoption choices that producers make could potentially facilitate such a strategy.

2. Method

2.1 Study Area and Sampling Procedure

The data for this study was gathered from a survey of 119 coffee producers in the southern state of Oaxaca. The survey related to activities undertaken during the 2013/2014 coffee season and was conducted in August-September, 2014. Field interviews with producers were carried out face to face using a structured survey questionnaire. During the initial sampling stage, a municipal district - San Agustin Loxicha - was selected (see Figure 1). The area in question was chosen due to its coffee growing potential and because it is located in one of the state's main coffee growing regions.

In the following stage, two villages within the municipality were randomly selected as field sites and the survey was pre-tested in a third village. Proportionate random sampling was carried out using a database of coffee producers in each village. The sample size was determined based on a reasonable estimate of the number of producers who could be interviewed during the field research period. Similar sample sizes have also been used in previous adoption studies (Boahene, Snijders & Folmer, 1999; Cramb, 2005; Ali et al., 2007). In addition, during field visits, semi-structured interviews were conducted with two groups of producers to obtain supplementary information on local issues related to coffee growing.

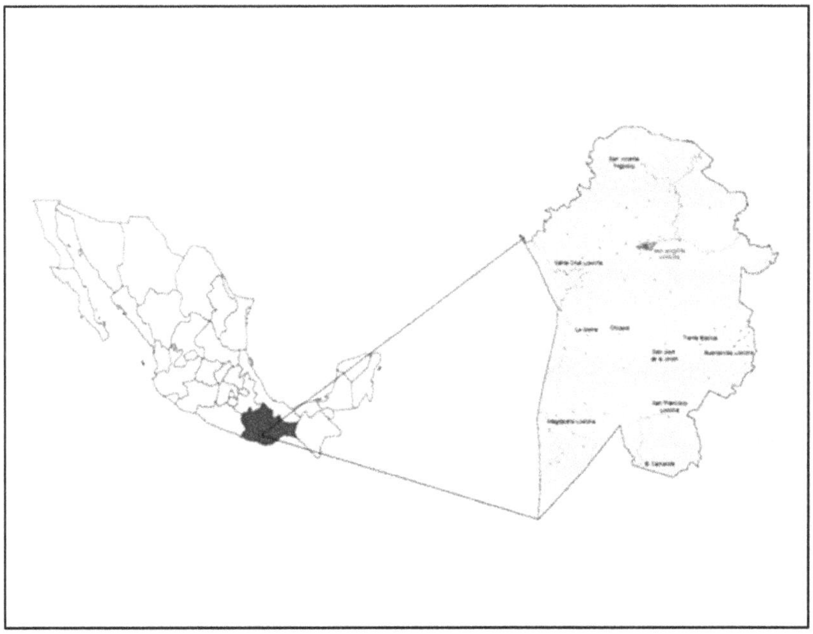

Figure 1. Map of surveyed area in Mexico

2.2 SMP Categories

Producers were asked a series of questions related to their use of SMPs. The practices mentioned in the survey are regarded as a high priority for all coffee growing regions in Mexico (AMECAFE, 2011). They are:

1) Renovating coffee tree stocks

- Replanting coffee trees
- Pruning branches

- Cutting off trunks

2) Applying fertilizer

- Organic fertilizer usage
- Chemical fertilizer usage

3) Implementing soil conservation measures

- Constructing terraces
- Building drainage ditches
- Building contour barriers
- Planting *Inga* trees or shrubs

4) Managing shade levels on coffee plots

- Pruning surrounding trees or shrubs
- Thinning sections of plot

For each of the four categories of SMPs, a producer was considered an adopter if at least one of the recommendations had been implemented. The first category lists three possible options producers have for improving coffee tree stocks. These include replanting plots by removing old or damaged trees, pruning branches (*podar*) or cutting off trunks (*recepar*) in order to stimulate new growth. Typically, coffee trees begin to produce berries three years after planting with peak production attained between the fifth and sixth year, after which, in the absence of pruning or cutting, yields begin to decline. The type of measure required depends on both the physical condition and age of the tree. For the second category of SMPs, both organic and chemical fertilizer adoption was accounted for although in Mexico coffee producers rarely use the latter (AMECAFE, 2011). Examples of organic fertilizer used in coffee production include chicken and cow manure, coffee pulp and household compost material.

The third category of SMPs outlines ways that producers can minimize soil erosion due to water runoff. Measures include building terraces, creating drainage ditches and erecting contour barriers. In addition, producers are encouraged to plant leguminous trees and shrubs of the genus *Inga*. Conservation strategies are vital since coffee is mostly cultivated in the river basins of the Pacific Ocean and the Gulf of Mexico. Given the mountainous topography of these regions, coffee plots are often vulnerable to soil erosion in addition to other environmental stressors caused by climate change (Castellanos et al., 2013).

Given that most coffee in Mexico is shade grown (Moguel & Toledo, 1999) adequate shade control measures are considered necessary. Excessive shading can create unfavourable phytosanitary conditions and negatively affect yields. Coffee farmers are advised to prune surrounding trees and shrubs, remove dead branches or thin out plot sections as needed (category 4).

The survey data indicates that SMP adoption varies considerably from one category to the next. For example, the number of producers who had renovated their stock of coffee trees was relatively high: 74% of those surveyed had followed at least one of the recommendations. By contrast, only 34% of producers had applied fertilizer and with one exception all of the adopters had used organic materials. Regarding soil conservation strategies, 58% of producers had implemented at least one measure while less than half (48%) of respondents had chosen to manage shade levels.

2.3 Econometric Strategy

A multivariate probit model (MVP) employs a system of simultaneous equations in order to determine the effect that explanatory variables have on adoption. The framework is based on the assumption that a producer's decision on whether or not to adopt a new practice is likely to be correlated with other adoption choices. By contrast, many studies develop univariate models to test which variables influence the adoption of a single agricultural practice. Such an approach however can be problematic since it fails to consider potential correlations among practices (Feder, Just, & Zilberman, 1985; Teklewold, Kassie, & Shiferaw, 2013). In many cases, adoption choices are interdependent meaning producers adopt new practices simultaneously or sequentially either as complements or substitutes. When adoption choices are correlated, a univariate model carries the risk of over or underestimating the significance of explanatory variables. By contrast, the MVP approach considers any potential correlations between the unobserved errors in the adoption equations as well as the relationship among adoption choices (Rodríguez-Entrena & Arriaza, 2013).

Formally, it is assumed that producers will evaluate a series of practices and choose the bundle that maximizes their expected utility. Following Teklewold et al. (2013), we consider the ith coffee producer confronted with a decision on whether to adopt a new practice. Let U_0 denote the benefits derived from traditional management practices and let U_n represent the benefits of adopting the nth practice. In our case, n denotes one of the four categories of SMPs. The producer chooses to adopt if $Y^*_{in} = U_n - U_0 > 0$. The net benefits (Y^*_{in}) from adoption is a latent variable determined by a vector of household and farm characteristics (X_{in}) as a well as non-observable disturbances captured by the error term ε_{in}:

$$Y^*_{in} = \beta_n \cdot X_{in} + \varepsilon_{in} \tag{1}$$

In this equation, the term β_n represents the vector of parameters to be estimated. By using the indicator function, we can derive an observed binary equation from the unobserved preferences in equation (1). Specifically, a producer will adopt a SMP (Y_{in}=1) if $Y^*_{in} >0$ but will forego adoption (Y_{in}=0) when $Y^*_{in} <0$. Since the MVP model takes into account the possibility of multiple adoptions, the error terms of the equations jointly follow a multivariate normal distribution. Estimates are obtained using the maximum likelihood approach and the model captures any correlations among the stochastic elements of the SMPs. As a result, the observed binary equation will jointly represent the choice to adopt a new practice. When SMPs are interdependent, the MVP solution allows for greater efficiency in estimating the parameters of interest.

2.4 Explanatory Variables

The explanatory variables (X_{in}) included in the model as well as their hypothesized effect on the adoption of SMPs are discussed below. Table 1 presents the descriptive statistics from the survey relating to the variables of interest.

Table 1. Definitions of variables and descriptive statistics

Dependant Variables	Name	Mean	Std. Err
Renovating coffee tree stock (1= yes; 0 = no)		0.74	0.04
Applying fertilizer (1= yes; 0 = no)		0.34	0.04
Implementing soil conservation measures (1= yes; 0 = no)		0.58	0.05
Controlling shade levels on coffee plots (1= yes; 0 = no)		0.48	0.05
Explanatory Variables			
Farm and Management Characteristics			
Total farm size (hectares)	Hectares	2.58	0.27
Producer has multiple plots (1= yes; 0 = no)	Mult_Plot	0.43	0.05
Average plot distance from home (minutes)	Plot_Dist	32.94	3.00
Producer uses hired labour (1= yes; 0 = no)	Hired_Lab	0.45	0.05
Producer receives coffee subsidies (1= yes; 0 = no)	Coff_Subs	0.42	0.05
Socio-economic Profile			
Age of producer (years)	Age	48.04	1.39
Producer is male (1= yes; 0 = no)	Gender	0.75	0.04
Education level of producer (years of schooling)	Education	4.72	0.33
Total family size (number)	House_Size	4.84	0.23
Household member has salaried employment (1= yes; 0 = no)	Off_Farm_Emp	0.23	0.04
Social Capital			
Membership in farmers' group or association (1= yes; 0 = no)	Farm_Member	0.25	0.04
Number of adult relatives in the village	King_Ties	7.71	0.60
Village Dummy			
Village of Chilapa (1= yes; 0 = no)	Village_Dum	0.63	

The adoption of agricultural innovations in developing countries has long attracted the attention of researchers (Feder et al., 1985; Doss, 2006). For the purposes of empirical analysis, studies have generally distinguished between farm and aggregate level adoption. At the farm level, the decision process is usually modelled as a series of intrinsic and extrinsic factors that determine individual adoption behaviour (Meijer, Catacutan, Ajayi, Sileshi, & Nieuwenhuis, 2015). A macro level approach by contrast measures the aggregate use of a practice within a given population or territory (Feder et al., 1985). Generally, studies seek ways of relating adoption choices to farm and socio-economic factors through the use of econometric models (Knowler & Bradshaw, 2007). While various factors are hypothesized to play a role, results from different studies also indicate that the effect of certain variables is not always uniform.

Among the possible determinants of adoption include factors relating to farm and management characteristics. For instance, having a larger plot can facilitate adoption since larger producers on average tend to have lower fixed costs and can expect higher yields after implementing a new practice (Boahene et al., 1999). On the other hand, producers with larger holdings might choose to practice more extensive forms of agriculture by not adopting intensive farming methods (Kassie, Jaleta, Shiferaw, Mmbando, & Mekuria, 2013). Working multiple plots is expected however to have a positive effect since conditions on different land parcels may increase the need for certain SMPs to be implemented (Nkegbe & Shankar, 2014). Access to hired labour also potentially facilitates the uptake of new practices by enabling producers to overcome workforce related production constraints (Yila & Thapa, 2008). On the other hand, the average distance from household to plot is expected to make adoption less likely due to the transaction costs involved in moving heavy materials and inputs (Bizimana, Nieuwoudt, & Ferrer, 2002). From a policy perspective, government support programs can give producers the confidence to invest in new practices despite uncertainty (Teklewold et al., 2013). The main coffee program in Mexico (*Fomento Productivo*) offers subsidies to producers who are able to document that they marketed coffee during each of the three years prior to applying for benefits. In 2013, 29% of producers were beneficiaries under the program (SAGARPA, 2014).

Studies have also suggested that certain socio-economic characteristics of producers and their households influence adoption choices. For instance, age and education are considered possible determinants although the direction of the effect is not always clear. Older producers or those with higher levels of education will have accumulated a greater stock of information, which could mitigate the risks involved in adopting new practices (Doss, 2006). On the other hand, age is associated with a shorter planning horizon and reduced physical capacities. As well, producers with higher levels of educational attainment often have greater off-farm employment opportunities, potentially limiting the range of plot work they choose to undertake. Several studies have confirmed that education has played a major role in enabling rural households in Mexico to pursue non-agricultural employment (Janvry & Sadoulet, 2001; Yúnez-Naude & Edward Taylor, 2001; Barham et al., 2011).

Family members are often the principal source of labour for smallholders (Netting, 1993). As a result, household size is expected to have a positive effect on the likelihood of adoption. With regard to gender, it has been argued that female producers often have less access to agricultural resources (credit, land, labour, inputs, information…) which could limit their ability to adopt new practices (De Groote & Coulibaly, 1998; Teklewold et al., 2013). In Mexico, 28% of coffee producers are women and female participation rates have increased in recent years (AMECAFE, 2012).

Access to off-farm employment can be a catalyst for adoption especially since coffee producing households in Mexico are increasingly diversifying their sources of income (Barham et al., 2011; Padrón & Burger, 2015). Nonetheless, the anticipated effect of this trend on adoption is ambiguous. Off-farm revenues allow for greater consumption smoothing, thereby reducing the perceived risks associated with implementing new practices. On the other hand, the opportunity cost of undertaking plot work increases.

Finally, different aspects of a producer's endowment in social capital have been shown in various contexts to influence adoption choices (Narayan & Pritchett, 1999; Isham, 2002; Cramb, 2005). In regions where markets and information sources are imperfect, local forms of association enable farmers to exchange knowledge and experiences (Kassie et al., 2013). As such, membership in rural institutions can encourage adoption by giving producers the opportunity to learn about new or unfamiliar practices. In addition, extended families in many developing countries share resources through informal insurance mechanisms which allow households to better manage risk (Fafchamps & Gubert, 2007). Coffee growers with greater kingship ties might therefore be more likely to adopt SMPs. On the other hand, family networks characterized by strong links between members ("bonding social capital") can be closed off to outside ideas in such a way that the adoption of new agricultural practices is discouraged (Warriner & Moul, 1992).

Certain variables of interest were not considered during the final analysis. For instance, tenure arrangements were not taken into account since households in the sampled area are governed by a communal land resources committee, an institution independent of the municipality. Under this system, producers hold usufruct rights over their coffee parcels. While a 1991 amendment to the Mexican constitution made the privatization of communal lands (*ejidos*) possible, it remains the dominant type of ownership, especially in the rural areas of the south (Brown, 2004).

In addition, the survey data indicated a perfect correlation between membership in a producer association and participation in extension training. Respondents who were members of an organization all belonged to organic coffee cooperatives and no other type of producer group was present in the area. Producers who were selected to participate in the semi-structured interviews confirmed that independent growers had not received technical assistance during the 2013/2014 coffee season. Generally speaking, outside of cooperatives, the availability of technical support depends largely on government funding. However, since 2009, public spending earmarked for the coffee sector has been declining (FUNDAR, 2015) and in 2013 no funds were allocated for the provision of extension services. Given the perfect correlation between participation in extension training and membership in a producer organization, it was decided that only the latter variable would be retained.

3. Results and Discussion

3.1 General Performance of the Model

The MVP model was used to estimate the effects of explanatory variables on adoption behaviour. The natural logarithm of certain variables was taken in order to correct for skewed distributions and allow for a better model fit. In addition, we tested for the presence of multicollinearity by running the variance inflation factors (VIF). The results of the test (VIF <2.30) did not suggest any potential bias. The Wald test [chi-square (56) $=98.570***$] also indicates that the data fit of our model performs reasonably well. In addition, a likelihood ratio test was carried out in order to compare four univariate probit models to the multivariate solution (Table 2). The test result was statistically meaningful [chi-square (6) $=14.216**$] indicating that the error correlations between practices are jointly significant. This justifies our use of the MVP model rather than separate probit regressions and supports the hypothesis that SMP adoption choices are interrelated. Specifically, coffee producers who used fertilizer were more likely to adopt soil conservation measures (rho $=0.642***$) as well as manage shade levels (rho $= 0.359*$). However, renovating coffee tree stocks was not correlated with other categories of SMPs and can therefore be considered a 'stand-alone' practice.

Table 2. Correlation coefficients for MVP adoption equations

Equations	Coefficients (rho)	Standard error	t-value
Renovating coffee tree stock vs Applying fertilizer	0.171	0.242	0.710
Renovating coffee tree stocks vs Implementing soil conservation measures	0.202	0.196	1.030
Renovating coffee tree stock vs Managing shade levels	0.217	0.204	1.060
Applying fertilizer vs Implementing soil conservation measures	0.642	0.244	2.630***
Applying fertilizer vs Managing shade levels	0.359	0.214	1.680*
Implementing soil conservation measures vs Managing shade levels	−0.015	0.187	−0.080
Likelihood ratio test chi-square (6) =14.216**			

$p*** <0.01, p** <0.05, p* <0.10$.

3.2 Farm Characteristics and Management Practices

Tables 3 summarizes the MVP estimates obtained for the four categories of SMPs. The results indicate that larger coffee producers are less likely to use fertilizer or manage shade levels. With one exception, surveyed producers who adopted fertilizer had chosen to use organic materials. However, this result is not unusual given that coffee growers in Mexico tend not to use chemical fertilizer due to its perceived high cost (AMECAFE, 2011). While organic fertilizer can be more cost-effective, its availability depends crucially on local materials. This makes it difficult for producers to increase the amounts that can be obtained over a certain threshold (Valkila, 2009). As a result, fertilizer use will be more attractive to smaller producers since the application of relatively limited levels of organic fertilizer will have a greater positive impact on soil fertility than what larger producers can achieve.

From a policy perspective, these results suggest that smallholder coffee growing does not in itself represent a barrier to the adoption of SMPs. Smaller-scale coffee producers have greater facility implementing certain practices by utilizing their own resources as compared to larger producers who are more likely to require access to off-farm inputs before making a decision to adopt. When the markets for these inputs (credit, labour, fertilizer…) are imperfect, larger producers will face greater constraints in terms of their ability to implement SMPs. Viewed in this light, targeted policy reforms aimed at increasing the efficiency of agricultural input markets could indirectly facilitate the adoption process for producers with larger coffee plots.

Table 3. Coefficient estimates of the MVP model

Variables	Renovating coffee tree stocks			Fertilizer use		
	Coefficients	Standard error	t-value	Coefficients	Standard error	t-value
Hectares (log)	−0.246	0.265	−0.930	−0.609	0.265	−2.300**
Mult_Plot	0.023	0.363	0.060	0.069	0.406	0.170
Plot_Dist (log)	−0.078	0.110	−0.710	0.032	0.116	0.270
Hired_Lab	0.498	0.409	1.220	0.478	0.414	1.150
Coff_Subs	0.127	0.345	0.370	−0.436	0.391	−1.120
Age	−0.198	0.079	−2.490**	0.136	0.085	1.610
Age (squared)	0.002	0.001	2.520**	−0.001	0.001	−1.300
Gender	−0.717	0.459	−1.560	0.868	0.467	1.860*
Education (log)	−0.070	0.295	−0.240	0.108	0.272	0.400
House_Size (log)	0.433	0.427	1.010	−0.489	0.419	−1.170
Salar_Emp	0.101	0.403	0.250	0.283	0.464	0.610
Farm_Member	1.858	0.649	2.860***	2.780	0.481	5.780***
King_Ties (log)	0.570	0.249	2.290**	0.578	0.261	2.220**
Village_Dum	−0.144	0.471	−0.310	−0.455	0.462	−0.980
Constant	4.010	2.122	1.890*	−5.703	2.194	−2.600***

Variables	Soil conservation measures			Managing shade levels		
	Coefficients	Standard error	t-value	Coefficients	Standard error	t-value
Hectares (log)	−0.183	0.230	−0.800	−0.402	0.219	−1.830*
Mult_Plot	0.341	0.347	0.980	0.017	0.317	0.050
Plot_Dist (log)	0.122	0.104	1.180	−0.091	0.098	−0.930
Hired_Lab	0.466	0.361	1.290	−0.273	0.351	−0.780
Coff_Subs	0.009	0.324	0.030	0.084	0.314	0.270
Age	0.136	0.073	1.860*	−0.108	0.068	−1.590
Age (squared)	−0.002	0.001	−2.070**	0.001	0.001	1.660*
Gender	0.225	0.400	0.560	0.362	0.357	1.010
Education (log)	−0.482	0.293	−1.650*	−0.329	0.244	−1.350
House_Size (log)	−1.032	0.394	−2.620***	−0.886	0.370	−2.390**
Salar_Emp	0.309	0.386	0.800	0.599	0.386	1.550
Farm_Member	2.085	0.543	3.840***	1.051	0.359	2.930***
King_Ties (log)	0.404	0.228	1.770*	0.252	0.204	1.230
Village_Dum	−0.337	0.433	−0.780	−0.154	0.406	−0.380
Constant	−1.790	1.889	−0.950	3.917	1.901	2.060**

Wald chi-square (56)   98.570***
Log pseudo-likelihood   −213.394
Number of observations   119

$p^{***} <0.01, p^{**} <0.05, p^{*} <0.10.$

Apart from the total size of coffee holdings, the farm and management variables included in the model were not significant. For instance, the number of plots as well as average distance did not have any noticeable effect on adoption rates. In addition, access to labour was not a determining factor. Producers who responded to the semi-structured interviews explained that workers are normally hired to help during the coffee harvest. Since plot improvement work is undertaken earlier in the season this explains why access to non-household labour did not translate into higher adoption rates. Moreover, hired labourers tend to come from neighbouring rural areas where coffee is not grown due to lower altitudes. During the pre-harvest period, these workers usually cultivate their own agricultural plots meaning the supply of available labour to coffee producers is highly seasonal.

Also of note is the fact that producers who benefited from government subsidies did not have higher adoption rates in any of the SMP categories. The absence of noticeable results derives from a general policy failure in Mexico to implement stable coffee support programs for producers. As a case in point, since 2003 there have been four major changes to the largest subsidy program (*Fomento Productivo*) in terms of both eligibility and funding criteria. Funding allocated to the coffee sector is also renegotiated on an annual basis and tends to fluctuate sharply (Fundar, 2015). The resulting uncertainty around subsidies makes it challenging for producers to plan their coffee investment strategies around the level of government support they can expect to receive. This in turn reduces their incentive to adopt SMPs. The question of whether or not non market incentives should be used to support coffee production remains of course a matter open to debate. However, the lack of results suggests that coffee policies in Mexico have not encouraged the uptake of SMPs and that producers seem to be passive beneficiaries of current programs.

3.3 Producers' Socio-Economic Profile

The results indicate that age is a determining factor although the effect is not uniform. Older producers are less likely to renovate their stock of coffee trees. This preference for drawing down capital stocks could be the result of shorter planning horizons associated with age. It might also reflect recent transformations in the Mexican economy, notably the increasing returns from education and off-farm employment over coffee as highlighted by Barham et al. (2011). Specifically, older producers might not renovate their stocks if they expect that younger household members will rely less on coffee growing for their livelihood. Such an expectation would reduce the incentive to improve coffee tree stocks since the benefits in terms of improved yields would only materialize over the long run. In this respect, new policies designed to promote the efficient sale and transfer of landholdings could encourage older producers to maintain their coffee trees. In contrast, the MVP model reveals that age is positively correlated with the adoption of soil conservation practices. This outcome is likely due to the fact that older producers on average have accumulated a greater stock of information on the importance and proper implementation of such measures.

The variable for gender indicated that male coffee producers are more likely to use fertilizer. As previous studies have shown, male producers often have social advantages when it comes to accessing agricultural inputs (Udry, 1996; De Groote & Coulibaly, 1998). The result might therefore indicate that women coffee growers have fewer resources for making fertilizer. Other factors might also be relevant such as the possibility that female producers face greater difficulties when it comes to transporting inputs and materials. Given the growing number of female coffee producers in Mexico, initiatives should focus on removing the constraints that women face when making adoption choices.

The results also indicate that a producer's level of education negatively influences the adoption of soil conservation measures. While education is generally regarded as helpful to adoption, producers with more schooling often have better opportunities for off-farm work and may limit the number of practices they choose to implement (Umatsu & Mishra, 2010). Moreover, several of the soil conservation recommendations in particular the building of ditches, terraces and contour barriers are especially labour-intensive. On average, the opportunity costs of implementing these measures will be greater for producers with higher levels of educational attainment.

Also of note is the fact that household size is negatively correlated with the adoption of soil conservation and shade control measures. In smallholder systems it is generally assumed that household size has a positive effect on adoption since it can be interpreted as a proxy for available farm labour (Netting, 1993). However, while household labour used to be an important factor of production, the growth in returns to education in Mexico over the last two decades has increased both school attendance and retention rates. Comparing the investments made by rural households in education and coffee production can provide some insight into the negative relationship observed between household size and adoption. In a study by Barham et al. (2011), the average educational costs (both direct and opportunity costs) per child for a coffee producing household were compared with the daily opportunity cost of making capital improvements to coffee plots. Due to the comparatively higher returns from schooling, it was estimated that rural households in Mexico invest on average twice as much in education as they do in coffee production. The MVP results appear to confirm this trade-off and suggest that larger households (with generally more children) tend to invest more in education and less in plot improvement work.

3.4 Social Capital

Finally, the estimates reveal that the adoption of SMPs depends crucially on different aspects of a producer's endowment in social capital. For every SMP category, membership in a local producer association increased the likelihood of adoption. As was previously mentioned, the only institutions present in the area visited were organic coffee cooperatives and all surveyed members had benefited from some form of extension support provided by the cooperative to which they belonged. The results therefore indicate that training provisions have a positive effect on the accumulation of coffee-related knowledge by producers and facilitate the adoption of new practices. Furthermore, coffee cooperatives arrange frequent member meetings, giving producers the opportunity to access information and share experiences that sensitize non-adopters to the benefits of SMPs. One of the main obstacles to adoption is that the costs of implementing new practices usually exceed the short-term profits, despite yielding important benefits over the long-term. In this respect, the social function of local institutions such as cooperatives is to provide a venue through which producers can learn from others during the adoption phase. These results corroborate what previous studies have highlighted regarding the impact of social capital on adoption choices in sustainable agriculture (Warriner & Moul, 1992; Swinton, 2000; Cramb, 2005; Rodríguez-Entrena & Arriaza, 2013).

In addition, the number of adult relatives in the village was positively correlated with the renovation of coffee tree stocks, fertilizer use and soil conservation measures. The evidence suggests that informal support mechanisms among family members allow producers to mitigate the risks they face when implementing SMPs. Additionally, in many coffee growing regions of Mexico, extended families participate in collaborative work exchanges (Lewis, 2005) which effectively increase the supply of labour available to producers. These exchanges seem to confer advantages on producers with greater kinship ties since they allow extended families to "self-exploit" their own labour in the sense of the term as used by Chayanov (1986). Given the communal land ownership system, it is also possible that producers with extended kinship networks face greater social pressures to adequately maintain their land holdings. As well, improving the productive state of coffee plots by adopting SMPs could confer on adopters a higher social status with the benefits derived being greater for producers with larger kinship ties.

4. Conclusion

Improving physical conditions on coffee plots through the adoption of SMPs is vital for increasing yields, reducing rural household poverty and ensuring the long-term sustainability of the coffee sector in Mexico. In this study, we analyzed the probability of adopting multiple SMPs by coffee producers using data obtained from field survey interviews. An MVP model was used to jointly analyze the adoption of multiple practices while accounting for possible correlations between them.

The results showed a relationship between the adoption of SMPs and the total size of coffee holdings, socio-economic variables (age, gender, education, household size) and a producer's endowment in social capital (association membership, kinship ties) the latter revealing itself to be a key factor in the overall adoption process. This confirms what previous studies have highlighted, namely that direct forms interaction among producers are one of the principal driving forces behind the adoption of sustainable agricultural practices. We therefore conclude that policy efforts aimed at encouraging coffee growers to adopt SMPs must account for the crucial role that social capital plays in smallholder coffee systems. From a development perspective, this implies strengthening the institutions that producers use for interacting with each other as well as encouraging greater organization at the local level. Additionally, since government subsidies had no effect on adoption rates, current coffee programs should be reformed with a view to encouraging the uptake of SMPs.

Regarding the limits of the model, it should be noted that the use of binary variables for classifying producers as adopters has been criticized as it leaves aside the issue of adoption *intensity*. In addition, the results obtained must account for the limited geographical coverage of the area sampled. While the model did capture certain trends, it is often the case with microstudies that variations between households with regard to the explanatory variables are not sufficiently large to produce meaningful results.

One promising option for further research would be to study the perceptions of coffee farmers as they relate to the adoption of SMPs. Indeed, a variety of methods could be drawn upon in order to identify and analyze which factors producers regard as being the principal barriers to adoption. Additional research in this area could contribute to further understanding the livelihood strategies of rural households and inform the design of more effective coffee policies in Mexico.

Acknowledgments

This study was supported by the Social Sciences and Humanities Research Council of Canada (SSHRC) through a General Research Fund.

References

Ali, L., Mangheni, N. M., Sanginga, P. C., Delve, R. J., Mastiko, F., & Miiro, R. (2007). Social capital and adoption of soil fertility management technologies in Tororo district, Uganda. In A. Bationo, B. Waswa, J. Kihara, & J. Kimetu (Eds.), *Advances in integrated soil fertility management in sub-Saharan Africa: challenges and opportunities* (pp. 947-954): Springer Netherlands.

AMECAFE (Asociación Mexicana de la Cadena Productiva del Café). (2011). *Plan de innovación en la cafeticultura de México.* Mexico City, Mexico.

AMECAFE (Asociación Mexicana de la Cadena Productiva del Café). (2012). Padron Nacional Cafetalero. http://amecafe.org.mx/padron-nacional-cafetalero/

Bacon, C. (2005). Confronting the coffee crisis: Can fair trade, organic, and specialty coffees reduce small-scale farmer vulnerability in Northern Nicaragua? *World Development, 33*(3), 497-511. http://dx.doi.org/10.1016/j.worlddev.2004.10.002

Barham, B. L., Callenes, M., Gitter, S., Lewis, J., & Weber, J. (2011). Fair Trade/organic coffee, rural livelihoods, and the "agrarian question": Southern Mexican coffee families in transition. *World Development, 39*(1), 134-145. http://dx.doi.org/10.1016/j.worlddev.2010.08.005

Barham, B. L., & Weber, J. G. (2012). The economic sustainability of certified coffee: Recent evidence from Mexico and Peru. *World Development, 40*(6), 1269-1279. http://dx.doi.org/10.1016/j.worlddev.2011.11.005

Beuchelt, T. D., & Zeller, M. (2011). Profits and poverty: Certification's troubled link for Nicaragua's organic and fairtrade coffee producers. *Ecological Economics, 70*(7), 1316-1324. http://dx.doi.org/10.1016/j.ecolecon.2011.01.005

Bizimana, C., Nieuwoudt, W. L., & Ferrer, S. R. (2002). Factors influencing adoption of recommended farm practices by coffee farmers in Butare, southern Rwanda. *Agrekon, 41*(3), 237-248. http://dx.doi.org/10.1080/03031853.2002.9523597

Blackman, A., & Naranjo, M. A. (2012). Does eco-certification have environmental benefits? Organic coffee in Costa Rica. *Ecological Economics, 83*, 58-66. http://dx.doi.org/10.1016/j.ecolecon.2012.08.001

Boahene, K., Snijders, T. A. B., & Folmer, H. (1999). An integrated socioeconomic analysis of innovation adoption: The case of hybrid cocoa in Ghana. *Journal of Policy Modeling, 21*(2), 167-184. http://dx.doi.org/10.1016/S0161-8938(97)00070-7

Brown, J. (2004). Ejidos and communidades in Oaxaca, Mexico: Impact of the 1992 reforms. *Reports on Foreign Aid and Development no. 120*. Seattle, Washington: Rural Development Institute

Castellanos, E. J., Tucker, C., Eakin, H., Morales, H., Barrera, J. F., & Díaz, R. (2013). Assessing the adaptation strategies of farmers facing multiple stressors: Lessons from the coffee and global changes project in Mesoamerica. *Environmental Science & Policy, 26*, 19-28. http://dx.doi.org/10.1016/j.envsci.2012.07.003

Chayanov, A. V. (1986). *The Theory of Peasant Economy.* Madison, Wisconsin: The University of Wisconsin Press.

CONAPO (Consejo Nacional de Población). (2010). *Índice de marginación por entidad federativa y municipio 2010.* Mexico City, Mexico.

Cramb, R. A. (2005). Social capital and soil conservation: Evidence from the Philippines. *Australian Journal of Agricultural and Resource Economics, 49*(2), 211-226. http://dx.doi.org/10.1111/j.1467-8489.2005.00286.x

De Groote, H., & Coulibaly, N. G. (1998). Gender and generation: An intra-household analysis on access to resources in southern Mali. *African Crop Science Journal, 6*(1), 79-95. http://dx.doi.org/10.4314/acsj.v6i1.27827

Doss, C. R. (2006). Analyzing technology adoption using microstudies: Limitations, challenges, and opportunities for improvement. *Agricultural Economics, 34*(3), 207-219. http://dx.doi.org/10.1111/j.1574-0864.2006.00119.x

Fafchamps, M., & Gubert, F. (2007). The formation of risk sharing networks. *Journal of Development Economics, 83*(2), 326-350. http://dx.doi.org/10.1016/j.jdeveco.2006.05.005

Feder, G., Just, R. E., & Zilberman, D. (1985). Adoption of agricultural innovations in developing countries: A survey. *Economic Development and Cultural Change, 33*(2), 255-298.

FUNDAR (Fundar Centro de Análisis e Investigación). (2015). Café 2005-2012. Retrieved from http://subsidiosalcampo.org.mx/analiza/padrones/cafe/

Isham, J. (2002). The effect of social capital on fertiliser adoption: Evidence from rural Tanzania. *Journal of African Economies, 11*(1), 39-60. http://dx.doi.org/10.1093/jae/11.1.39

Janvry, A. d., & Sadoulet, E. (2001). Income strategies among rural households in Mexico: The role of off-farm activities. *World Development, 29*(3), 467-480. http://dx.doi.org/10.1016/S0305-750X(00)00113-3

Kassie, M., Jaleta, M., Shiferaw, B., Mmbando, F., & Mekuria, M. (2013). Adoption of interrelated sustainable agricultural practices in smallholder systems: Evidence from rural Tanzania. *Technological Forecasting and Social Change, 80*(3), 525-540. http://dx.doi.org/10.1016/j.techfore.2012.08.007

Knowler, D., & Bradshaw, B. (2007). Farmers' adoption of conservation agriculture: A review and synthesis of recent research. *Food Policy, 32*(1), 25-48. http://dx.doi.org/10.1016/j.foodpol.2006.01.003

Lewis, J. (2005). Strategies for survival: Migration and fair trade-organic coffee production in Oaxaca, Mexico. University of California, San Diego The Center for Comparative Immigration Studies CCIS.

Meijer, S. S., Catacutan, D., Ajayi, O. C., Sileshi, G. W., & Nieuwenhuis, M. (2015). The role of knowledge, attitudes and perceptions in the uptake of agricultural and agroforestry innovations among smallholder farmers in sub-Saharan Africa. *International Journal of Agricultural Sustainability, 13*(1), 40-54. http://dx.doi.org/10.1080/14735903.2014.912493

Moguel, P., & Toledo, V. M. (1999). Biodiversity conservation in traditional coffee systems of Mexico. *Conservation Biology, 13*(1), 11-21. http://dx.doi.org/10.1046/j.1523-1739.1999.97153.x

Narayan, D., & Pritchett, L. (1999). Cents and sociability: Household income and social capital in rural Tanzania. *Economic Development and Cultural Change, 47*(4), 871-897. http://dx.doi.org/10.1086/452436

Netting, R. (1993). *Smallholders, householders: Farm families and the ecology of intensive, sustainable agriculture.* Stanford, CA: Stanford University Press.

Nkegbe, P. K., & Shankar, B. (2014). Adoption intensity of soil and water conservation practices by smallholders: Evidence from Northern Ghana. *Bio-based and Applied Economics, 3*(3), 159-174. http://dx.doi.org/10.13128/BAE-13246

Padrón, B. R., & Burger, K. (2015). Diversification and labor market effects of the Mexican coffee crisis. *World Development, 68*, 19-29. http://dx.doi.org/10.1016/j.worlddev.2014.11.005

Rodríguez-Entrena, M., & Arriaza, M. (2013). Adoption of conservation agriculture in olive groves: Evidences from Southern Spain. *Land Use Policy, 34*(0), 294-300. http://dx.doi.org/10.1016/j.landusepol.2013.04.002

Ruben, R., & Fort, R. (2012). The impact of Fair Trade certification for coffee farmers in Peru. *World Development, 40*(3), 570-582. http://dx.doi.org/10.1016/j.worlddev.2011.07.030

SAGARPA (Secretaría de Agricultura Ganadería Desarrollo Rural Pesca y Alimentación). (2014). Padrones de Beneficiarios. Retrieved from http://www.sagarpa.gob.mx/agricultura/Paginas/Padrones.aspx

Swinton, S. M. (2000). *More social capital, less erosion: Evidence from Peru's Altiplano.* Paper presented at the Annual Meeting of the American Agricultural Economics Association, Tampa, FL.

Teklewold, H., Kassie, M., & Shiferaw, B. (2013). Adoption of multiple sustainable agricultural practices in rural Ethiopia. *Journal of Agricultural Economics, 64*(3), 597-623. http://dx.doi.org/10.1111/1477-9552.12011

Udry, C. (1996). Gender, agricultural production, and the theory of the household. *Journal of Political Economy, 104*(5), 1010-1046. http://dx.doi.org/10.1086/262050

Umatsu, H., & Mishra, A. (2010). *Net effect of technology adoption by US farmers.* Paper presented at the Southern Agricultural Economics Association Annual Meetings, Orlando, FL.

Valkila, J. (2009). Fair Trade organic coffee production in Nicaragua — sustainable development or a poverty trap? *Ecological Economics, 68*(12), 3018-3025. http://dx.doi.org/10.1016/j.ecolecon.2009.07.002

Warriner, G. K., & Moul, T. M. (1992). Kinship and personal communication network influences on the adoption of agriculture conservation technology. *Journal of Rural Studies, 8*(3), 279-291. http://dx.doi.org/10.1016/0743-0167(92)90005-Q

Yila, O. M., & Thapa, G. B. (2008). Adoption of agricultural land management technologies by smallholder farmers in the Jos Plateau, Nigeria. *International Journal of Agricultural Sustainability, 6*(4), 277-288. http://dx.doi.org/10.3763/ijas.2008.0374

Yúnez-Naude, A., & Edward Taylor, J. (2001). The determinants of nonfarm activities and incomes of rural households in Mexico, with emphasis on education. *World Development, 29*(3), 561-572. http://dx.doi.org/10.1016/S0305-750X(00)00108-X

7

Baseline Farmer Survey of Smallholder Cocoa Farming Systems in Ghana

F. Aneani[1] & F. Padi[1]

[1]Cocoa Research Institute of Ghana, P. O. Box 8, New Tafo-Akim, Ghana

Correspondence: Francis Aneani, Cocoa Research Institute of Ghana, P. O. Box 8, New Tafo-Akim, Ghana. E-mail: faneani@gmail.com

Abstract

The effects of the prevalent low-input systems of production, over-aged farms, and unstable climate are worsened by weaknesses inherent in the formal system of production and distribution of recommended cocoa varieties. Generally, the purpose of the baseline survey was to obtain perception of the farmers in the target communities on the possibility of re-introduction of cocoa in denuded and marginal areas which were previously cropped to cocoa, but now food crops; determine farmers' interest in planting new cocoa varieties; and determine farmer behavior in the use of technologies of cocoa farm establishment and maintenance. The survey was conducted in the period starting from 10th December, 2013 to 5th July, 2014 at Asesewa (Konkoney) in the Eastern Region, Akomadan and Afrancho in the Ashanti Region, as well as Kenyasi (Atwidie), Bechem (Breme) and Acherensua (Kokontreso) in Brong-Ahafo Region. The project sites and 192 respondents were purposively sampled. A standard questionnaire was employed to interview the respondents. Data analysis indicated that 40.0% of the respondents would want their farms to be rehabilitated whereas 60.0% indicated they would not. Also, 98.4% of them reported a higher possibility of re-introduction of cocoa in the denuded and marginal areas whilst 1.6% indicated that it was impossible. Additionally, 79.6% of the respondents expressed interest in testing any new cocoa varieties on their farms as part of the project while 20.4% were disinterested. In conclusion, the survey has indicated that re-introduction of cocoa in marginal and denuded area is highly probable.

Keywords: baseline survey, sustainable intensification, farming systems, cocoa establishment, rehabilitation, Ghana

1. Introduction

Cocoa is an understorey tree and, in Ghana, the crop is traditionally cultivated under the shade of selectively thinned forest. The forest shade trees contribute to the build-up of organic matter, nutrient recycling and the maintenance of biodiversity. Poor farmland management and excessive timber extraction have resulted in the deforestation and degradation of most of the natural forest sites suitable for cocoa cultivation (World Bank, 1987).

Sustainability of the smallholder cocoa production systems in West and Central Africa is threatened by a myriad of constraints. The effects of the prevalent low-input systems of production, over-aged farms, and unstable climate are exacerbated by weaknesses inherent in the formal system of production and distribution of recommended varieties (Asare et al., 2010). Sustainable production can be achieved by increasing farmers' access to cultivars improved for tolerance to stress, and by providing a stimulating framework for rejuvenating aged farms with genetically improved varieties.

Three major constraints have largely accounted for the current low productivity and extensive systems of cocoa farming in West and Central Africa. First, access to, and adoption of improved varieties are between 10% and 40% across the sub-region (Gockowski, 2011). In spite of 75 years of development of improved varieties, inefficiencies in the formal system of seed delivery and lack of appreciation of improved varieties have accounted for this low adoption rates (Asare et al., 2010). Second, a large proportion of the commercial plantations have aged beyond their economic lifespan of 30 years. With lack of capacity and incentives to rejuvenate these farms, and the difficulties that are inherent in re-establishment of cocoa on previously used land, extensive systems of production have been adopted with a consequent destruction of available secondary forests. The effects of these challenges on productivity are exacerbated by a climate change phenomena that have created an unstable production environment, with increasing frequency of droughts, increasing temperatures and reducing humidity (Anim-Kwapong & Frimpong, 2006). The rapidly changing climate is also increasing the susceptibility of current varieties to the major pests, and previously considered minor pests which are now gaining prominence. The most notable effect of the climate change phenomena on production is high mortality of tree stocks within the first two years of establishment (Gockowski, 2011).

To sustain the cocoa industry, a programme of re-establishing cocoa in denuded and degraded forest areas which previously had carried cocoa was started by Cocoa Research Institute of Ghana (CRIG) in 1986 (Anim-Kwapong & Teklehaimanot, 2001). The program was underpinned by the supply of improved hybrid varieties (MASDAR, 1998).

In Ghana, improved hybrid cocoa varieties were introduced for the past 25 years or more (MASDAR, 1998). There is evidence that hybrid varieties outperform the older varieties in two ways by producing trees that bear fruits in three years compared with at least five years for the older varieties and by producing more pods per tree (MASDAR, 1998). But hybrid cocoa trees require optimal weather conditions and good farming practices such as the application of chemical inputs, adoption of new planting procedures, pruning, and spraying. Hybrid varieties also require frequent harvesting throughout the year (MASDAR, 1998; Business & Financial Times, 2014).

The current project aimed at stimulating the adoption of intensive systems of production, and the attainment of sustained, increased productivity in the predominantly smallholder cocoa farming systems in Ghana. Specific objectives of the baseline survey were to:

- Determine the perception of the respondents on cocoa farm rehabilitation and the inherent constraints;
- Obtain perception of the farmers in the target communities on the possibility of re-introduction of cocoa in marginal areas which were previously cropped to cocoa, but now food crops;
- Determine farmers' interest in planting new cocoa varieties.

2. Materials and Methods

2.1 Study Areas

The study areas were Akomadan and Afrancho in the Ashanti Region; Asesewa (Konkoney) in the Eastern Region; as well as Kenyasi (Atwidie), Bechem (Breme) and Acherensua (Kokontreso) in Brong-Ahafo Region (Figure 1)

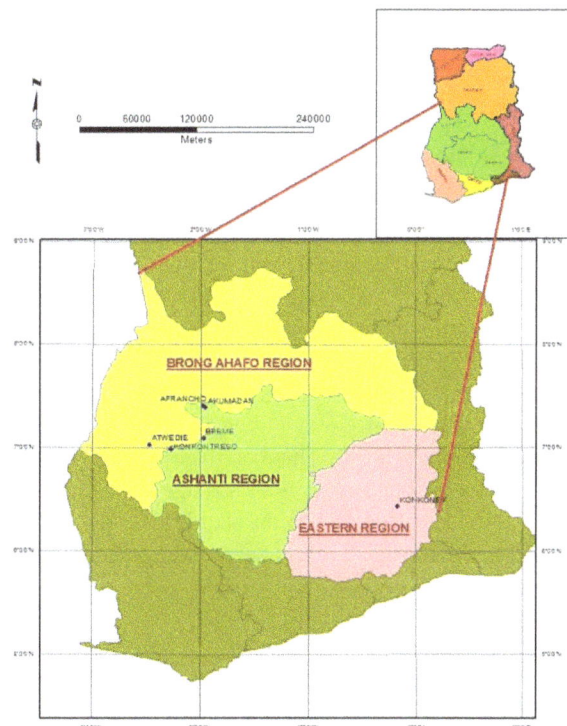

Figure 1. A map indicating the study areas

2.2 Sampling Procedure

The project sites were purposively selected based on the criteria that they are located in denuded and marginal cocoa areas while a sample size of 192 respondents was selected. In each community, the respondents were assembled at a meeting place and each was randomly interviewed. The number of respondents in each community is presented in Table 1.

Table 1. Study areas and number of respondents selected for the interview.

Region	District	Location	No. of farmers
Ashanti	Offinso	Akumandan	30
Ashanti	Offinso	Afrancho	60
Eastern	Oyoko-Nankese	Konkoney	14
Brong Ahafo	Bechem	Breme	32
Brong Ahafo	Goaso	Atwedie	43
Brong Ahafo	Goaso	Konkontreso	13
Total			192

2.3 Data Collection

The survey was conducted in the period starting from 10th December, 2013 to 5th July, 2014 at Asesewa (Konkoney) in the Eastern Region; Akomadan and Afrancho in the Ashanti Region; as well as Kenyasi (Atwidie), Bechem (Breme) and Acherensua (Kokontreso) in Brong-Ahafo Region. The project team interacted with a group of farmers who were first briefed on the purpose of the gathering, the objectives of the project and the expected contribution of the farmers to the project objectives. This was done to create awareness of the project activities in the target areas. Feedback was also obtained from the gathering in the form of questions and answers with some explanations to ensure the clarification of doubtful issues. Then respondents were individually interviewed with a questionnaire to obtain data on their demographic and farm characteristics, perception on possibility of rehabilitation of old cocoa and re-introduction of cocoa in the denuded areas, farmers' interest in planting new cocoa varieties, and willingness and ability of respondents to establish and maintain the project farms.

2.4 Data Analysis

Data were analyzed, after processing and computer entry, with Statistical Package for Social Science software (SPSS, Version 19) using quantitative analytical techniques such as frequencies, percentages, cross-tabulations, etc.

3. Results

3.1 Demographic and Farm Characteristics

From Table 2 which describes the demographic and farm features of the respondents, 98.4% were cocoa farmers whereas 1.6% also food crop farmers. The respondents consisted of males (74.9%) and females (25.1%). In terms of ownership of the farm, 58.1% of the respondents were owners while 41.9% were caretakers. More of the respondents had one cocoa farm (67.4%) while those having two farms accounted for 22.1% (Table 2). Also, generally the educational level of the respondents was low. About 46% of them had middle/JHS education and 22.4% had no education at all (Table 2). From Table 3 which presents the summary of additional demographic and farm characteristics of the respondents, the average age of the respondents was 51 years, with a range of 23 – 95 years. The average year of farming experience was 24 years, having a range of 1 – 65 years. The mean farm size in hectares (ha) of the cocoa producers was 3.0. The mean age of the cocoa farms was 13 years. The cocoa farms produced a mean output of 614.4 kg per farmer with a range of 192.0 kg to 2 560.0 kg per farmer. The food crop farm characteristics were not reported due to unreliable data given by the respondents. It should be noted that the number of respondents in the results section keeps changing because of non-responses in the data.

Table 2. Distribution of demographic and farm characteristics of respondents in the study areas

Characteristic	Cases	Percentage (%)
Gender:		
Male	140	74.9
Female	47	25.1
Total	187	100.0
Educational Level:		
Non-formal	15	7.8
Primary	20	10.4
Middle/JHS	88	45.8
SHS	20	10.4
Tertiary	6	3.1
No education	43	22.5
Total	192	100.0
Cocoa farm ownership:		
Owner	68	58.1
Caretaker	49	41.9
Total	117	100.0
Status of farmer:		
Cocoa farmer	184	98.4
Non-cocoa farmer	3	1.6
Total	187	100.0
Number of cocoa farms:		
1	58	67.4
2	19	22.1
3	8	9.3
4	1	1.2
Total	86	100.0

Table 3. Summary of farmer and farm characteristics of the respondents in the study areas

Characteristics	Mean	Minimum	Maximum	Sample Size (n)
Farmer age (years)	51	23	95	191
Working experience (years)	24	1	65	185
Cocoa farm size (hectares)	3.0	0.4	16.0	182
Cocoa farm age (years)	13	1	56	187
Cocoa output (kg/farmer)	614.4	192.0	2 560.0	68

3.2 Perception on Rehabilitation (Planting Cocoa in Marginal Areas Which Were Previously Cropped To Cocoa, But Now Food Crops)

3.2.1 Knowledge about Rehabilitation

In inquiring about the knowledge status of the respondents relating rehabilitation of cocoa farms (Table 4), 92.9% indicated that they knew something about rehabilitation whereas 7.1% had no information of it. From those respondents who had information, 71.3% said that cocoa farm rehabilitation meant re-planting of a farm which is over 30 years, 38.8% stated re-planting of cocoa farm with new cocoa variety (hybrid), 26.3% indicated re-planting of cocoa farm infected with Cocoa Swollen Shoot Virus Disease (CSSVD), 12.5% said re-planting of cocoa farm of Amelonado trees, 7.5% also indicated re-planting of cocoa farm of abandoned cocoa trees while 2.5% of the respondents specified re-planting of cocoa farm with other types of cocoa varieties.

3.2.2 Willingness of Farm Rehabilitation

When asked whether they would want their farms to be rehabilitated, 40.0% said "yes" whereas 60.0% indicated "no". Also, 66.7% of the respondents said they were willing to pay for the rehabilitation of their farms whilst 33.3% said "no". The specific characteristics of cocoa the respondents prefer for the rehabilitation of their farms (Table 4) include early yielding cocoa (73.3%), high yielding cocoa (56.7%), big cocoa beans (8.3%), less susceptible to disease (8.3%), small cocoa trees (6.7%), less susceptible to pests (5.0%), and big cocoa pods (1.7%).

Information on whether respondents had plans for rehabilitation of their existing cocoa farms was obtained (Table 4). Only 53.6% said 'yes' and 46.4% 'no'. Of those who indicated 'yes', 71.0% planned to rehabilitate their farms by infilling the open spaces with young plants; 16.1% by decreasing or increasing the shade in the plantation, etc.

The respondents gave their reasons for the preference of rehabilitation (Table 4) and 36.0% said it was because of low yields, 44.0% indicated old cocoa trees, and 8.0% reported that they want to plant hybrid because of obtaining higher incomes. However, some respondents rejected farm rehabilitation because the cocoa farm was young (83.2%), and that the farm size was too small for rehabilitation (2.1%).

3.2.3 Problem/Constraints in Farm Rehabilitation

For the problems/constraints that a cocoa farmer is likely to encounter in farm rehabilitation in the districts (Table 4), 22.7% of the respondents indicated loss of cocoa income, 29.5% reported the need of money to hire labour and purchase other inputs, 34.1% indicated initial financial hardship whilst 2.3% reported non-availability of inputs.

Table 5 summarizes the number of farms and farm size to be rehabilitated, and the costs of rehabilitation. Those who were willing to pay for the rehabilitation indicated an average amount of GH₵ 273.08 (n = 13), having a range of GH₵ 50.00 to GH₵ 1 000.00, as payment for an acre of farm to be rehabilitated. Also, they were prepared to offer one farm, on average, for rehabilitation with a range 1-3 farms. The average farm size they were willing to provide for rehabilitation was 3.9 acres (n = 32; 1 acre = 0.4 hectare), with a range of 1-5 acres.

Table 4. Distribution of respondents on perception on cocoa farm rehabilitation (planting cocoa in marginal areas which were previously cropped to cocoa, but now food crops)

Item	Frequency	Percentage
Knowledge of farm rehabilitation		
Yes (have knowledge)	79	92.9
No (have no knowledge)	6	7.1
Total	85	100
Meaning of cocoa farm rehabilitation*		
Replanting cocoa farm which is over 30 years	57	71.3
Replanting cocoa farm with new cocoa variety (hybrid)	31	38.8
Replanting cocoa farm infected with cocoa swollen shoot virus disease (CSSVD)	21	26.3
Re-planting cocoa farm of Amelonado trees	10	12.5
Re-planting cocoa farm of abandoned cocoa trees	6	7.5
Re-planting cocoa farm having other types of cocoa varieties	2	2.5
Total	80	158.9
Do you want cocoa farm rehabilitated?		
Yes	34	40
No	51	60
Total	85	100
Willingness to pay for rehabilitation		
Yes (willing)	20	66.7
No (unwilling)	10	33.3
Total	30	100.0
Specific characteristics of cocoa preferred for rehabilitation*		
Early yielding cocoa	44	73.3
High yielding cocoa	34	56.7
Big cocoa beans	5	8.3
Less susceptible to diseases	5	8.3
Less susceptible to pests	3	5.0
Small cocoa trees	4	6.7
Big cocoa pods	1	1.7
Total	60	160
Plans for rehabilitation of existing cocoa farms		
Yes (have plans)	37	53.6
No (have no plans)	32	46.4
Total	69	100.0

Item	Frequency	Percentage
For those who have plans to rehabilitate their farms		
Planned to rehabilitate farms by infilling of the open spaces with young plants	22	57.9
Planned to rehabilitate farms by decreasing or increasing shade in the plantation	5	13.2
Planned to rehabilitate their farms by drastic pruning of the old cocoa trees	3	7.9
Planned to rehabilitate their farms by spraying pesticides	4	10.5
Planned to rehabilitate their farms by applying fertilizer	3	7.9
Others	1	2.6
Total	38	100.0
Reasons given for cocoa farm rehabilitation		
Low yields	9	36.0
Old cocoa trees	11	44.0
Want to plant hybrid cocoa	2	8.0
Death of cocoa trees due to pest and disease attack	2	8.0
Higher income	1	4.0
Total	25	100.0
Reasons given against cocoa farm rehabilitation		
Because of young cocoa farm	40	83.2
Because the farm has already undergone rehabilitation	2	4.2
Because the farm still gives higher yields	3	6.3
Because the farm is too small for rehabilitation	1	2.1
Because the farm has no pest and disease attacks	2	4.2
Total	48	100.0
Problems/constraints likely to be encountered in the district		
Loss of income	10	22.7
Capsid infestation	1	2.3
The need of money to hire labour and purchase other inputs	13	29.5
The problem of farm maintenance	2	4.5
Initial financial hardship	15	34.1
Unfavourable weather conditions	1	2.3
No problem	1	2.3
Non –availability of inputs	1	2.3
Total	44	100.0

*multiple choice items (total of the percentage figures is not equal to 100%)

Table 5. Summary of number of farms and farm size to be rehabilitated, and the costs of rehabilitation

Characteristics	Mean	Minimum	Maximum	Sample Size (n)
Number of farms to be rehabilitated	1	1	3	23
Farm size to be rehabilitated (ha)	1.6	0.4	2	32
Amount to be paid for farm rehabilitation (GH₵)	273.0	50.00	1 000.00	13

3.3 Possibility of Re-Introduction of Cocoa

The respondents were questioned on the possibility of re-introduction of cocoa on their land (Table 6). Only 98.4% said 'yes' whilst 1.6% said 'no'. The reasons stated for the possibility of cocoa re-introduction by the respondents consisted of the land was already/currently planted to cocoa (35.5%), the land/soil was good for cocoa production (33.5%), the land was previously planted to cocoa (19.0%), the observation of indicator crops such as plantain showed that cocoa could be grown in the study areas (1.2%), etc. For possible chance of success after re-introduction of cocoa (Table 6), 93.3% of the respondents indicated that it could grow very well whereas 6.7% reported that cocoa could grow well. Also, the respondents indicated how successful cocoa establishment could be achieved in the area (Table 6). Their views included planting of temporary shade such as plantain, cocoyam, etc. for successful cocoa establishment (32.0%), planting of permanent shade trees (1.6%), and other activities (1.0%) such as providing assistance in the form of credit, forming association of farmers to combat bush fires, etc. On condition that cocoa could be successfully grown in the area, they were asked to indicate what percentage of their land holdings they would be happy to plant solely to cocoa. They reported an average of 46.98% (n = 154), with range from 10% to 100% (Table 7).

Table 6. Distribution of respondents on perception on the possibility of re-introduction of cocoa in the study areas

Item	Frequency	Percentage
Possibility of re-introduction of cocoa		
Yes (possible)	184	98.4
No (impossible)	3	1.6
Total	187	100
Reasons for possibility of cocoa re-introduction		
The land is already/currently planted to cocoa	61	35.5
The soil is good for cocoa production	58	33.5
Cocoa is additional source of income	13	7.4
The increase in incentives (e.g. increase in cocoa producer price) of cocoa production	1	0.6
The reduction of bushfire outbreaks	2	1.2
The land was previously planted to cocoa	16	9.0
The weather conditions are good	2	1.2
The observation of indicator crops such as plantain shows that cocoa can be grown in the study areas	2	1.2
The education received on the new cocoa varieties	4	2.4
Others	14	8.0
Total	173	100.0
Possible chance of success after re-introduction of cocoa		
Cocoa can grow very well	166	93.3
Cocoa can grow well	12	6.7
Total	178	100.0
How successful cocoa establishment could be achieved in the study areas*		
Planting of temporary shade trees	17	8.9
Planting of permanent shade trees	3	1.6
Providing extension officer to educate cocoa farmer on the use of good agronomic practices	15	7.8
Planting cocoa varieties that are easy to establish	10	5.2
Others	2	1.0
Total	192	24.5

*multiple choice items (total of the percentage figures is not equal to 100%)

Table 7. Summary of the percentage of land holdings the respondents are willing to offer for planting solely cocoa

Characteristics	Mean	Minimum	Maximum	Sample Size (n)
Percentage of land holdings to be planted to solely cocoa (%)	46.98	10	100	154

3.4 Perception on Farmers' Interest in New Cocoa Varieties

Concerning the interest the respondents had in testing any new cocoa varieties on their farms as part of the project (Table 8), 79.6% of them indicated "yes" and 20.4%, "no". Upon enquiring whether they had plans for extending their farms, that is, establishing new farms, 94.4% indicated 'yes' whereas 5.6% said 'no'. Of those respondents who opted for infilling to rehabilitate their farms or establishment of new farms with hybrid varieties, 81.9% chose to plant seeds from cocoa stations as seedlings, 16.1% preferred to plant seeds from cocoa station for direct planting at stake, and 2.0% decided to use clones from cocoa stations.

Upon asking them whether the new varieties they were interested in were available (Table 8), 77.6% said they were available whilst 22.4% indicated they were not. The respondents appeared to source their planting materials from the seed gardens (83.1%), that is, from Seed Production Unit (SPU) of Ghana Cocoa Board (COCOBOD); from Cocoa Research Institute of Ghana (CRIG) (3.2%); from other farmer's farm (11.3%); from their own farms (1.6%); and other sources (0.8%).

Giving reasons for non-availability of the new cocoa variety (Table 8), 26.3% of the respondents reported inadequate supply of pods from the cocoa stations, 2.6% reported inadequate number of cocoa stations, 60.5% indicated difficulty in getting pods from the cocoa stations while 10.5% indicated unawareness of the place to get the new variety.

Of those who were willing to test the new cocoa varieties (Table 8), 10.8% desired to infill the open spaces in existing cocoa farms whereas 89.2% wished to establish new cocoa farms. Data on willingness and ability of respondents to establish and maintain the project farms were elicited (Table 8). Those who agreed to clear the land for the new cocoa plantings accounted for 98.9% of the respondents, to plant temporary shade trees in the cocoa farm formed 99.4%, to plant permanent shade trees in the cocoa represented 99.4%, to harvest and break the cocoa pods constituted 99.4%, to count the total pods harvested per tree accounted for 98.9%, and to count healthy and diseased pods represented 97.7%. The difference between these percentage figures and 100% gives you the percentage of respondents who did not agree to perform the activities. Further, the percentage figures would not add up to 100% because of multiple choice items. In addition, 89.8% of the respondents were willing to pay for the new varieties whereas 10.2% were unwilling.

Table 8. Distribution of respondents on perception on interest in planting new cocoa varieties

Item	Frequency	Percentage
Interest of respondents in cocoa varieties		
Yes (interested)	129	79.6
No (disinterested)	33	20.4
Total	162	100
Plans for extending their farms (i.e. new establishments)		
Yes (have plans)	157	94.4
No (have no plans)	9	5.6
Total	161	100
For those who opted for infilling to rehabilitate their farms or make new plantings with any type of planting materials		
Plant seeds from cocoa stations as seedlings	24	81.9
Plant seeds from cocoa stations for direct planting at stake	122	16.1
Use clones from the cocoa stations	3	2.0
Total	149	100.0
Availability of interesting new varieties		
Yes (available)	118	77.6
No (unavailable)	34	22.4
Total	152	100.0
Source of planting materials		
Seed gardens (cocoa stations)	103	83.1
Cocoa Research Institute of Ghana (CRIG)	4	3.2
Other farmers' farms	14	11.3
Own farm	2	1.6
Others	1	0.8
Total	124	100.0
Reasons for non-availbility of the new cocoa varieties		
Inadequate supply of pods from the cocoa stations	10	26.4
Inadequate number of cocoa stations	1	2.6
Difficulty in getting pods from the cocoa stations	23	60.5
Unawareness of the place to get the new variety	4	10.5
Total	38	100.0
Interest in testing new cocoa variety as part of the project		
Yes (interested)	98	98
No (disinterested)	2	2
Total	100	100
Type of planting method for testing the new cocoa varieties		
Desired to infill open spaces in existing cocoa farms	9	10.8
Wished to establish new cocoa farms	74	89.2
Total	83	100.0
Willingness and ability to establish and maintain the project farms*		
To clear the land for the new plantings	173	98.9
To plant temporary shade trees in the cocoa farm	173	99.4
To plant permanent shade trees in the cocoa farm	173	99.4
To weed, spray and prune cocoa trees	173	99.4
To harvest and break the cocoa pods	173	99.4
To count the total pods harvested per tree	172	98.9
To count healthy and diseased pods	169	97.1
Total	174	692.5
Willingness to pay for the new varieties		
Yes (willing)	141	89.8
No (unwilling)	16	10.2
Total	157	100.0

*Multiple choice items (total of the percentage figures is not equal to 100%)

The possible pod and seedling prices, and number of trees for infilling are summarized in Table 9. The respondents were willing to pay an average price of GH₵ 0.47 per pod or cuttings (n = 73), ranging from GH₵ 0.20 to GH₵ 10.00 and GH₵ 0.36 per seedling (n = 56), ranging from GH₵ 0.10 to GH₵ 2.00. In addition, the average number of new cocoa trees to plant for filling of the open spaces in the old farm, or newly established farm was 2,097 (n = 81) with a range of 100-18,000 trees (450 trees per acre of land).

Table 9. Summary of possible prices of pods and seedlings, and number of cocoa seedlings for infilling

Characteristics	Mean	Minimum	Maximum	Sample Size (n)
Possible price to be paid for new cocoa varieties (GHs per pod or cuttings)	0.47	0.20	10.00	73
Possible price to be paid for new cocoa varieties (GHs per seedlings)	0.36	0.10	2.00	56
Number of cocoa seedlings for infilling	2 097	100	18 000	81

4. Discussion

The findings have demonstrated that the majority of the respondents have some information or knowledge of cocoa farm rehabilitation and were willing to offer their farms for rehabilitation. They were also willing to bear the expenses on the rehabilitation in which they would want to use seedlings of hybrid cocoa varieties for infilling of the open spaces in the existing cocoa farms. The main reasons given for the preference of rehabilitation were that the cocoa farms were old with low yields. However, some respondents rejected the rehabilitation of their farms because of initial loss of cocoa income which can lead to financial hardship that would limit the hiring of labour and purchasing of other non-labour inputs. This implies that farmers in the project areas may be interested in the COCOBOD's cocoa rehabilitation programme. Also, it is likely that farmers may encounter financial constraints in labour hiring and other non-labour input purchasing. Additionally, this might affect the adoption of the recommended cocoa establishment technologies of CRIG. The current project is not capable in providing credit to the interested farmers as was done in the Eastern Region Cocoa Rehabilitation Project (World Bank, 1970), but can only assist in supplying free hybrid seedlings to the project farmers for the cocoa farm establishment. According to MASDAR (1998), in 82% of cases replanting has been undertaken on a 'spot' basis of replacing individual dead or low yielding trees. Older and younger trees therefore get mixed together and exact proportions become difficult to determine. This is also an agronomic problem in that it leads to an uneven canopy structure that attracts capsids and weeds. The existing shade may be inappropriate for young trees and there is evidence that older trees also transfer diseases to the young ones (Petithuegenin, 1995).

Most of the respondents indicated that it was highly possible to re-introduce cocoa in the study areas mainly because the land/soil is good for cocoa production and the fact that the land is already or currently planted to cocoa. They also reported that successful cocoa establishment could be achieved in the area after re-introduction, and this could be made possible by planting temporary shade trees such as plantain as well as the control of insect pests and, if necessary, fertilizer application among other practices (World Bank, 1970; Adebiyi & Okunlola, 2013). Despite the perception of good soils and cocoa being already planted in the areas, the soils in the project sites should be tested for its real suitability for cocoa. This is because the soils might be exhausted of nutrients having been under prolonged cultivation for food and other perennial crops after the bush fire disasters which caused cocoa farm abandonment. Effects of degraded soils and soil acidity on germination and root development of seedlings have been reported in the literature (Turner et al., 1988; Marschner, 1991; Anim-Kwapong & Teklehaimanot, 2001). Temporary and permanent shade trees should be planted to ensure good seedling establishment and subsequent development of the cocoa trees as anticipated in the COCOBOD's cocoa rehabilitation programme, and as indicated by some studies (Anim-Kwapong & Teklehaimanot, 1995; Osei-Bonsu & Anim-Kwapong, 1998).

Majority of the respondents were interested in testing new cocoa varieties on their farms as part of the project since they have plans for extending the farms by planting seedlings from the cocoa stations, that is, the seed gardens of Seed Production Unit (SPU) of Ghana Cocoa Board (COCOBOD). However, some respondents stated that it was difficult in obtaining hybrid pods from the cocoa stations due to inadequate supply. Most of them wished to establish completely new cocoa farms with the new cocoa hybrid varieties instead of infilling of the open spaces in existing cocoa farms. The implication of this is that the current project must establish cocoa nurseries at the project sites to serve as ready sources of hybrid seedlings for the rehabilitation and re-introduction efforts to be successful. Good government policies to increase incentives for cocoa rehabilitation and re-introduction are very important. The decline of cocoa production in Ghana from the mid 1960s to the early 1980s was blamed on poor government policies including: lack of support funds to stabilize producer prices in the face of falling world prices; the establishment of a new internal marketing system which disrupted the efficient marketing system it replaced; and disbanding of extension and disease control service. These factors among others led to the neglect of existing cocoa farms and loss of interest in replanting (World Bank, 1970; Ghana Government &World Bank, 1978).

5. Conclusion

The conclusions drawn from the results are as follows:

- The respondents appeared to have knowledge about cocoa farm rehabilitation, and majority of them were unwilling to rehabilitate the cocoa farms. The respondents reported rehabilitation problems/constraints such as initial financial hardship, the need of funds to hire labour and purchase other farm inputs etc.

- The survey revealed that majority of the respondents reported that re-introduction of cocoa in marginal or denuded land was highly probable since the land/soil was already good for cocoa production, etc.

- Majority of the respondents were observed to have interest in testing any new cocoa varieties on their farms as part of the project. They were also willing to provide land, pay for and test the new hybrid cocoa to be supplied on their farms, and prepared to perform all the farm establishment, maintenance and harvesting activities in support of the project.

Acknowledgement

The support provided by the technical staff, Mr. E. Badger and Mr. D. Agyapong, all of the Social Science and Statistics Unit (SSSU) of Cocoa Research Institute of Ghana (CRIG) is gratefully acknowledged. Special thanks go to CORAF/WECARD and CRIG for funding the project. This paper/publication, CRIG/002/2015/045/008, is published by kind permission of the Executive Director of CRIG, New Tafo-Akim.

References

Adebiyi, S. & Okunlola, J. O. (2013). Factors affecting adoption of cocoa farm rehabilitation techniques in Oyo State of Nigeria. *World Journal of Agricultural Science, 9*(3), 258-265.

Anim-Kwapong, G. J., & Teklehaimanot, Z. (2001). Albizia zygia (DC) Macbride, a shade tree for cocoa: The effect of duration of acid scarification and substrate acidity on the germination of seeds. *Forests Trees and Livelihoods, 11*, 47-55. http://dx.doi.org/10.1080/14728028.2001.9752370

Anim-Kwapong, G. J., & Frimpong, E. B. (2006). *Vulnerability and adaptation assessment under the Netherlands Climate Change Assistance Programme Phase 2 (NCAP2):Vulnerability of agriculture to climate change-impact of climate on cocoa production.* Final report submitted to Environmental Protection Agency, Ghana. 122pp.

Asare, R., Afari-Sefa, V., Gyamfi, I., Okafor, C., & Mva Mva, J. (2010). Cocoa seed multiplication: an assessment of seed gardens in Cameroon, Ghana and Nigeria. STCP Working Paper Series 11 (Version August 2010). Sustainable Tree Crop Program, International Institute of Tropical Agriculture, Accra, Ghana.

Business & Financial Times (2014). Is cocoa still our economic backbone? Wednesday, March 6th, 2014.

Ghana Government /World Bank (1978). *Evaluation of Ghana Government /World Bank Cocoa Rehabilitation Projects in the Eastern and Ashanti Regions.* A report submitted to the Commissioner of Cocoa Affairs, Ghana Government/World Bank Cocoa Project Evaluation Committee. 200pp.

Gockowski, J. (2011). Agricultural intensification as a strategy for climate mitigation in Ghana: an evaluative study of the COCOBOD high tech program, rural incomes, and forest resources in the Bia (Juaboso) district of Ghana.

Sustainable Tree Crop Program, International Institute of Tropical Agriculture, Accra, Ghana.

Marschner, H. (1991). Mechanisms of adaptation of plants to acid soils. *Plant and Soils, 134*, 1-2. http://dx.doi.org/10.1007/978-94-011-3438-5_78

MASDAR (1998). *Socio-economic Study.* A Final Report. Accra, Ghana: Ghana Cocoa Board (COCOBOD)/ MASDAR Consultants, 384.

Osei-Bonsu, K., & Anim-Kwapong, G. J. (1998). Preliminary evaluation of some forest trees for cocoa cultivation. *Journal of Ghana Science Association, 1*, 141-150. http://dx.doi.org/10.4314/jgsa.v1i1.17796

Petithuguenin, P. (1995). *Regeneration of cocoa cropping systems: The Ivoirian and Togolese experience.* In: Ruf F. & Siswotputrano, P. S. Cocoa Circle-the Economics of cocoa supply. Abington Woodhead, 1995. http://dx.doi.org/10.1016/b978-1-85573-215-5.50008-9

Turner, G. D., Lau, R. R., & Young, D. R. (1988). Effect of acidity on germination and seedling growth of *Paulownia tomentosa. Journal of Applied Ecology, 25*(2), 561-567. http://dx.doi.org/10.2307/2403844

World Bank (1970). *Ghana Eastern Region Cocoa Project.* A Project Report, Agriculture Projects Department, World Bank Association.

8

Water Use Efficiency of Maize Varieties under Rain-Fed Conditions in Zambia

Ethel Mudenda[1], Elijah Phiri[1], Lydia M. Chabala[1] & Henry M. Sichingabula[2]

[1]Department of Soil Science, School of Agricultural Sciences, University of Zambia, P.O. Box 32379, Lusaka, Zambia

[2]Department of Geography and Environmental Studies, School of Natural Sciences, University of Zambia, P.O. Box 32379, Lusaka, Zambia

Correspondence: Ethel Mudenda, Department of Geography, Mazabuka Girls Secondary School, P.O. Box 670211, Mazabuka, Zambia. E-mail: muleyanam@gmail.com

Abstract

This study evaluated water use efficiency (WUE) of selected hybrid maize (*Zea Mays L.*) varieties in Zambia under rain-fed conditions. A randomized complete block field experiment was carried out during the 2014/2015 rainy season at the University of Zambia Agricultural Demonstration Centre. Treatments were 30 maize varieties from the early, medium and late maturity classes. WUE was calculated as the ratio of yield to evapotranspiration (ET) and transpiration (T). Results showed significant differences in WUE dry matter ($_{DM}$) for transpiration ($_T$) of early maturing varieties. However, no significant differences were observed in WUE_{DM} for evapotranspiration ($_{ET}$), WUE grain yield ($_{GY}$)$_{,T}$ and $WUE_{GY, ET}$. $WUE_{DM, T}$, $WUE_{DM, ET}$, $WUE_{GY, T}$, and $WUE_{GY, ET}$ were statistically the same among medium maturing varieties. Results further showed that among the late maturing varieties, $WUE_{DM, T}$, $WUE_{DM, ET}$ and $WUE_{GY, T}$ showed significant differences but no significant differences were observed in $WUE_{GY, ET}$. It was concluded that maize varieties from the same maturity classes have different WUEs. The study thus provided options in variety selection based on which varieties performed better, particularly SC 525, SC 513 and PAN 4M 21 from the early maturity class; PHB 30G19, ZMS 606, MRI 634 and SC 637 from the medium maturity class; and PAN ZM 83, SC 709, PAN 8M 93 and SC 719 from the late maturity class. It was recommended however, that repeated experiments over time should be done to validate the findings given that the trial was only conducted in one season.

Keywords: maize varieties, rain-fed, water use efficiency

1. Introduction

Maize (*Zea mays* L.) is a C4 plant, which potentially has more efficient use of carbon dioxide (CO_2), solar radiation, water and nitrogen in photosynthesis than C3 crops, resulting in higher production of dry matter (DM) (Huang, Birch & George, 2006). Even though maize makes efficient use of water, it is considered more susceptible to water stress than other crops because of its unusual floral structure with separate male and female organs and the near-synchronous development of florets on a usually single ear borne of each stem. The production of maize, as is for all other crops is directly related to the capture of resources, such as water and sunlight, and the efficiency with which it converts these physical resources into biological materials. Water Use Efficiency (WUE) is one of the ways to analyze the response of crops to different conditions of water availability as it relates to the production of dry biomass with the amount of water applied or evapotranspired. It is aptly defined as the amount of yield produced per unit of water evapo-transpired or transpired (Doorenbos et al., 1979).

Evapotranspiration (ET) is composed of soil evaporation (E) and plant transpiration (T) and the two are dependent on wetness of the surface soil and canopy size. The actual ET in any field experiment will therefore vary considerably depending on the frequency of the rains and the crop canopy dynamics. Since ET is plant specific, different crops and different varieties from the same crops evapotranspire different amounts of water under similar climatic conditions. These differences provide opportunities to select for appropriate varieties with an efficient use of water. To assess such options, ET estimates of different varieties are required for a specific region. Since allowance is made for differences in atmospheric evaporative demand among varieties, T in general becomes a reasonably stable quantity for most green crops having a closed canopy. This is to be expected because of the close link between CO_2 usage for photosynthesis and plant water use (Gardner, Laryea & Unger, 1999). The advantage of WUE_T over WUE_{ET} is that it avoids the compounding effect of the non-productive soil E and weed T losses (Nyakudya & Stroosnijder, 2014).

Notwithstanding the efforts of breeders to develop high yielding varieties, the yield of maize throughout the production regions in Zambia are generally low, averaging < 1.5 ton ha^{-1} for the majority of small scale farmers (Jayne et al., 2007; Ministry of Agriculture and Livestock [MAL], 2015). Most of the maize that the country produces each year is grown under rain-fed conditions. In areas such as the semi-arid and dry sub-humid environments, the amount of rainfall is not only the limiting factor of rain-fed maize production but also its erratic nature. Therefore, a major focus of rain-fed cropping systems is increasing efficiency of water utilization by crops. One strategy to reduce the effect of water stress on crop yield is to use water efficient species (Stewart & Nielsen, 1991). Bibi, Sadaqat, Akram & Mohammed (2010) also noted that crop plants are usually under stress conditions at one time or another and the plant species able to withstand stresses have great economic potential. Thus rain-fed maize production could be enhanced by adopting varieties that efficiently use soil moisture for biomass and grain production.

Studies on maize WUE have been conducted in some parts of Africa (Frimpong, Amoatey, Ayeh & Asare, 2011; Asare, Frimpong, Ayeh & Amoatey, 2011). However, there is limited information on maize varieties that use water more efficiently under rain-fed agriculture in Zambia with the exception of the work by Phiri, Verplancke, Kwesiga & Mafongonya (2003) who investigated WUE of rain-fed maize in eastern Zambia under different fallow systems with only one variety evaluated. There is need therefore to locally determine WUE of different maize varieties under rain-fed conditions in Zambia under the conventional type of farming. Quantifying WUE of crops is important to identify and subsequently disseminate the best suitable varieties for a specific region. The reason for this is to enhance maize productivity through improved selection of maize varieties which are efficient in water use and to generate evidence not currently existing of the maize varieties for promotion among small scale farmers. This would greatly help in information and decision guides. Given the many varieties readily available on the market, variety selection is a cost effective way of maximizing WUE. It is against this background that the study was initiated. The overall purpose of the current study was to evaluate 30 hybrid maize varieties for their efficiency in use of water for DM and grain yield (GY) production under rain-fed conditions in agro-ecological region IIa of Zambia. Identified maize cultivars, when adopted by farmers could assist in enhancing sustainable maize production in areas where rain-fed maize production is mostly practiced, particularly in areas that experience low and erratic rainfall.

2. Materials and Methods

2.1 Area Description

The field experiment was conducted at the University of Zambia Agricultural Demonstration Centre. The site is located at latitude 15° 21′ 25″ South and longitude 28° 27′ 25″ East. The elevation is 1 160 m above sea level. The site is in Chongwe District of Lusaka Province in Zambia. The soil of the study site is of a sandy loam texture, taxonomically belonging to the Chromic Luvisol category (Jones et al., 2013). The pre-planting analysis of the soil of the experimental field indicated generally low total nitrogen and exchangeable potassium contents of 0.06% and 0.231 cmol kg^{-1}, respectively. The amount of available phosphorous was found to be 16.20 mg kg^{-1}, sufficient enough to prevent phosphorous deficiency. Other soil parameters analyzed are shown in Table 1. The study site falls in Agro-Ecological region IIa of Zambia.

2.2 Treatments and Experimental Design

Treatments were 30 hybrid maize varieties consisting of 10 early, 10 medium and 10 late maturity varieties. These varieties were selected based on availability from seed companies during the season (here and hereinafter, mention of brand names of maize varieties is for identification only and does not constitute endorsement of the product(s) by the author(s) or the institution(s) mentioned herein). Maturity of maize hybrids is a genetic characteristic and is generally defined as the period from germination to when the kernel ceases to increase in weight (Brown, Zuber, Darrah & Glover, 1985). The experiment was set up in a Randomized Complete Block Design (RCBD) with three replications. Each subplot measured 5 m x 5 m separated from each other by a border space of 2 m.

2.3 Agronomic Management

The experiment was conducted in the 2014/2015 rainy season and maize varieties were subjected to conventional agricultural practices which included no irrigation. The study site was prepared using a moldboard plough and harrowed to a fine tilth with a tractor before the onset of the rains. Maize varieties were sown on the 17th of December, 2014 at a spacing of 75 cm between rows and 30 cm within rows to give a population of 44, 444 plants ha[-1]. Compound D (10% N: 20% P_2O_5: 10% K_2O: + 6% S) and Urea (46% N) fertilizers were applied at a rate of 200 kg ha[-1] to provide 20 kg N ha[-1], 17 kg P ha[-1], 17 kg K ha[-1] from Compound D and 92 kg N ha[-1] from Urea. Glyphosate herbicide was applied before planting at a rate of 125 g ha[-1] as pre-emergence weed control.

2.4 Biomass and Grain Yield Measurement

Measurements of biomass or DM and GY were done from the net plots and determined at harvest time. A bordered area of 9 m[2] (3 m x 3 m) consisting of 40 plants in each plot was hand-harvested for the measurement of yield. In terms of total biomass, maize plants were cut just above the ground using sickles and weighed. In terms of GY, the harvested ears were counted, weighed, manually shucked and the grain weighed. The grain was then tested for moisture using a grain analysis meter and GY adjusted to 12.5% moisture content. The results of the harvest were expressed in ton ha[-1].

Table 1. Soil Characterization of the Study Site

Depth (cm)	pH (CaCl₂)	EC (mS cm⁻¹)	OC %	OM %	Sand %	Silt %	Clay %	Texture (USDA)	ρb g cm⁻¹	FC % vol	WP % vol	AWC mm m⁻¹
0-10	4.21	0.20	0.25	0.50	73.4	17.5	9.1	SL	1.43	13.8	5.5	83
10-20	4.05	0.23	0.15	0.30	73.5	15.4	11.1	SL	1.45	14.6	6.5	81
20-30	4.04	0.15	0.13	0.26	70.1	13.5	16.4	SL	1.53	17.8	9.6	82
30-40	4.08	0.44	0.14	0.28	70.8	12.1	17.1	SL	1.65	17.3	10.3	70
40-50	4.15	0.23	0.26	0.52	66.8	14.1	19.1	SL	1.70	19.0	11.7	73
50-60	4.10	0.19	0.14	0.28	64.8	12.1	23.1	SCL	1.70	21.6	13.9	77
60-70	4.22	0.11	0.14	0.28	63.5	10.8	25.7	SCL	1.59	24.4	15.7	87
70-80	4.34	0.12	0.10	0.20	60.8	12.8	26.4	SCL	1.57	25.0	15.7	93
80-90	4.53	0.26	0.07	0.14	62.1	11.5	26.4	SCL	1.59	24.6	15.6	90
90-100	4.52	0.26	0.18	0.36	60.1	14.1	25.8	SCL	1.51	25.8	15.8	100

pH = potential of hydrogen (i.e. soil reaction), EC = electrical conductivity, OC = organic carbon, OM = organic matter, USDA = United States Department of Agriculture, ρb = bulk density, FC = field capacity, WP = wilting point and AWC = available water content

2.5 Measurement of Weather Data and Calculation of Water Use Efficiency (WUE)

Weather data during the study was collected from an automated weather station located at latitude 15° 19′ 9″ and longitude 28° 26′ 25″ and altitude 1 149 m above sea level and 4.5 km north of the site. Weather elements recorded included rainfall (mm), minimum and maximum temperatures (°C), wind speed (m s[-1]), relative humidity (%) and solar irradiance (MJ m[-2] day[-1]). This data, together with the soil and crop management and growth data were input into the AquaCrop model (Steduto, Hsiao, Raes & Fereres, 2009; Raes, Steduto, Hsiao & Fereres, 2009 and Hsiao et al., 2009) to help estimate the water balance model for each variety from seed emergence to crop maturity as follows:

$$\Delta S = P + I - R - D - E - T \tag{1}$$

Where in ΔS = change in root zone soil moisture storage, P = precipitation, in this case, rainfall, I = irrigation, R = runoff, D = downward drainage out of the root zone, E = direct evaporation from the soil surface, and T = transpiration by plants. Irrigation was not considered as the study was conducted under rain-fed conditions. All measurements were in mm.

WUE was calculated as the ratio of yield to ET and T as follows:

$$WUE_{GY,ET} = \frac{GY}{ET} \tag{2}$$

$$WUE_{GY,T} = \frac{GY}{T} \tag{3}$$

$$WUE_{DM,ET} = \frac{DM}{ET} \tag{4}$$

$$WUE_{DM,T} = \frac{DM}{T} \tag{5}$$

Where WUE_{GY} = water use efficiency for grain yield (kg ha[-1] mm), WUE_{DM} = water use efficiency for dry matter or total biomass (kg ha[-1] mm[-1]), ET = evapotranspiration (mm), T = transpiration (mm).

2.6 Data Analysis

Data were analyzed by analysis of variance (ANOVA) using GenStat Statistical Software. The least significant difference test (LSD) was used to compare variety differences. Differences were declared significant at $P < 0.05$.

3. Results

3.1 Weather Conditions

The total amount of rainfall received during the experimental year was 1 031.8 mm (Table 2). The highest monthly total rainfall was recorded in December (326.6 mm), with the lowest rainfall occurring in May (0.4 mm). The highest minimum and maximum temperatures recorded at the experimental site were 19.3 °C and 33.4 °C respectively, in the month of November. The highest mean relative humidity (RH) at the experimental site was 86.28%, which occurred in January. The mean RH of the entire period was 64.55%. The highest solar radiation and wind speed were 24.77 MJ m^{-2}day^{-1} in November and 2.4 m s^{-1} in September, respectively.

Table 2. Monthly Weather Variables Recorded at the Experimental Site during the Study Period

Year	Month	Rainfall	T_{min}	T_{max}	RH $_{average}$	Solar $_{radiation}$	Wind $_{speed}$
		mm	°C		%	MJ m^2 day^{-1}	m s^{-1}
2014	June	0.0	9.7	24.8	61.98	17.31	1.9
	July	0.0	8.9	25.0	57.40	18.53	1.9
	August	0.8	8.7	28.0	46.10	20.10	2.2
	September	0.0	14.3	30.2	40.98	22.74	2.4
	October	0.0	17.2	33.4	39.54	24.47	2.2
	November	76.0	19.3	33.4	46.30	24.47	1.9
	December	326.6	18.9	29.4	74.29	18.56	0.9
2015	January	243.6	18.3	27.6	86.28	17.90	1.1
	February	140.0	18.2	28.4	85.80	20.12	1.1
	March	56.4	16.9	28.7	79.23	21.81	1.7
	April	188.0	15.9	25.5	85.74	15.50	1.6
	May	0.4	11.3	26.2	71.01	20.35	1.7
		1 031.8	**14.8**	**28.4**	**64.55**	**20.16**	**1.7**

T_{min} = minimum temperature, T_{max} = maximum temperature and RH = relative humidity

3.2 Yield, Transpiration and Evapotranspiration

3.2.1 Yield, Transpiration and Evapotranspiration of Early Maturing Maize Varieties

Results on total DM, GY, T and ET of early maturing maize varieties are presented in Table 3. DM varied from 4.8 ton ha^{-1} to 13.5 ton ha^{-1}. The average DM was 8.96 ton ha^{-1}. The lowest was with maize variety SC 303 while the highest was with maize variety SC 513. There were no significant differences observed in DM (P > 0.05). GY varied from 2.1 ton ha^{-1} to 5.1 ton ha^{-1}. The average grain yield was 3.26 ton ha^{-1}. The lowest GY was found to be with SC 303 while the highest GY was found to be with SC 525. No significant differences were found in GY (P > 0.05).

T ranged from 98.7 mm to 232.0 mm with an average of 173.54 mm. The lowest T was observed with SC 303 while the highest was observed with SC 525. There were significant differences observed in T among early maturing maize varieties (P < 0.05). ET varied from 288.7 mm to 427.0 mm with an average of 372.20 mm. The lowest ET was observed with SC 303 while the highest was observed with SC 525. Very highly significant differences were observed in ET (P < .001).

Table 3. Yield, Transpiration and Evapotranspiration of Early Maturing Maize Varieties

Variety	Biomass ton ha^{-1}	Grain ton ha^{-1}	Transpiration mm	Evapotranspiration mm
SC 303	4.8[a]	2.1[a]	98.7[a]	288.7[a]
ZMS 402	7.4[a]	2.4[a]	160.0[abc]	344.3[bc]
GV 409	5.6[a]	2.5[a]	131.7[ab]	336.3[ab]
P 3253	6.9[a]	2.6[a]	178.7[bcd]	393.3[cd]
PAN 413	10.7[a]	2.5[a]	149.0[abc]	343.0[bc]
SC 403	7.7[a]	3.2[a]	192.0[bcd]	403.0[d]
MRI 514	10.5[a]	3.5[a]	183.0[bcd]	391.7[cd]
PAN 4M 21	11.1[a]	3.8[a]	206.3[cd]	377.0[bcd]
SC 513	13.5[a]	4.9[a]	204.0[cd]	417.7[d]
SC 525	11.6[a]	5.1[a]	232.0[d]	427.0[d]
Mean	**8.98[ns]**	**3.26[ns]**	**173.54[*]**	**372.20[***]**
LSD	5.586	2.249	69.5	50.6
CV (%)	36.3	40.3	23.4	7.9
P-value	0.058	0.116	0.028	<.001

ns = not significant, *= significant, ***= very highly significant, means followed by the same letter(s) are not significantly different

3.2.2 Yield, Transpiration and Evapotranspiration of Medium Maturing Maize Varieties

Results on DM, GY, T and ET of medium maturing maize varieties are presented in Table 4. DM varied from 6.2 ton ha^{-1} to 13.4 ton ha^{-1}. The mean DM was 10.50 ton ha^{-1}. Maize variety SC 647 had the lowest DM while maize variety SC 637 had the highest DM yield. No significant differences were observed in DM (P > 0.05). GY varied from 2.3 ton ha^{-1} to 4.8 ton ha^{-1}. The mean GY was 3.52 ton ha^{-1}. Maize variety SC 647 had the lowest GY while maize variety PHB 30G19 had the highest GY. No significant differences were observed in the GY (P > 0.05).

T ranged from 151.3 mm with PAN 53 as the lowest to 217.0 mm with PHB 30G19 as the highest. The average T was 193.06 mm. Yet again, no significant differences were observed (P > 0.05). Maize variety PAN 53 had the lowest amount of ET (380.7 mm) while maize variety PHB 30G19 had the highest (424.0 mm). The average ET was 407.96 mm. However, there were no significant differences observed in ET of medium maturing maize varieties (P > 0.05).

Table 4. Yield, Transpiration and Evapotranspiration of Medium Maturing Maize Varieties

Variety	Biomass ton ha^{-1}	Grain ton ha^{-1}	Transpiration mm	Evapotranspiration mm
SC 647	6.2^a	2.3^a	189.3^a	408.3^a
PAN 53	9.7^a	2.7^a	151.3^a	380.7^a
MRI 694	9.3^a	2.9^a	181.3^a	405.0^a
P 3812W	10.4^a	3.4^a	181.0^a	403.3^a
MRI 624	11.4^a	3.5^a	198.7^a	404.7^a
MRI 634	10.9^a	3.6^a	191.3^a	400.3^a
ZMS 616	10.1^a	3.6^a	198.3^a	413.0^a
SC 637	13.4^a	4.2^a	211.7^a	423.0^a
ZMS 606	11.3^a	4.2^a	210.7^a	417.3^a
PHB 30G19	12.3^a	4.8^a	217.0^a	424.0^a
Mean	**10.50^{ns}**	**3.52^{ns}**	**193.06^{ns}**	**407.96^{ns}**
LSD	3.833	2.722	80.71	44.58
CV (%)	44.8	44.6	24.4	6.4
P-value	0.846	0.702	0.855	0.692

ns = not significant, means followed by the same letter are not significantly different

3.2.3 Yield, Transpiration and Evapotranspiration of Late Maturing Maize Varieties

Results on DM, GY, T and ET of late maturing maize varieties are presented in Table 5. Total DM varied from 5.3 ton ha^{-1} to 22.5 ton ha^{-1}, with a mean of 14.29 ton ha^{-1}. MRI 724 had the lowest DM while SC 709 had the highest. The DM of late maturing maize varieties showed significant differences (P < 0.05). On the other hand, late maturing maize varieties exhibited considerable but not statistical differences in GY which varied from 1.5 ton ha^{-1} to 6.0 ton ha^{-1}, with an average of 4.24 ton ha^{-1}. MRI 724 had the lowest GY while PAN ZM-83 had the highest GY.

Table 5. Yield, Transpiration and Evapotranspiration of Late Maturing Maize Varieties

Variety	Biomass ton ha^{-1}	Grain ton ha^{-1}	Transpiration mm	Evapotranspiration mm
MRI 724	5.3^a	1.5^a	174.3^a	402.3^a
ZMS 702	10.4^{ab}	2.9^a	203.3^a	415.0^a
PAN ZM 81	12.7^{abc}	3.9^a	197.3^a	416.0^a
GV 635	12.8^{abc}	3.9^a	217.3^a	424.4^a
ZMS 720	13.8^{abcd}	4.1^a	236.0^a	437.0^a
MRI 744	13.4^{abc}	4.4^a	237.0^a	435.0^a
PAN 8M 93	13.9^{abcd}	4.9^a	223.7^a	425.3^a
SC 719	18.5^{bcd}	5.2^a	265.7^a	459.7^a
SC 709	22.5^d	5.6^a	267.0^a	461.3^a
PAN ZM 83	19.6^{cd}	6.0^a	261.7^a	455.3^a
Mean	**14.29***	**4.24^{ns}**	**228.33^{ns}**	**433.13^{ns}**
LSD	8.785	3.452	96.6	59.15
CV (%)	35.8	47.3	24.7	8.0
P-value	0.033	0.294	0.528	0.446

*= significant, ns = not significant, means followed by the same letter(s) are not significantly different

T ranged from 174.3 mm to 267.0 mm with an average of 228.33 mm. The lowest T was with MRI 724 while the highest was with SC 709. However, no significant differences were observed in T (P > 0.05). ET ranged between 402.3 mm and 461.3 mm. The average was 433.13 mm. Following the pattern of T, maize varieties MRI 724 and

SC 709 had the lowest and highest ET amounts, respectively but no significant differences were observed among the varieties (P > 0.05).

3.3 Water Use Efficiency

3.3.1 Water Use Efficiency of Early Maturing Maize Varieties

Among early maturing maize varieties, WUE for dry matter yield for evapotranspiration ($WUE_{DM, ET}$) varied from 16.67 kg ha^{-1} mm^{-1} to 32.23 kg ha^{-1} mm^{-1} (Table 6) with a mean of 23.36 kg ha^{-1} mm^{-1}. The lowest and highest $WUE_{DM, ET}$ were found to be with GV 409 and SC 513, respectively. WUE for dry matter for transpiration ($WUE_{DM, T}$) varied from 36.54 kg ha^{-1} mm^{-1} to 68.51 kg ha^{-1} mm^{-1}. The average was 50.52 kg ha^{-1} mm^{-1}. The lowest value was observed to be with P 3253, followed by SC 403 and ZMS 402 while the highest was observed to be with PAN 413, followed by SC 513 and the MRI 514.

WUE for grain yield of the water evapo-transpired ($WUE_{GY, ET}$) for the entire growing season was in the range of 6.38 kg ha^{-1} mm^{-1} to 11.86 kg ha^{-1} mm^{-1}. The average was 8.39 kg ha^{-1} mm^{-1}. The lowest value was observed with P 3253 while the highest was observed to be with SC 525. WUE for GY for transpiration ($WUE_{GY, T}$) varied from 13.42 kg ha^{-1} mm^{-1} to 23.83 kg ha^{-1} mm^{-1}. The average was 17.99 kg ha^{-1} mm^{-1}. The lowest value was observed to be with P 3253 while the highest was observed to be with SC 513. The ANOVA of the WUE of early maturing maize varieties (Table 7) indicated that there were no significant differences observed among varieties (P > 0.05) in $WUE_{DM, ET}$, $WUE_{GY, ET}$ and $WUE_{GY, T}$. However, there were significant differences observed in $WUE_{DM, T}$ (P < 0.05).

Table 6. Water Use Efficiency of Early Maturing Maize Varieties

Variety	$WUE_{DM, ET}$	$WUE_{DM, T}$	$WUE_{GY, ET}$	$WUE_{GY, T}$
	kg ha^{-1} mm^{-1}			
GV 409	16.67[a]	43.63[ab]	7.26[a]	18.45[a]
SC 303	16.72[a]	50.29[abc]	7.11[a]	20.91[a]
P 3253	17.19[a]	36.54[a]	6.38[a]	13.42[a]
SC 403	18.93[a]	39.58[ab]	7.92[a]	16.46[a]
ZMS 402	20.46[a]	42.79[ab]	6.45[a]	13.84[a]
MRI 514	26.29[a]	55.89[bcd]	8.68[a]	18.41[a]
SC 525	27.22[a]	50.07[ab]	11.86[a]	21.82[a]
PAN 4M 21	28.51[a]	51.39[abc]	9.57[a]	16.81[a]
PAN 413	29.44[a]	68.51[d]	7.01[a]	15.97[a]
SC 513	32.23[a]	66.55[cd]	11.63[a]	23.83[a]
Mean	23.36[ns]	50.52*	8.39[ns]	17.99[ns]
LSD	11.88	16.39	5.049	7.031
CV%	29.6	18.9	34.9	22.8

ns = not significant, *= significant, means followed by the same letter(s) are not significantly different

Table 7. ANOVA of the WUE of Early Maturing Maize Varieties

Source of variation	df	s.s	m.s	v.r	F pr.
Variate: $WUE_{DM, ET}$					
Replication stratum	2	568.25	284.13	5.93	
Variety	9	961.85	106.87	2.23	0.071
Residual	18	862.74	47.93		
Total	29	2392.84			
Variate: $WUE_{DM, T}$					
Replication stratum	2	790.65	395.32	4.33	
Variety	9	30.97	344.22	3.77	0.008
Residual	18	1642.73	91.62		
Total	29	5531.34			
Variate: $WUE_{GY, ET}$					
Replication stratum	2	69.138	34.569	3.99	
Variety	9	106.821	11.869	1.37	0.271
Residual	18	155.916	8.662		
Total	29	331.875			
Variate: $WUE_{GY, T}$					
Replication stratum	2	120.38	60.19	3.58	
Variety	9	311.05	34.56	2.06	0.092
Residual	18	302.36	16.80		
Total	29	733.79			

df = degrees of freedom, s.s = sum of squares, m.s = mean square, v.r = variance ratio

3.3.2 Water Use Efficiency of Medium Maturing Maize Varieties

$WUE_{DM, ET}$ for medium maturing maize varieties varied from 15.10 kg ha^{-1} mm^{-1} to 30.88 kg ha^{-1} mm^{-1} (Table 8). The mean was 25.25 kg ha^{-1} mm^{-1}. The lowest and highest values were observed with SC 647 and SC 637, respectively. $WUE_{DM, T}$ was in the range 32.30 kg ha^{-1} mm^{-1} to 63.52 kg ha^{-1} mm^{-1}. The mean was 52.85 kg ha^{-1} mm^{-1} with maize variety PAN 53 having the highest and SC 647 having the lowest values.

$WUE_{GY, ET}$ varied from 5.62 kg ha^{-1} mm^{-1} to 11.29 kg ha^{-1} mm^{-1} with average of 8.56 kg ha^{-1} mm^{-1}. The lowest among the medium maturing varieties was found to be with SC 647. The highest was found to be with PHB 30G19. In terms of $WUE_{GY, T}$ values varied from 12.03 kg ha^{-1} mm^{-1} to 22.07 kg ha^{-1} mm^{-1} with average of 17.79 kg ha^{-1} mm^{-1}. The lowest was observed with SC 647 and the highest was observed with PHB 30G19. Results of the ANOVA showed that statistically no significant differences were observed ($P > 0.05$) among medium maturing varieties in $WUE_{DM, ET}$, $WUE_{DM, T}$, $WUE_{GY, ET}$ and $WUE_{GY, T}$ (Table 9).

Table 8. Water Use Efficiency of Medium Maturing Maize Varieties

Variety	$WUE_{DM, ET}$	$WUE_{DM, T}$	$WUE_{GY, ET}$	$WUE_{GY, T}$
	kg ha^{-1} mm^{-1}			
SC 647	15.10^a	32.30^a	5.62^a	12.03^a
MRI 694	22.65^a	49.91^a	7.28^a	15.89^a
ZMS 616	24.07^a	48.42^a	8.67^a	17.27^a
PAN 53	24.47^a	63.52^a	6.92^a	17.17^a
P 3812W	25.52^a	57.08^a	8.25^a	18.20^a
ZMS 606	26.91^a	52.82^a	10.44^a	20.65^a
MRI 634	27.08^a	56.94^a	9.03^a	18.95^a
MRI 624	27.37^a	52.24^a	8.44^a	16.49^a
PHB 30G19	28.50^a	54.42^a	11.29^a	22.07^a
SC 637	30.88^a	60.88^a	9.69^a	19.22^a
Mean	25.25^{ns}	52.85^{ns}	8.56^{ns}	17.79^{ns}
LSD	16.82	22.37	5.791	7.929
CV%	38.8	24.6	39.4	26.0

ns = not significant, means followed by the same letter are not significantly different

Table 9. ANOVA of the WUE of Medium Maturing Maize Varieties

Source of variation	df	s.s	m.s	v.r	F pr.
Variate: $WUE_{DM, ET}$					
Replication stratum	2	410.29	205.14	2.13	
Variety	9	429.94	54.77	0.57	0.805
Residual	18	1731.56	96.20		
Total	29	2634.79			
Variate: $WUE_{DM, T}$					
Replication stratum	2	693.0	346.5	2.04	
Variety	9	2001.9	222.4	1.31	0.299
Residual	18	3062.0	170.1		
Total	29	5757.0			
Variate: $WUE_{GY, ET}$					
Replication stratum	2	17.70	8.85	0.78	
Variety	9	76.77	8.53	0.75	0.662
Residual	18	205.12	11.40		
Total	29	299.58			
Variate: $WUE_{GY, T}$					
Replication stratum	2	11.72	5.86	0.27	
Variety	9	207.58	23.06	1.08	0.423
Residual	18	384.60	21.37		
Total	29	603.90			

df = degrees of freedom, s.s = sum of squares, m.s = mean square, v.r = variance ratio

3.3.3 Water Use Efficiency of Late Maturing Maize Varieties

The range of $WUE_{DM, ET}$ of late maturing maize varieties varied from 13.11 kg ha^{-1} mm^{-1} to 47.91 kg ha^{-1} mm^{-1} with a mean of 32.20 kg ha^{-1} mm^{-1}. The lowest value was observed with MRI 724 while the highest value was observed with SC 709 (Table 10). In terms of $WUE_{DM, T}$ values varied from 30.26 kg ha^{-1} mm^{-1} to 84.44 kg ha^{-1} mm^{-1}. The mean was 60.43 kg ha^{-1} mm^{-1} and MRI 724 had the lowest while the highest was found to be with SC 709.

The efficiency with which soil water was used for GY production varied from 3.69 kg ha^{-1} mm^{-1} to 12.85 kg ha^{-1}

mm^{-1}. The mean $WUE_{GY, ET}$ was 9.56 kg ha^{-1} mm^{-1}. The lowest value was found to be with MRI 724 while the highest was found to be with PAN ZM-83. $WUE_{GY, T}$ was in the range 8.36 kg ha^{-1} mm^{-1} to 22.54 kg ha^{-1} mm^{-1}. The mean was 17.77 kg ha^{-1} mm^{-1}. Maize varieties MRI 724 and PAN ZM 83 had the lowest and highest values, respectively. The ANOVA of late maturing maize varieties (Table 11) showed that $WUE_{DM, ET}$ and $WUE_{GY, T}$ were significantly different ($P < 0.05$) among the varieties while $WUE_{DM, T}$ revealed very highly significant differences ($P < .001$) among varieties. However, $WUE_{GY, ET}$ was not significantly different among the varieties ($P > 0.05$).

Table 10. Water Use Efficiency of Late Maturing Maize Varieties

Variety	$WUE_{DM, ET}$	$WUE_{DM, T}$	$WUE_{GY, ET}$	$WUE_{GY, T}$
	kg ha^{-1} mm^{-1}			
MRI 724	13.11[a]	30.26[a]	3.69[a]	8.36[a]
ZMS 702	24.79[ab]	49.82[b]	6.90[a]	13.56[ab]
GV 635	29.74[bc]	56.63[bc]	9.29[a]	17.79[bc]
PAN ZM 81	30.30[bc]	63.69[bcd]	9.26[a]	19.17[bc]
MRI 744	30.76[bc]	56.63[bc]	10.16[a]	18.56[bc]
ZMS 720	31.23[bc]	57.03[bc]	9.19[a]	16.68[bc]
PAN 8M 93	32.24[bc]	62.33[bcd]	11.38[a]	21.74[c]
SC 719	39.98[bcd]	69.11[cde]	11.24[a]	19.40[bc]
PAN ZM 83	41.92[cd]	74.31[de]	12.85[a]	22.54[c]
SC 709	47.91[d]	84.44[e]	11.70[a]	19.90[bc]
Mean	**32.20***	**60.43***	**9.56[ns]**	**17.77***
LSD	15.46	16.59	6.360	7.692
CV%	28.0	16.0	38.8	25.2

*= significant, ***= very highly significant, ns= not significant, means followed with the same letter(s) are not significantly different

Table 11. ANOVA of the WUE of Late Maturing Maize Varieties

Source of variation	df	s.s	m.s	v.r	F pr.
Variate: $WUE_{DM, ET}$					
Replication stratum	2	117.00	58.50	0.72	
Variety	9	2502.23	278.03	3.42	0.013
Residual	18	1462.90	81.27		
Total	29	4082.12			
Variate: $WUE_{DM, T}$					
Replication stratum	2	353.09	176.55	1.89	
Variety	9	5773.26	641.47	6.85	< .001
Residual	18	1684.57	93.59		
Total	29	7810.92			
Variate: $WUE_{GY, ET}$					
Replication stratum	2	10.30	5.15	0.37	
Variety	9	191.30	21.26	1.55	0.206
Residual	18	247.41	13.75		
Total	29	449.02			
Variate: $WUE_{GY, T}$					
Replication stratum	2	13.62	6.81	0.34	
Variety	9	467.24	51.92	2.58	0.041
Residual	18	361.90	20.11		
Total	29	842.77			

df = degrees of freedom, s.s = sum of squares, m.s = mean square, v.r = variance ratio

4. Discussion

During the crop growing season, there was a continuous 23 days-dry spell that occurred in the month of March. As a result, plants experienced water stress and this was evidently visible on the plants especially through wilting of leaves to near permanent point; therefore, optimum crop growth was not achieved. Limitation of transpiration, which is the process that ensures use of water for plant growth as well as for cooling purposes, as was observed with the early maturing maize variety SC 303, caused plants to pay, sooner or later, in terms of reduced growth since the same stomates that transpired water also served to absorb CO_2 that was needed in photosynthesis. Additionally, reduced transpiration often results in warming of the plants and hence in increased respiration and further reduction of net photosynthesis (Hillel, 2004), and this was evidenced by the low GY (see also Ludlow, 1975). Whereas no variety tested in this experiment gave a yield of < 1.5 ton ha^{-1}, the Zambian Central Statistical Office (CSO) reported that unit yields stood at 1.5 - 2.0 ton ha^{-1} each year in most provinces of Zambia (CSO,

2014). Factors that accounted for the yield expansion in this study could have included changing hybrid seed use and also the influence of the ever changing weather pattern. In this study, it was not possible to detect significant varietal differences in GY, especially of late maturing maize varieties. Although significant differences in GY could not be detected, it cannot be concluded that varietal differences did not exist, since C.Vs were very high. One of the effects of drought on maize is delay in silking. The trial was planted in mid-December, 2014. Tasselling and silking coincided with the drought in March, 2015. This caused poor synchrony between silking and tasselling. This is because the anthesis-silking interval is a prediction of seed set in many maize varieties when under stress at flowering (Edmeades, Bolanos, Elingo, Banziger & Westgate, 2000). Slatyer (1969) explains that water reduction especially at anthesis can markedly reduce fertilization and grain set in most cereals, especially maize; with reductions of over 50% in yield being caused by relatively brief periods of wilting.

Water extracted from the soil by the roots of plants has been shown to depend on the plant characteristics such as leaf surface area and rooting density; and soil physical properties such as the water holding capacity and hydraulic conductivity (Kamara, Kling, Ajala & Menkir, 2004). Many authors are of the consensus view that often, over 98% of water taken up by the plant is lost as vapor in the process of T (Hillel, 2004). However, the results of this study showed that contrary to that view, of the total 433.1 mm of seasonal water use by late maturing maize varieties, only 53% was lost as T. Although maize is susceptible to water deficit, research has shown that there is a marked genotypic variation in root density, morphological and physiological characteristics in the crop thereby causing differences in ET under identical environmental conditions (Farhad, Cheema, Saleem & Saqib, 2011; Kamara et al., 2004). These characteristics include resistance to transpiration, plant height, plant leaf roughness and reflection and ground cover. Maize varieties PAN 53 and P 3812w had lower ET (380.0 and 403.0 mm, respectively) yet produced greater GY (2.7 and 3.4 ton ha^{-1}) and had higher $WUE_{GY, ET}$ (6.92 and 8.25 kg ha^{-1} mm^{-1}, respectively) compared to that of hybrid SC 647 and MRI 694 (5.62 and 7.28 kg ha^{-1} mm^{-1}) with GY of 2.3 and 2.9 ton ha^{-1}, respectively yet had higher ET (408.0 and 405.0 mm, respectively). Thus it seems that increases in yield and WUE by PAN 53 and P 3812w hybrids were not due to a better ability to take up water from the soil, but a better ability to create yield per unit of water.

DM and GY depend on photosynthesis; and photosynthesis involves the uptake of carbon dioxide through stomata. However, open stomata required for carbon dioxide uptake are an open gate for water loss. Thus, there is a tight trade-off between uptake of carbon dioxide and water loss, and this explains the close link between crop production and water use. The comparatively higher seasonal $WUE_{DM, T}$ for maize variety PAN 413 (68.51 kg ha^{-1} mm^{-1}) to that of ZMS 402 (42.79 kg ha^{-1} mm^{-1}) and P 3253 (36.54 kg ha^{-1} mm^{-1}) was due to higher biomass accumulated (10.7 ton ha^{-1} for PAN 413 compared to 7.4 and 6.9 ton ha^{-1} for ZMS 402 and P 3253, respectively) at relatively low T (149.0 mm for PAN 413) as opposed to the T of ZMS 402 and P 3253 of 160.0 mm and 178.7 mm which was higher. Maize varieties SC 525 and SC 513 resulted in higher GY compared to SC 303 and ZMS 402. This may have been due to their longer growth cycle (about 30 days difference), allowing the crops to obtain more water resources for plant growth and grain production. It means that yield increase in this case was not dependent on greater WUE but on the more rain water received from the longer growing season. The mean seasonal $WUE_{GY, ET}$ of 8.42 kg ha^{-1} mm^{-1} of early maturing maize varieties was close to 8.80 kg ha^{-1} mm^{-1} reported by Phiri et al. (2003) for rain-fed continuous maize with fertilizer kind of farming in eastern Zambia and also fell within the range that Sadras, Grassini & Steduto (2011) reported for maximum yield per unit seasonal ET for maize as 6 - 23 kg ha^{-1} mm^{-1}. The relationship between WUE and yield was direct, meaning that the higher the yield, the higher the WUE. Even though no significant differences were observed in WUE_{GY}, however, it turns out that high producing varieties were more efficient water users. Several previous studies have indicated that relationships between resource capture and use efficiency are affected by the temporal distribution of captured resources in different development stages (Mwale, Azam-Ali & Massawe, 2007 a,b). Notably, crop biomass production and grain yield appear to be most strongly correlated to resource capture during the reproductive stage since grain yield especially is directly associated with the current rates of assimilation and translocation. Since inadequate water will produce no grain, WUE expressed on the basis of grain rises rapidly as water availability increases.

High C.Vs were observed in the statistical analysis of all maturity classes. The C.V expresses the standard deviation per experimental unit as a percentage of the general mean of the experiment. The high C.Vs were attributed to crop failure in some plots due to the severe rain reduction (drought) observed during flowering and yield formation stages of growth. Drought effects make plot to plot error differences high. Although the total rainfall received during the experimental year (1 031.8 mm) should have satisfied the crop water requirements, plant water deficits originated due to the erratic rainfall distribution during the crop growing period. The 23-days dry spell that occurred during the critical phase of crop growth (flowering and grain formation) affected efficiency of water utilization by the crop varieties. Yield production was thus critically affected and consequently WUE too. Maize generally requires well rainfall distribution during the growing period. Any inconsistent or irregular rainfall distribution affects the efficiency of water utilization by the crop.

Generally, the mean seasonal $WUE_{DM, ET}$ for medium maturing maize varieties were comparable to the value of 20.7 kg ha^{-1}mm^{-1} reported by Phiri et al. (2003) for rain-fed maize in eastern Zambia. Efficient use of water in the production of dry matter depends on many factors. ET is essentially dependent on the energy available from net radiation and advection. DM production or net photosynthesis is largely determined by crop and soil characteristics that maximize use of net radiation. When a grain crop like maize, runs out of water at the critical period, yields and WUE are drastically reduced without lowering much of the total seasonal ET which then explains why WUE of medium maturing maize varieties at the end of the season would not be significantly different among varieties. In fact, maize was shown to prioritize radiation interception by maintaining leaf area expansion at the cost of nitrogen concentration per unit leaf area which causes photosynthetic rates to be reduced (Lamaire, van Oosterom, Jeuffroy, Gastal & Massignam, 2008). Similar $WUE_{GY, ET}$ values ranging from 11.0 – 18.0, 9.3 – 13.8, and 11.4 – 14.4 kg ha^{-1} mm^{-1} have been reported by Tijani, Oyedele and Aina (2008), El-Tantawy, Ouda and Khalil (2007) and Meena, Meena and Bhimavat (2009) respectively, for maize grown under rain-fed conditions in Africa.

With regard to late maturing maize varieties, the significant differences observed in WUE were a direct result of differences in total aboveground dry matter produced. The higher the dry matter produced, the higher the WUE. The lower total dry matter production was generally due to the less soil water extraction by the varieties. The prolonged crop growing period of long maturity maize varieties may have led to greater vapor flow and ET, which in theory would be harmful to the sustainable use of dry land agricultural production. Considering that hybrids were compared under the same environment and had the same length to maturity, yet their yields were different, the difference was simply due to superior hybrid(s) resulting in higher WUE (Ritchie & Basso, 2007). Thus, whenever water becomes limited during development or maturation, then yield differences may be partly or wholly due to plants WUE. When water supply is limited, the assimilation rate, plant growth and consequently crop yield are all related quantitatively to the water supply. Additionally, although photosynthate accumulated prior to anthesis contribute to grain filling, and in some cases may provide a significant portion of grain yield, the greatest contribution is usually from photosynthate after anthesis by the ear, leaves and stem (Eastin, Haskins, Sullivan & van Bavel, 1969). In another study, Liu et al. (2009) found that under water limited conditions, crop yields appear to be strongly related to water resource use; thus GY can be dramatically reduced by water resource deficits. Blum (2009) also noted that effective use of water implies maximum capture for T and minimal loss through E. Seasonal rainfall therefore has an impact on biomass partitioning for GY in maize and consequently has effects on WUE_{GY}. Sadras et al. (2011) reported a maximum yield per unit seasonal T for maize of 30 – 37 kg ha^{-1} mm^{-1}. Arguably, the $WUE_{GY, T}$ for all the varieties tested in the current study irrespective of the maturity class fell far below the reported values (13.42 – 23.83, 12.03 – 22.07 and 8.36 – 22.54 kg ha^{-1} mm^{-1} for early, medium and late maturing maize varieties, respectively). This clearly indicated that WUE is variety and location specific and thus justified the need to have locally determined WUE with regional varieties under local climate conditions.

5. Conclusion

The sustainability of agriculture depends on the improvement of WUE, which is more production with less water. This study evaluated the WUE of 30 maize varieties within the early, medium and late maturity classes under rain-fed conditions in Zambia's agro-ecological region IIa. As plant reactions were affected by the amount of water directly or indirectly, efficient use of soil water varied among maize varieties. It was concluded that maize varieties from the same maturity classes have different WUEs. Therefore, breeding for maximal water capture and use for increased transpiration would be important targets for yield improvement under unfavorable water conditions. However, because of the dry spell that occurred during the flowering and grain formation stages of plant growth and the trial only having been evaluated in one season, these results are not conclusive. Nevertheless, the results provided an indication that some varieties tested used water more efficiently than others. The study thus provided options in variety selection for high WUE based on which varieties performed better, particularly SC 525, SC 513 and PAN 4M 21 from the early maturity class; PHB 30G19, ZMS 606, MRI 634 and SC 637 from the medium maturity class; and PAN ZM 83, SC 709, PAN 8M 93 and SC 719 from the late maturity class. It was recommended that repeated experiments over time should be done to validate the findings given that the trial was only conducted in one season.

Acknowledgements

The authors are greatly indebted to the Southern African Science Service Centre for Climate Change and Adaptive Land Use (SASSCAL Task 109), sponsored by the German Federal Ministry of Education and Research under promotion number 01LG1201M and Agricultural Productivity Program for Southern Africa (APPSA) for the financial support towards this study. Both the technical and academic staff of the Departments of Soil and Plant Sciences in the School of Agricultural Sciences and the Department of Geography and Environmental Studies at the University of Zambia are much appreciated for the assistance they rendered during data collection and analysis, and for their valuable contributions during initial preparation of this paper.

References

Asare, D. K., Frimpong, J. O., Ayeh, E. O., & Amoetey, H. M. (2011). Water use efficiencies of maize cultivars grown under rain-fed conditions. *Agricultural Sciences, 2*(2), 125-130. http://dx.doi.org/10.4236/as.2011.22018

Bibi, A., Sadaqat, H. A., Akram, H. M., & Mohammed, M. I. (2010). Physiological markers for screening Sorghum (*Sorghum bicolor*) germ-plasm under water stress conditions. *Int. J. Agric. Biol., 12*, 451-455. Retrieved from https://www.researchgate.net/.../2284423635_Physiological_Makers_for_So

Blum, A. (2009). Effective use of water (EUW) and not water use efficiency (WUE) is the target of crop yield improvement under drought stress. *Field Crops Research, 112,* 119-123. http://dx.doi.org/10.1016/j.fcr.2009.03.009

Brown, W. L., Zuber, M. S., Darrah, L. l., & Glover, D. V. (1985). Origin, adaptation and types of Corn. In: *National Corn handbook – 10.* Iowa: Cooperative Extension Service, Iowa State University of Science and Technology. Retrieved from www.corn.agronomy.wisc.edu/Management/pdfs.NCH10.pdf

Doorenbos, J., Kassam, A. H., Bentvelson, C. L. M., Branschield, V., Plusje, J. M. G. A., Uittenbogaard, G. O., & van Der Wal, H. K. (1979). Yield response to water. *FAO irrigation and drainage paper No 33.* Rome: FAO.

Eastin, J. D., Haskins, F. A., Sullivan, C. Y., & van Bavel, C. H. M. (Eds.). (1969). *Physiological aspects of crop yield.* Madison: American Society of Agronomy.

Edmeades, G. O., Bolanos, J., Elings, A., Banziger, M., & Westgate, M. E. (2000). The role and regulation of anthesis-silk interval. In: Westgate, M. E., & Boote, K. J. (Eds.), *Physiology and modelling kernel set in Maize. CSSA Special Publication, 29,* 43-73. http://dx.doi.org/10.2135/cssaspecpub29.c4

El-Tantawy, M. M., Ouda, A. S., & Khalil, A. F. (2007). Irrigation scheduling for maize grown under Middle Egypt conditions. *Research Journal of Agriculture and Biological Sciences, 3*, 456-462.

Farhad, W., Cheema, M. A., Saleem, M. F., & Saqib, M. (2011). Evaluation of drought tolerant and sensitive maize hybrids. *Int. J. Agric. Biol., 13*(4), 523-528.

Frimpong, J. O., Amoatey, H. M., Ayeh, E. O., & Asare, D. K. (2011). Productivity and soil water use by rain-fed maize genotypes in a coastal Savanna environment. *International Agrophysics, 25*, 123-129.

Gardner, C. M. K., Laryea, K. B., & Unger, P. W. (1999). *Soil physical constraints to plant growth and crop production.* Rome: Land and Water Development Division: Food and Agricultural Organization of the United Nations.

Hillel, D. (2004). *Introduction to environmental soil physics.* San Diego: Elsevier Academic Press.

Hsiao, T. C., Heng, L., Steduto, P., Rojas-Lara, B., Raes, D., & Fereres, E. (2009). AquaCrop – the FAO crop model to simulate yield response to water: III. Parameterization and testing for maize. *Agronomy Journal, 101*(3), 448-459. http://dx.doi.org/10.2134/agronj2008.0218s

Huang, R., Birch, C. J., & George, D. L. (2006). Water use efficiency in Maize production-The challenge and improvement strategies. *6th Triennial Conference.* Maize Association of Australia.

Jayne, T. S., Goverch, J., Chilonda, P., Mason, N., Chapota, A., & Haantuba, H. (2007). Trends in agricultural and rural development indicators in Zambia. *Working paper No. 24: Food security research project.* Lusaka. Retrieved http://www.aec.msu.edu/agecon/fs2/zambia/index.htm

Jones, A., Breuning-Madsen, H., Brossard, M., Dampha, A., Deckers, J., Dewitte ... Zougmoré R., (Eds.), (2013). *Soil atlas of Africa.* Luxembourg: Publications Office of the European Union.

Kamara, A. Y., Kling, J. G., Ajala, S. O., & A. Menkir. (2004). Vertical root pulling resistance in maize is related to nitrogen uptake and yield. *7th Eastern and Southern Africa regional Maize conference, Kenya, Nairobi,*

11- 15 February, 2001 pp 228-232.

Lamaire, G., van Oosterom, E., Jeuffroy, M. H., Gastal, F., & Massignam, A. (2008). Crop species present different qualitative types of nitrogen deficiency during their vegetative growth. *Field Crops Research,* 105(3), 253-265. http://dx.doi.org/10.1016/j.fcr.2007.10.009

Liu, C. A., Jin, S. L., Zhou, L. M., Jia, Y., Li, F. M., Xiong, Y. C., & Li, X. G. (2009). Effects of plastic film mulch and tillage on maize productivity and soil parameters. *European Journal of Agronomy,* 31, 241-249. http://dx.doi.org/10.1016/j.eja.2009.08.004

Ludlow, M. M. (1975). Effects of water stress on the decline of leaf net photosynthesis with age. In: Mcelle, R (Ed.), *Environmental and biological control of photosynthesis,* 123-134. http://dx.doi.org/10.1007/978-94-010-1957-6_13

Meena, R. P., Meena, R. P., & Bhimavat, B. S. (2009). Moisture use functions and yield of rain-fed maize as influenced by indigenous technologies. *Asian Agri-History, 13*(2), 155-158.

Ministry of Agriculture and Livestock. (2015). *Investment opportunities in agriculture.* Lusaka: Ministry of Agriculture and Livestock.

Mwale, S. S., Azam-Ali, S. N., & Massawe, F. J. (2007a). Growth and development of bambara groundnut (*Vigna subterranea*) in response to soil moisture: 1. Dry matter and yield. *Eur. J. Agron., 26*(4), 345-353. http://dx.doi.org/10.1016/j.eja.2006.09.007

Mwale, S. S., Azam-Ali, S. N., & Massawe, F. J. (2007b). Growth and development of bambara groundnut (*Vigna subterranea*) in response to soil moisture: 2. Resource capture and conversion. *Eur. J. Agron., 26*(4), 354-362. http://dx.doi.org/10.1016/j.eja.2006.12.008

Nyakudya, I. W., & Stroosnijder, L. (2004). Effect of rooting depth, plant density and planting density on maize (*Zea mays* L.) yield and water use efficiency in semi-arid Zimbabwe: modeling with AquaCrop. *Agricultural Water Management, 146,* 280-296. http://dx.doi.org/10.1016/j.agwat.2014.08.024

Phiri, E., Verplancke, H., Kwesiga, F., & Mafongonya, P. (2003). Water balance and maize yield following Sesbania fallow in Eastern Zambia. *Agroforestry Systems, 59*(3), 197-205. http://dx.doi.org/10.1023/b:agfo.0000005220.67024.2c

Raes, D., Steduto, P., Hsiao, T. C. & Fereres, E. (2009). AquaCrop-the FAO crop model to simulate yield response to water: II. Main algorithms and software description. *Agronomy Journal, 101*(3), 438-447. http://dx.doi.org/10.2134/agronj2008.0140s

Ritchie, J. T., & Basso, B. (2007). Water use efficiency is not constant when crop supply is adequate or fixed: the role of agronomic management. *Eur. J. Agron,* http://dx.doi.org/10.1016/j.eja.2007.08.003.

Sadras, V. O., Grassini, P., & Steduto, P. (2011). Status of water use efficiency of main crops. *SOLAW background thematic report-07.* Rome: FAO.

Slatyer, R. O. (1969). Physiological significance of internal water relations to crop yield. In: Eastin, J. D., Haskins, C. Y., Sullivan, C. H. M., & Van Bavel (Eds.), *Physiological aspects of crop yield* (pp. 53-85). Madison: American Society of Agronomy.

Steduto, P., Hsiao, T. C., Raes, D., & Fereres, E. (2009). AquaCrop – the FAO crop model to simulate yield response to water: I. Concepts and underlying principles. *Agronomy Journal, 101*(3), 426-437. http://dx.doi.org/10.2134/agronj2008.0139s

Stewart, B. A., & Nielson, D. R. (1991). Irrigation of agricultural crops. *Soil Science, 152*(2), 137 http://dx.doi.org/10.1097/00010694-199108000-00013

Tijani, F. O. Oyedele, D. J., & Aina, P. O. (2008). Soil moisture storage and water use efficiency of maize planted in succession to different fallow treatments. *International Agrophysics, 22,* 81-87.

9

Farmers' Own Research: Organic Farmers' Experiments in Austria and Implications for Agricultural Innovation Systems

Susanne Kummer[1], Friedrich Leitgeb[1] & Christian R. Vogl[1]

[1]Division of Organic Farming, Department of Sustainable Agricultural Systems, University of Natural Resources and Life Sciences Vienna, Austria

Correspondence: Susanne Kummer, University of Natural Resources and Life Sciences Vienna (BOKU), Gregor-Mendel-Strasse 33, A-1180 Vienna, Austria. E-mail: susanne.kummer@boku.ac.at

Abstract

Farmers' experiments can be defined as the autonomous activities of farmers to try or introduce something new at the farm, and include evaluation of success or failure with farmers' own methods. Experiments enable farmers to adapt their farms to changing circumstances, build up local knowledge, and have resulted in countless agricultural innovations. Most research on the topic has been conducted in countries of the south. In this paper, however, we present experiments of randomly sampled organic farmers in Austria, and we discuss implications for agricultural innovation systems. In 76 structured questionnaire interviews we investigated topics, motives, methods and outcomes of farmers' experiments, and factors related to the frequency of experimentation. From the interviewed farmers, 90% reported experiments, and the majority of experiments (94%) involved monitoring and evaluation strategies. Farmers who reported a high frequency of experimentation showed a significantly higher propensity to plan, document and repeat their experiments, and had a more positive attitude towards experimenting than farmers that rarely experimented. We conclude that experimenting is a common activity among organic farmers in Austria, and that farmers have their own methods to conduct and assess their experiments. The most significant outcome is the creation of new knowledge, stressing the importance of experimentation for learning and adaptive farm management. Farmers' experiments are significant on two levels, i.e. at individual farm level and at the level of agricultural innovation systems. Taking full advantage of this innovative potential requires a better involvement of farmers as co-researchers into the development of agricultural innovations.

Keywords: adaptive management, European Innovation Partnership, farmer innovation, farmer learning, knowledge exchange, local innovation, local knowledge, organic agriculture, participatory research

1. Introduction

The historical development of locally adapted farming systems worldwide can be ascribed to continuous autonomous experimentation activities of farmers (Hoffmann, Probst, & Christinck, 2007). Experimenting enables farmers to adapt to constantly changing conditions (Bentley, 2006; Darnhofer, Bellon, Dedieu, & Milestad, 2010), is a means to generate local knowledge (Sumberg & Okali, 1997), and builds the base for countless agricultural innovations. The potential of farmers' experiments to contribute to agricultural development has not been taken into consideration by agricultural scientists for a long time. Only a small group of anthropologists and agricultural historians have shown interest in the topic in the past (Johnson, 1972; Sumberg & Okali, 1997). With the relatively recent interest in rural development including the concepts of participation, empowerment and sustainability, farmers' experiments and local knowledge began to attract more attention within research, especially in the context of development studies (Bentley, van Mele, & Acheampong, 2010; Sumberg, Okali, & Reece, 2003).

Scientific research about farmers' experiments mainly focused on case studies set in development contexts in countries of the South, especially in Asia, Africa and Latin America (Chambers, Pacey, & Thrupp, 1998; Haverkort, van der Kamp, & Waters-Bayer, 1991; Reij & Waters-Bayer, 2001b; Sturdy, Jewitt, & Lorentz, 2008; Laurens van Veldhuizen, Waters-Bayer, Ramírez, Johnson, & Thompson, 1997), and little has been written about the situation in so called industrialized countries (Kandel, Porter, Carr, & Zwinger, 2008; Kummer, 2011; McKenzie, 2013). Furthermore, few research activities investigated farmers' experiments in a systematic way, by studying the entire process of experimentation and the applied methods. Literature on farmers' experiments mainly focused on few examples (Bentley, 2006) of active experimenters within the farming community (Haverkort et al., 1991; Reij & Waters-Bayer, 2001b), outstanding 'research-minded farmers' (Biggs, 1990) and 'farmer innovators' (Critchley, 2000; Reij & Waters-Bayer, 2001a), and most presented experiments concerned plant production issues (Sumberg & Okali, 1997; van Veldhuizen, Waters-Bayer, Ramirez, Johnson, & Thompson, 1997). Furthermore, most sources refer to experiments carried out in the context of participatory research (Bentley et al., 2010; Kandel et al., 2008; Wortmann et al., 2005), i.e. experiments together with researchers and extensionists and not independent farmers' experimentation. Consequently, our research aims at filling the knowledge gap on farmers' autonomous experiments in a 'non-development' context.

Case studies of smallholder farmers in the Global South emphasize the need and creative capacity of finding appropriate solutions within conditions of resource scarcity (Leitgeb, Kummer, Funes-Monzote, & Vogl, 2014) and poverty (Reij & Waters-Bayer, 2001a). It is a powerful motive to try a new idea when *farmers are driven by the need to feed their families'* (Reij & Waters-Bayer, 2001a, p.83). Although the situation in less constrained conditions will be less threatening and so the urgency involved in experimenting is presumably lower, the overarching significance of experimentation is still the same: Farmers, no matter in which part of the earth, have always lived in changing environments where uncertainty and disturbances are inevitable. Therefore, farmers need the ability to adapt to change in order to be able to maintain their farms. Experimenting is one way for farmers to learn, and is a key strategy to adapt to change and thus enhance adaptive management of a farm. Conducting and monitoring experiments allows a better understanding of system dynamics, widens the range of options in case of change, and enables farmers to improve their management practices based on the knowledge gained (Darnhofer et al., 2010). The outcome of the iterative process of adaptive management is learning about the farming system, i.e. an ongoing reconsideration of the efficiency of measures taken, the accuracy of the consequences of actions, the relationship between actions and indicators, and learning about trade-offs (Milestad, Dedieu, Darnhofer, & Bellon, 2012).

Besides the significance of experimentation for adaptive management of farms in general, there are two reasons why it is particularly interesting to explore farmers' experiments in the context of organic agriculture. First, sustainable land use practices are more knowledge-intensive (Röling & Brouwers, 1999). While conventional farmers can use external inputs such as synthetic pesticides and synthetic fertilizers to handle adverse dynamics in their agro-ecosystem, organic farmers need to develop knowledge about the agro-ecosystem to a larger extent to be able to manage their farms successfully without these inputs. Second, organic agriculture was developed by farmers' grassroots organizations, where farmers themselves were responsible for advances and innovations. Official research and extension only played a minor role in the development of organic agriculture (Brunori et al., 2013; Padel, 2001), and so organic agriculture developed by practical experiments and trials of farmers and practical researchers. The lack of advice and formal research in the initial phase of organic agriculture leads to the assumption that organic farmers have developed a culture of experimentation (Gerber, Hoffmann, & Kügler, 1996).

1.1 Defining Farmers' Experiments

Farmers' experiments can be defined as the activity of trying or introducing something totally or partially new at the farm, including evaluation of the success or failure of this introduction (Quiroz, 1999), or as the comparison of something known with something unknown (Stolzenbach, 1999). Sumberg and Okali (1997) consider two conditions necessary for an activity to be labeled an experiment: the creation or initial observation of conditions, and the observation or monitoring of subsequent results.

A common concept of on-farm experimentation is 'on-farm research', which means research conducted, and usually also controlled, by scientists on farms, involving the farmer more or less actively (Lawrence, Christodoulou, & Whish, 2007). Another term used in literature is 'farmer-initiated research', which refers to *research conducted by farmers for discovery or production of information'* (Wortmann et al., 2005, p.244) in cooperation with research and extension. In this study, however, we focus on experiments carried out by farmers on their own initiative, and we explicitly do not refer to on-farm research. Farmers themselves often do not use the term 'experiment' to refer to their practical on-farm experiments (Stuiver, Leeuwis, & van der Ploeg, 2004), but relate this term more to a scientific and formal procedure. In various empirical studies on the topic, the term 'trying' instead of 'experimenting' has been perceived as being more appropriate (Sumberg & Okali, 1997), while in other cases local terms were used to address the subject in the field (Stolzenbach, 1999).

We conceptualize farmers' experiments as individual research processes with a certain problem or topic as starting point, the experiment itself involving methods to conduct and monitor the process, and different possible outcomes that can be classified into: i) adaptations of a method or practice, ii) local innovations (i.e., innovations that are not new in general but to the specific area or context), iii) inventions and iv) failures (i.e., experiments that do not lead to satisfactory results). Experiments are influenced by various intervening factors such as environmental, social and personal conditions (Leitgeb et al., 2014; Sumberg & Okali, 1997), and are embedded into the wider agricultural communication and innovation system (Leitgeb, Funes-Monzote, Kummer, & Vogl, 2011).

1.2 Positioning Farmers' Experiments within Agricultural Innovation Systems

An agricultural innovation system (AIS) consists not only of actors directly involved in the agricultural production chain and the agricultural research, extension and education system, but of a diversity of stakeholders within and outside the agricultural sector that are involved in the development of agricultural innovations (Hermans, Klerkx, & Roep, 2015). In contrast to the overcome linear 'transfer of technology' approach where innovations were seen as being exclusively developed by science and then transferred to farmers that were expected to adopt them, the AIS perspective considers the development of innovation as co-evolutionary process shaped by all actors involved (Klerkx, van Mierlo, & Leeuwis, 2012), and includes institutional and political dimensions (Schut, Rodenburg, Klerkx, van Ast, & Bastiaans, 2014). The main focus is on learning as a means of developing new arrangements specific to local contexts, and on strengthening the capacity of actors to create, diffuse and use knowledge and enable innovation (Rivera & Sulaiman, 2009), and so consequently the role of farmers as innovators and the value of local knowledge receives more attention (Brunori et al., 2013).

The shift to a systemic perception on innovation development also displays within the Common Agricultural Policy (CAP) of the European Union: A current program of the European Commission is the 'European Innovation Partnership on Agricultural Productivity and Sustainability' (EIP-AGRI). The proposed EIP-AGRI stresses the importance of innovation and knowledge exchange in the agrarian sector to meet societal challenges of food production, and emphasizes the need for systemic feedback from practice to science, for enhancing knowledge exchange, and for joint efforts to invest in sustainable innovation (European Commission [EC], 2012).

In the context of this current EIP-AGRI it is relevant to investigate, which experiments and innovations take place at the farm level, and to make this innovative potential visible that could be relevant for such political initiatives. To be able to draw conclusions on farmers' innovative potential, we investigate whether organic farmers in Austria experiment, the topics, motives, methods and outcomes of farmers' experiments, and the factors related to the propensity to experiment. By doing so, our research aims at filling the knowledge gap on farmers' autonomous experiments in a 'non-development' context. We then discuss the significance of these experiments for adaptive farm management and for policy initiatives such as EIP-AGRI proposing possibilities to support farmers' experiments.

2. Methods

Research on farmers' experiments so far mainly used case studies of well-known experimenters. To be able to investigate the topic among 'average' organic farmers, we set up a structured questionnaire and applied it to a random sample of 10% of organic farmers in four regions of Austria. The questionnaire confronted all interviewees with the same set and sequences of questions. The pre-defined answer categories in the questionnaire were based on results and preliminary analysis of 47 semi-structured interviews carried out previously (Kummer, 2011), and on results from literature research. Answer categories included one- to multiple-choice answers, open-ended questions, and ratings along Likert scales (Bernard, 2006). We conducted two pre-test interviews to check if the questionnaire was comprehensible, and implemented final adaptations regarding wording and sequence of questions. Altogether, 76 organic farmers were interviewed with the questionnaire (Table 1).

Table 1. Characteristics of interviewees (n=76)

Characteristics	Definition	%	Arith. mean	Median	Max	Min
Sex	Male	71%				
	Female	29%				
Age (years)			45.3	46	70	23
Farm environment	Farm in mountainous region	50%				
	Farm in (predominantly) flat region	50%				
Farm operation type	Regular (full-time)	46%				
	Sideline (part-time)	54%				
Farm size (hectare)			41.1	28	230	2
Farming experience	Farmer since (years)		20.5	20	46	2
	Organic since (years)		13.2	13	27	2

We generated the random sample based on a complete list of all Austrian organic farmers that was provided by the Austrian Ministry of Agriculture (Bundesministerium für Land- und Forstwirtschaft, Umwelt und Wasserwirtschaft) after signing a confidentiality statement. We selected four regions that represented different climatic and agricultural production zones in Austria (Table 2). Region 1 (N=135; n=13) is located in the south-east of Austria, a flat to hilly area dominated by crop production. Region 2 (N=146; n=13) comprises an alpine region in the north-west of Austria, characterised by grassland farming and animal husbandry (milking cows and suckler cows for meat production). Region 3 (N=285; n=25) is characterized by flat arable land of high productivity, comparably large farm sizes and intensive crop production, mainly without livestock. Region 4 (N=248; n=25) is located in the north of Austria, characterized by continental climate and an emphasis on fodder crop production. The selected regions represent four of the eight agricultural production areas in Austria.

Table 2. Characteristics of the four study regions

Characteristics	Region 1	Region 2	Region 3	Region 4
Climatic zone	Temperate lowlands	Alpine region	Pannonian basin	Continental highland
Altitude above sea level	200-400 m	800-1500 m	140-200m	500-600m
Average annual temperature	10.6 °C	7.2 °C	10.1 °C	8.3 °C
Annual precipitation	830 mm	1370 mm	540 mm	780 mm
Proportion of organic farms	9%	17%	11%	24%
Average farm size	18 ha	7 ha	49 ha	34 ha
Main agricultural activities	Cropping, fruits, wine and vegetables	Grassland, milk, beef	Cropping, vegetables, sugar beet	Fodder crops, forestry

Data sources:

Data on average annual temperature and annual precipitation for the years 2009-2014 provided by Zentralanstalt für Meteorologie und Geodynamik [ZAMG] (http://www.zamg.ac.at).

Data on proportion of organic farms and average farm size for the year 2012 provided by Agrarmarkt Austria [AMA].

All data have been processed and compiled by the first author.

We are aware that the sample is not representative for all organic farmers in Austria, but still we can draw conclusions from this sample to the overall situation of organic farmers' experiments in the country, as we selected four contrasting regions representing major differences within agricultural production in Austria, and applied a random sample of 10% of the total organic farmers in each region. The relatively small sample size is mainly due to our decision to conduct personal interviews on the respondents' farms, and not to send out the questionnaire per mail or e-mail. We consciously decided to apply the questionnaire in 'face-to-face' settings, as we had the experience from the semi-structured interviews conducted previously that the topic of farmers' experimentation is not self-explanatory and needs a qualitative explanation and the possibility to ask questions to get into the topic and find a common language and wording.

In each interview setting one researcher and the farmer were sitting together on a table with the questionnaire positioned in front of them in a way that both persons were able to read the text. The researcher read the questions out loud in the sequence of the questionnaire, and the farmer read the answer categories and decided which answers applied. In this face-to-face setting, interviewees were able to ask questions or make qualitative comments and explanations to the questionnaire. Each interview started with a short introduction in the research topic by reading the following definition to the interviewee: *'When we use the terms trying or experimenting we refer to how YOU test and monitor if and how something works, and if it is suitable for your farm. We explicitly do not refer to scientific experiments, but to practical experiments of organic farmers taking place at their farms.'* After this introduction, farmers were asked to freely list experiments they had conducted on their farms. To learn about the individual experimentation process in a detailed way, one of the mentioned topics was then selected together with the interviewee, and systematic questions about topics, motives, methods and outcomes of these experiments followed. At the end of each interview, socioeconomic data about the farmer and the respective farm was inquired.

For data analysis, we applied descriptive and statistical analysis (frequencies and Spearman correlations), using the software SPSS and Microsoft Excel. Regarding significance levels for correlations, we defined $p<0.01$ as highly significant, and $p<0.05$ as significant. In the results section, we present quantitative data from 76 structured questionnaire interviews. As eight of the interviewees stated not to experiment ('non-experimenters'), most results are based on answers from 68 interviewees, if not indicated differently.

3. Results

From all 76 interviewed farmers, 89.5% reported at least one activity in the past where they had experimented on their farms, and eight farmers (10.5%) stated that they had never carried out any experiment. Regarding the frequency of experimenting, 18.4% stated that they 'very often' experimented (i.e. frequently during each season or year), 34.2% stated to experiment 'sometimes' (i.e. at least once every season or year), and 36.9% stated to experiment 'rarely' (i.e. not regularly and not every year). Farmers were asked to freely list experiments they had carried out. Between one and ten experiments were mentioned, with an arithmetic mean of 3.1 experiments per farmer. In total, the interviewees mentioned 239 individual experiments, and 68 experiments were discussed in detail.

3.1 Topics of Farmers' Experiments

Farmers were asked for all kinds of experiments they conducted on their farm in order to assess the full range of experimental activities of organic farmers. The 239 experiments mentioned by interviewees were clustered into 13 topics (Table 3).

Table 3. Frequency of topics for farmers' experiments according to thematic clusters (239 experiments mentioned by interviewees, n=68)

Topics	%
Plant production	**51.0%**
Cropping, plant production	23.8%
Tillage, soil management	13.8%
Weed and pest management	6.3%
Fertilization	5.4%
Vegetable, fruit and wine growing	1.7%
Animal husbandry	**16.7%**
Processing and commercialization	**15.5%**
Processing	8.8%
Commercialization	6.7%
Other areas	**16.8%**
Alternative remedies and supplements	6.3%
Tools, machinery, construction	6.3%
Labor management / reduction	1.3%
Social issues	1.3%
Others	1.6%

Of all experiments, 51% were conducted in the context of plant production, cropping and tillage, and included testing of

- new crops and varieties;

- different tillage tools and systems, including systems such as ploughless tillage, reduced tillage and direct sowing;

- different alternatives within plant production, e.g. intercropping and undersowing;

-   possibilities for optimizing the crop rotation;

-   different methods of weed control, e.g. different tools and machines for tillage, methods and time schedules for mechanical weed control, or introducing new crops into the crop rotation to suppress weeds;

-   different fertilizers, e.g. commercial organic fertilizers, farm manure, compost or mulching.

Experiments in the area of animal husbandry (17%) included testing of

-   new breeds and species on the farm;

-   different feedstuffs and optimization of feed composition;

-   different ways to handle animals, e.g. rearing animals and young animals (assistance before, during and after birth, handling and feeding of suckling animals);

-   new forms of housing and pasturing;

-   converting to alternative husbandry systems, e.g. from dairy farming to suckler cow systems.

Experiments regarding processing and commercialization (16%) included testing of

-   recipes, new ingredients, development of new products, establishment of product ranges, and improvements in processing procedures;

-   different marketing systems, e.g. implementation of direct selling (farm shops, self-harvest systems, web shops, catering).

A range of further experimentation activities (17%) were found in the interviews and included

-   technical experiments, i.e. testing or modifying and adapting tools and machinery on the farm, or experiments in the context of farm constructions;

-   testing of different alternative remedies, preparations and supplements to improve plant or animal health, or to improve compost, manure and soil quality, e.g. testing of effective microorganisms, homeopathy, biodynamic preparations, and other alternative remedies, or testing the lunar influence and farming according to the moon's cycle;

-   experiments to reduce farm labor;

-   implementation of social activities on the farm, e.g. offering educational activities or holidays on the farm.

3.2 Motives and Information Sources for Farmers' Experiments

The majority of the interviewees considered personal reasons (i.e. intrinsic motives like interest or curiosity) as an important motive to start an experiment. Other important motives were confronting challenges and problem solving (i.e. extrinsic motives) (Figure 1).

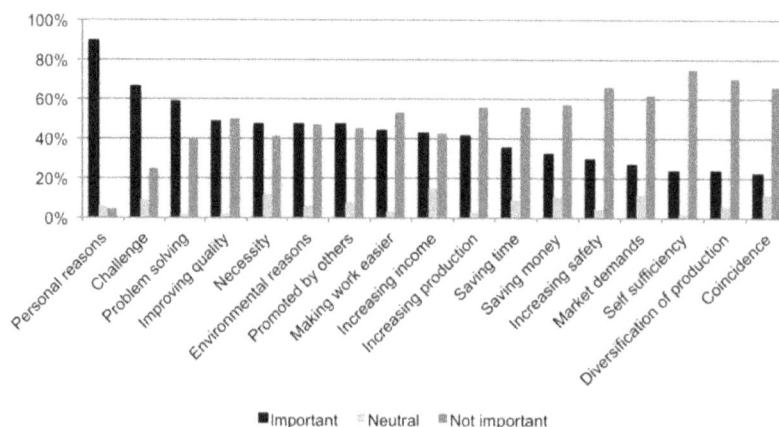

Figure 1. Importance of motives for farmers' experiments (n=68)

These results suggest that by experimenting farmers not only respond to external incentives or problems, but to the same extent have proactive motives for experimenting, developing the farm into a desired direction according to personal values and aims. Both types of motives are important features to enhance the adaptability of farms.

Eighty-four percent of the experimenting farmers had searched for information before or during their experiments. Literature was rated as most important information source, and also other farmers and advisors were considered important. Scientists were rated as least important information sources, indicating the well-known gap between farmers' and scientists' knowledge systems (Figure 2).

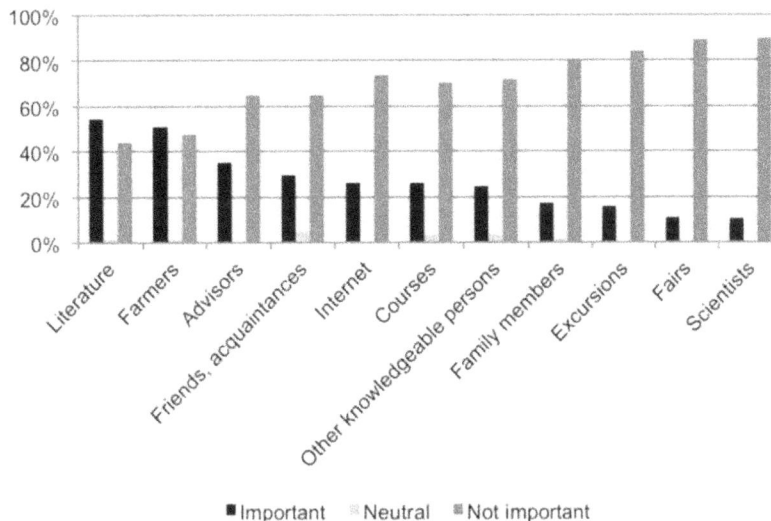

■ Important   ▨ Neutral   ■ Not important

Figure 2. Importance of information sources for farmers' experiments (n=68)

3.3 Methods and Outcomes of Farmers' Experiments

Two thirds of the interviewees had an explicit mental or written plan before starting an experiment (Table 4). The majority (63%) of the experimenting farmers stated that they set up their experiments first on a small scale and enlarged them if the outcome of the experiment was satisfactory. By doing so, farmers were able to first try a new method or practice with little risk. Thirty-seven per cent started the experiment on a large scale, either because they were convinced the new method would work satisfactorily, or because it was difficult or impossible to run a small-scale test. The impracticability of a small-scale test was often cited in the case of experiments involving technical constructions such as buildings or machinery that were implemented at once for the entire production unit. Setting up a test version in these cases would have been more costly than the construction of the entire production unit.

The majority (94%) of experimenting farmers monitored the progress of their experiments, mainly through observation, but also by comparisons. A small share of interviewees employed some kind of measurements, e.g. yield quantity, counting (e.g. of plants affected by a certain disease), or economic measurements (e.g. price calculations for different processed products for marketing). To evaluate the success or failure of an experiment, most interviewees compared their experiments with experiences from former experiments (historical comparison), and with units or experiences of other farmers. Comparisons with other units or practices on their own farm (including side-by-side comparisons), with results described in literature, and with information from advisors or other experts in the subject were employed less frequently (Table 4).

Of the experimenting farmers, 57% stated that they had documented their experiments by taking individual notes, but also obligatory records that farmers had to provide to comply with requirements of the regulation on organic farming and the agricultural subsidy system were seen as a possibility to document experiments. Less frequent documentation strategies involved taking photos, taking samples, or making a video (Table 4).

Table 4. Frequency of different methods used in farmers' experiments (n=68; multiple answers possible)

Methods	%
Planning	69.1%
Mental plan	55.9%
Written plan	13.2%
No explicit plan	30.9%
Monitoring	94.1%
Observation	88.2%
Comparisons	82.4%
Measurements	13.2%
No monitoring	5.9%
Comparisons	82.4%
With own experiences	78.6%
With other farmers	64.3%
With other unit on the farm	26.8%
With results from literature	23.2%
With information from advisors	10.7%
No comparison	17.6%
Documentation	57.4%
Taking notes	33.8%
Obligatory records	33.8%
Photographs	16.2%
Taking sample	4.4%
Video	2.9%
No documentation	42.6%

Obtaining more knowledge was rated as most important outcome of experiments. The pre-defined answer categories that applied least for the interviewed farmers were: reduction of labor, increasing income, and increasing production (Figure 3). Regarding labor reduction, several farmers stated that the outcome of the specific experiment (e.g. new product, new or additional marketing channel) resulted in even more work, or that the overall outcome (e.g. new working method) did not increase the income directly, or may even have caused additional costs.

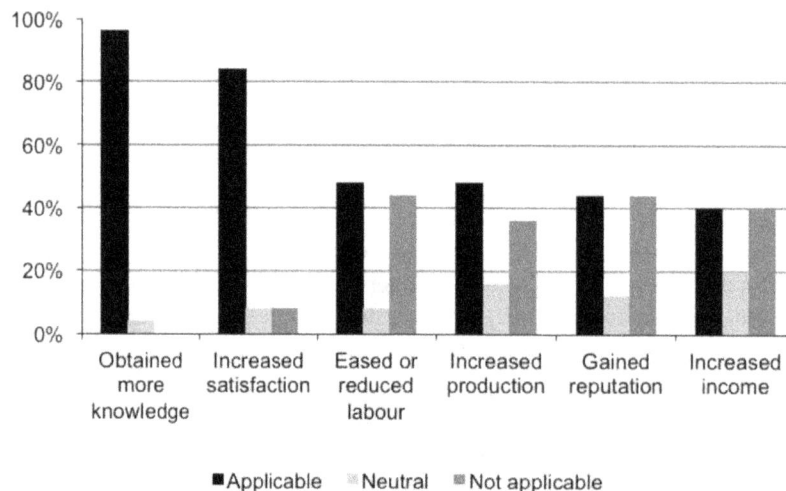

Figure 3. Frequency of different kinds of outcomes of farmers' experiments (n=68, 1 missing case, 67=100%)

About half of the experimenting farmers (47%) reported that other persons had used the results of their experiments, most often other farmers (90%) and friends (31%), but rarely advisors or scientists (6% in each case), indicating that the dissemination of experiments' outcomes was mainly limited to the personal network of the farmers and received little attention by the official research and advisory system.

3.4 Factors related to the frequency of experimentation

Some methodological differences in the experimentation process were significantly related to the frequency of experimentation: Farmers with a high frequency of experimentation showed a higher propensity to set up written or mental plans before starting an experiment, and they more often documented and repeated their experiments (Table 5).

Table 5. Correlation between frequency of experimentation and methods used in farmers' experiments (n=68; % within frequency of experimentation categories)

| | Frequency of experimentation | | | Spearman correlation | |
| | Very often | Sometimes | Rarely | | |
Methods	%	%	%	r	p
Written plan	28.6%	15.4%	3.6%	0.321**	0.008
Mental plan	50.0%	65.4%	50.0%		
No plan	21.4%	19.2%	46.4%		
Documentation	92.9%	57.7%	39.3%	0.384**	0.001
No documentation	7.1%	42.3%	60.7%		
Repetition	85.7%	53.8%	42.9%	0.259*	0.015
No repetition	14.3%	46.2%	57.1%		

Socio-economic factors that were not found to have an influence on the propensity to experiment were mode of farm operation (full-time or part-time) (r=0.177; p=0.125), age (r=0.136; p=0.240), sex (r=0.072; p=0.534), level of education (r=0.003; p=0.982), or years of farming experience (r=0.032; p=0.783). On the other hand, farmers who owned bigger farms reported a higher experimentation frequency, although the correlation between farm size and frequency of experimentation is slightly below the significance level (r=0.215; p=0.062).

Significant correlations were found between the frequency of experimentation and personal habits and characteristics of the interviewees, such as travel habits: Farmers who stated that they often experimented showed a higher travel activity outside Austria (r=0.253*, p=0.028) and outside Europe (r=0.410**, p=0.000). To gain an insight how personal attitudes were related to the propensity to experiment, we provided farmers with pre-defined statements. Farmers who reported a high frequency of experimentation agreed significantly more with the statement that they 'like to do things differently than others do', and that 'change and challenges make my life interesting'. In contrast, farmers who reported a low experimentation frequency agreed significantly more often with the statement that they 'like it when things are stable', and that they 'only change things when necessary'. Interestingly, farmers with a low experimentation frequency also agreed significantly more often to the statements 'I am well structured and organized' and 'I pass on my ideas and experiences' than frequent experimenters, indicating that some frequent experimenters did not want to communicate their findings, and that experimentation tends to involve a certain degree of chaos (Table 6). These results suggest that farmers reporting a high frequency of experimentation consciously define themselves as experimenters and innovators, whereas less frequent experimenters adhere more to values of tradition and stability, indicating that personality may have a significant influence on the propensity to experiment.

Table 6. Correlation between frequency of experimentation and farmers' attitudes to experimenting (n=76; % within frequency of experimentation categories for each statement)

Statement about farmers' attitude to experimenting		Frequency of experimentation				Spearman correlation	
		Very often	Some-times	Rarely	Never	r	p
I like to do things differently than others do.	Agree	**92.9%**	80.8%	50.0%	37.5%	0.384**	0.001
	Neutral	0.0%	7.7%	0.0%	0.0%		
	Don't agree	7.1%	11.5%	50.0%	62.5%		
Change and challenges make my life interesting.	Agree	**92.9%**	80.8%	50.0%	62.5%	0.353**	0.002
	Neutral	7.1%	15.4%	21.4%	12.5%		
	Don't agree	0.0%	3.8%	28.6%	25.0%		
I only try or change things when it is necessary.	Agree	14.3%	42.3%	42.9%	**62.5%**	-0.288*	0.012
	Neutral	0.0%	11.5%	14.3%	12.5%		
	Don't agree	85.7%	46.2%	42.9%	25.0%		
I like it when things are stable.	Agree	14.3%	57.7%	82.1%	**87.5%**	-0.478**	0.000
	Neutral	28.6%	23.1%	7.1%	0.0%		
	Don't agree	57.1%	19.2%	10.7%	12.5%		
I pass on my ideas and experiences to others.	Agree	71.4%	80.8%	92.9%	**100.0%**	-0.267*	0.020
	Neutral	21.4%	15.4%	7.1%	0.0%		
	Don't agree	7.1%	3.8%	0.0%	0.0%		
I have my things well structured and organized.	Agree	35.7%	57.7%	60.7%	**75.0%**	-0.227*	0.049
	Neutral	21.4%	34.6%	21.4%	25.0%		
	Don't agree	42.9%	7.7%	17.9%	0.0%		

4. Discussion

The majority of the interviewed farmers reported that they had carried out activities of experimental character on their farms, and most of them listed several topics of experimentation. The capacity of farmers to experiment is widely accepted within the scientific community (e.g. Bentley, 2006; Hoffmann et al., 2007; Johnson, 1972; Maat, 2011; Reij & Waters-Bayer, 2001b; Sumberg & Okali, 1997; Tambo & Wünscher, 2014; Wortmann et al., 2005). Literature in the context of participatory research mainly focuses on cases of active experimenters (Haverkort et al., 1991; Reij & Waters-Bayer, 2001b), 'research-minded farmers' (Biggs, 1990) and 'farmer innovators' (Critchley, 2000; Tambo & Wünscher, 2014), and thus little is known about less active experimenters, or the relative proportions of active and less active experimenters. In this random sample, 18.5% of the interviewees declared themselves as very active experimenters, 71% reported to experiment sometimes or rarely ('less active experimenters'), and 10.5% of the interviewees reported not to experiment at all.

Frequent experimenters in this study significantly more often had an explicit plan for their experiments, and more often documented and repeated their experiments. This may be due to the fact that these farmers are more aware of their experiments and define themselves as experimenters, in contrast to the majority of farmers that presumably do not see experiments as particular research processes but rather as a normal part of every day farming activities (Saad, 2002). Characteristics such as replication and documentation are seen as crucial for evaluating the success of experiments and for fine-tuning production systems (Wortmann et al., 2005).

Socio-economic factors that influence the propensity for farmers' experiments mentioned in literature are age, sex, education level, farming occupation (full-time or part-time), socio-economic status, and political, social or ecological constraints (Critchley & Mutunga, 2003; Saad, 2002; Sumberg & Okali, 1997). In this study, no socio-economic factors were found to have a significant influence on the frequency of experimentation. This is in line with Sumberg and Okali (1997), who found that there were no strong relationships between the socio-economic characteristics they assessed and either the propensity to experiment or the characteristics of the experiments. However, some similarities can be found among experimenting farmers described in the literature, and these refer mainly to personal characteristics of the farmers. For example, many farmer experimenters have travelled and experienced other areas (Critchley & Mutunga, 2003) and many are devoted to full-time farming and are flexible enough to experiment (Reij & Waters-Bayer, 2001a). In this study, frequent experimenters reported higher travelling activities to distant places and reported a significantly more positive attitude towards experimenting.

Altogether our results suggest that most farmers do experiment, but only a smaller group of farmers define themselves as active experimenters. The higher frequency of experimentation is mainly related to personal characteristics and positive attitude towards experimenting. Stimulating a positive attitude of farmers towards experimenting may therefore have a significant influence on the propensity to experiment.

4.1 Topics, Motives and Information Sources for Farmers' Experiments

Most of the experiments discussed in the interviews concerned agronomic topics, but also non-agronomic topics such as processing and commercialization, alternative remedies, labor management, or social issues. Literature mainly concentrates on experiments in the area of crop production and related activities such as fertilization or tillage (Leitgeb et al., 2014; Sumberg & Okali, 1997; van Veldhuizen et al., 1997). Sumberg and Okali (1997) catalogued 155 examples of farmers' experiments in three African countries and found that only 5% of the experiments were about non-agronomic topics such as labor management and marketing. This points to a certain 'blind spot' within agricultural research, and so integrating farmers' experiments better into the respective agricultural innovation systems could help to broaden the perception of which topics or problems have relevance and priority for farmers, and how they could be adequately addressed.

The frequency of non-agronomic topics in comparison to research findings of studies carried out in a development context may also be due to the fact that farming in industrialized countries is partly moving from the production of agricultural raw products to more multifunctional farm activities (Björklund & Milestad, 2006; Hubert, Ison, & Röling, 2000) and rural development activities (Darnhofer, 2005), a development that is also driven by the Common Agricultural Policy (CAP) of the EU (Morgan, Marsden, Miele, & Morley, 2010). Another factor that influences agricultural activities is decreasing prices due to market liberalization (Hubert et al., 2000). Decreasing agricultural income motivates farmers to experiment with alternative marketing strategies such as direct marketing, or with the production of alternative goods and services such as composting of organic residues for the community, or social services such as education or leisure time activities (parties, catering, holidays) on farm. Decreasing income in agriculture and changed social conditions, such as off-farm work of farmers or the attempt to separate farm work from family life, also causes farmers to search for time saving measures (Cournut & Dedieu, 2006).

Both reactive and proactive motives drive farmers to experiment and by doing so enhance the adaptability of their farms (Darnhofer et al., 2010). In the context of rapid and constant changes affecting farming activities, it is particularly important to adequately and actively respond to change, rather than reacting to the adverse impact of change. This requires the ongoing development of a range of alternative activities and resource use patterns that can be implemented quickly if needed (Milestad et al., 2012), and these alternatives can be developed and tested through experimenting.

As most important information sources to carry out experiments, the interviewed farmers ranked literature, farmer colleagues and advisors. Scientists were ranked as least important information sources, indicating that the interviewed farmers in Austria are not well connected to academic agricultural research, or that scientists fail to provide research results in a way that farmers want to use it. Results of academic research such as academic papers are generally complex and theoretical, and therefore inappropriate information sources for practical farming problems (Sewell et al., 2014). Or putting it more drastically: many farmers consider 'expert knowledge' as being of limited practical value (Lyon, Bell, Gratton, & Jackson, 2011; Stuiver et al., 2004), and prefer insider information coming from other farmers (Hoffmann et al., 2007; Schneider, Fry, Ledermann, & Rist, 2009). The situation can be different in other contexts, e.g. in Cuba, where efforts to enhance information exchange between scientists, farmers and extensionists were successful, and farmers' experiments are institutionalized and well integrated into the agricultural innovation system (Leitgeb et al., 2011). Similar approaches also exist in the European context, for example 'farmer-to-farmer'-approach (Schneider et al., 2009) or farmer stable schools (Vaarst et al., 2007).

4.2 Methods and Outcomes of Farmers' Experiments

When analyzing farmers' individual experiments, we found similar characteristics like comparable studies: most experiments involved planning (Sumberg & Okali, 1997), small-scale setup (Saad, 2002; Sturdy et al., 2008), monitoring strategies such as frequent observation (Stolzenbach, 1999) and comparisons (both historical and direct comparisons) (Sumberg & Okali, 1997), and resulted in incremental improvements and minor modifications of established practices (Sumberg et al., 2003). In this sense, farmers' experiments share communalities with experiments of formal agronomic research (Maat, 2011; Sumberg et al., 2003). Beyond these similarities, the particularity and significance of farmers' experiments becomes evident when evaluating less the products but rather the process involved (Saad, 2002). These processes of experimentation are characterized by flexibility and adaptive performance, and are significantly different from the standardized experimental design generally employed in academic research (Vogl, Kummer, Leitgeb, Schunko, & Aigner, 2015). Spontaneous variation during experimentation is considered a valid source of information itself, and it can be the essence of success for an experiment (Stolzenbach, 1999).

Farmers have their own methods for carrying out and evaluating experiments (Bentley, 2006), and these methods are adapted to the needs and reality of the respective farm. Strengthening farmers' experiments is sometimes interpreted as the need to formalize them by including e.g. replications, standardizations and quantifications (Wilbois et al., 2004), but there is little if any evidence that training farmers in more formal research methods would make their experiments more effective (Sumberg et al., 2003). Additionally, such a formalization is likely to increase the cost and risk of experimentation (Sumberg & Okali, 1997), and requires additional time (Wortmann et al., 2005). Some authors even suspect that introducing scientific methods may sidetrack farmers into pseudo-scientific trials that do not take advantage of their own knowledge, especially when formal research is seen as more valid and relevant (Saad, 2002). Academic and farmers' research may have different approaches, but both are relevant for the further development of sustainable farming systems. Exchange between these two areas of research could be beneficial for both areas (Hoffmann et al., 2007). Farmers could use formal research results as an inspiration and source of information in their experiments, and some of the farmers in our study were found to do exactly this. In the same vein, outcomes of farmers' experiments could be spread to other farmers, incorporated into advisory programs, and act as an incentive for researchers.

Interviewees stressed the importance of learning and development of knowledge through experimenting. By monitoring the process and outcomes of experiments, farmers widen the repertoire of options to deal with and confront change, and gain context specific knowledge to actively adapt their farm management. In this sense, farmers' experiments are powerful tools to build up farmers' knowledge and to strengthen the adaptability and resilience of farms (Darnhofer et al., 2010; Kummer, Milestad, Leitgeb, & Vogl, 2012), but this is not necessarily the case. In some cases, an experiment may put the farm at risk (e.g. experimenting on too large scale), may make the farm more dependent on external resources (e.g. experiments that increase off-farm purchases), or may reduce diversity (e.g. experiments that aim to maximize output of one product at the expense of others). When supporting or advising farmers, it is therefore important to raise awareness for possible risks of experimentation. A resilience framework can help to distinguish between risky experimentation and experiments that support sustainable development of the farm (Kummer et al., 2012).

4.3 Integrating Farmers' Experiments into AIS

Farmers' experiments can be a driving force for agricultural development when integrated into the agricultural innovation system, as it is the case in Cuba (Leitgeb et al., 2011). There is potential to make more active use of farmers' experiments and innovations by raising awareness for the topic within the farming community, the respective political and institutional landscape, and the agricultural research and advisory services. But raising awareness will probably not be enough, as e.g. even extensionists working intensively with farmers in participatory research projects are often not aware of the experimental capacity of farmers (Bentley et al., 2010). It will be crucial to find appropriate ways of participatory research and joint learning between the actors within the AIS, and lately a considerable number of promising concepts evolved, such as adaptive co-management (Armitage, Marschke, & Plummer, 2008), networks of practice (Oreszczyn, Lane, & Carr, 2010), or learning and innovation networks (Brunori et al., 2013; Moschitz, Roep, Brunori, & Tisenkopfs, 2015). Concrete examples how to make more active use of farmers' experimental capacities include experiential learning groups of Danish dairy farmers that developed concrete solutions to improve the health of their herds based on mutual advice, group induced experiments and common evaluation of the results (Vaarst et al., 2007). In Switzerland, farmers' experiments with soil protection measures were used in films to inspire other farmers, but finally the films were found to have a far-reaching impact on various actors and institutions involved in soil protection, induced social learning and helped to overcome traditional conflicts between the involved actors (Schneider et al., 2009).

A possibly promising pathway to integrate farmers' research better into the agricultural innovation system is formulated in the European Innovation Partnership (EIP) of the European Commission (EC) on 'Agricultural Productivity and Sustainability'. The EIP stresses the importance of innovation and knowledge exchange in the agrarian sector to meet societal challenges of food production. *'Increased and sustainable agricultural output will be achievable only with major research and innovation efforts at all levels. Repeatedly, researchers and stakeholders have highlighted the gap between the provision of research results and the application of innovative approaches to farming practice. New approaches take too long to arrive on the ground, and the needs of practical farming are not communicated sufficiently to the scientific community. Thus, important innovations are not implemented on the necessary scale, (...).'* (European Commission [EC], 2012, p.3). Although the EC paper stresses the necessity to *'help translating research results into actual innovation'* (EC, 2012, p.5), which could be understood as yet another example of the overcome 'transfer of technology' approach, the EC paper also emphasizes the need for systemic feedback from practice to science, for enhancing knowledge exchange, and for joint efforts to invest in sustainable innovation. Furthermore it is suggested that the EIP aims at *'enhancing communication and cooperation between science and practice. It will help sharing experience, including failures, lessons learned, and good practice.'* (EC, 2012, p.7). Within the context of agricultural innovation development, a shift from linear technology-oriented approaches to systems-oriented approaches to innovation has taken place over the last decades (Klerkx et al., 2012; Schut et al., 2014), and the current EIP seems to acknowledge this development (Moeskops, Blake, Tort, & Torremocha, 2014). The results of our research suggest that a certain group of farmers interested and actively engaged in experiments and the development of local innovations have the relevant knowledge and skills to contribute to the pursued European partnership.

To take advantage of the potential of farmers' experiments, it is important to develop conditions that support farmers in their experimentation activities (Chikozho, 2005; Johnson, 1972; Quiroz, 1999). National regulations, subsidies and support payments could be used to give farmers room for creativity within the regulatory frameworks for farming. Another possibility would be to engage farmers more actively in the advisory system and in technology development (Maat, 2011; Sumberg et al., 2003), and make active use of the outcomes of farmers' experiments for the development of local agricultural systems (Leitgeb et al., 2011). Studies that investigated the introduction of new farming practices in Europe, USA or Australia conclude that especially advisors are sometimes not open to new developments and even prevent farmers from engaging in those practices, and frequently the farmers themselves are drivers to establish novel farming practices and innovations (Ingram, 2010; McKenzie, 2013). With all the enthusiasm for participatory and social learning approaches, we also have to consider their disadvantages, as they require considerable resources such as time and budget. It will therefore be crucial to respect these opportunity costs (Hoffmann et al., 2007), e.g. by establishing effective compensation programs and political incentives (Armitage et al., 2008), or making sure that the benefits exceed the costs of participation (Home & Rump, 2015).

5. Conclusion

Based on our findings, we conclude that farmers' experimentation is significant on two levels or scales. Firstly, experiments have concrete significance on the individual farm level, as they involve learning processes, and the knowledge developed through experimenting helps to enhance the adaptive capacity of farms. Outcomes of experiments can provide possible strategies to adapt to changing conditions, and to actively take up emerging opportunities. Secondly, we conclude that the outcomes of experiments frequently have an outreach on the regional level and/or into the wider agricultural innovation system: some experiments result in local innovations that can be useful in the regional context or for similar conditions. Frequent experimenters developing these innovations have in-depth knowledge on the subject, and have monitored and tested their developments over time. They are possible advisors for other farmers and partners, e.g. for EIP-AGRI initiatives.

Both levels of experiments are important within the AIS, but on different scales and within different networks. To stimulate experiments on farm level, it will be helpful to make farmers more aware of their experimenting potential and encourage them to use it, e.g. by providing relevant information and offer possibilities of knowledge exchange. Presenting positive examples of farmers' experiments and stressing their significance may also have a positive influence on farmers' general attitude towards experimenting and motivate farmers to experiment. For the level of agricultural innovation systems, it could be supportive to encourage the dissemination of outcomes of farmers' experiments, to integrate interested farmers into participatory research or initiatives such as EIP-AGRI, and to support already existent learning and innovation networks. Investigating concrete possibilities and tools to integrate farmers' experiments into the agricultural knowledge and innovation system is the next logical step building on this basic research, so currently our suggestions are still on a general level. Research and policy initiatives that relate to the topic of farmers' experiments and innovations, such as the EIP-AGRI, provide a possible framework to further this research into a more applied direction.

Concerning the current European Innovation Partnership on Agriculture (EIP-AGRI), we can draw two conclusions: Firstly, EIP-AGRI states that research results do not reach the farmers, and at the same time the needs of practical farming do not reach the scientific community. Our results support this observation, as the interviewed farmers ranked scientists as least important information source for their experiments, and only in few cases were outcomes of farmers' experiments taken up by scientists and advisors. Secondly, EIP-AGRI claims the need to enhance communication and cooperation and share experiences between science and practice. We conclude that in order to tackle current challenges of farming and food production, it will be crucial to involve interested farmers as co-researchers into the development of innovative solutions. The concept of co-learning and co-production of knowledge within agricultural innovation systems could be a suitable framework to benefit from the specific skills and knowledge of the involved actors. Experimenting farmers have relevant experiences to share with other stakeholders in the agricultural system. When implemented in an appropriate way, the current European Innovation Partnership could be a base for making more active use of the innovative potential of farmers' experiments.

Acknowledgements

The authors thank the Austrian Science Foundation (FWF) for financing the project 'Organic Farmers' Experiments' (project P 19133). We are especially thankful to all the persons that took their valuable time for conducting interviews with us. We also thank persons and organizations that supported this work by providing data, information and inspiration: people from the organic farmers' association 'Bio Austria', the Chamber of Agriculture in the respective research regions, the organization 'Steirisches Vulkanland', the Ministry for Agriculture, and from FiBL Austria. Finally, we thank Andrea Moldaschl for her collaboration in data collection, Christoph Schunko for support in quantitative data analysis and for his valuable comments on earlier paper drafts, and our colleagues from the Working Group of Knowledge Systems and Innovations for the fruitful discussions on the subject of our research.

Note

This paper is partly based on the dissertation of the first author:
Kummer, S. (2011). *Organic farmers' experiments in Austria. Learning processes and resilience building in farmers' own experimentation activities.* (Doctoral thesis), University of Natural Resources and Life Sciences, Vienna. Retrieved from http://permalink.obvsg.at/bok/AC07810093

References

Armitage, D., Marschke, M., & Plummer, R. (2008). Adaptive co-management and the paradox of learning. *Global Environmental Change, 18*, 86-98. http://dx.doi.org/10.1016/j.gloenvcha.2007.07.002

Bentley, J. W. (2006). Folk Experiments. *Agriculture and Human Values, 23*(4), 451-462. http://dx.doi.org/10.1007/s10460-006-9017-1

Bentley, J. W., van Mele, P., & Acheampong, G. K. (2010). Experimental By Nature: Rice Farmers in Ghana. *Human Organization, 69*(2), 129-137. https://doi.org/10.17730/humo.69.2.r078vjvqx23675g1

Bernard, H. R. (2006). *Research Methods in Anthropology. Qualitative and Quantitative Approaches* (Fourth edition ed.). Oxford: Altamira Press.

Biggs, S. D. (1990). A multiple source of innovation model of agricultural research and technology promotion. *World Development 18*(11), 1481-1499. https://doi.org/10.1016/0305-750X(90)90038-Y

Björklund, J., & Milestad, R. (2006). Multifunctional farms and rural development: A study of four Swedish rural areas. In H. Langeveld & N. Röling (Eds.), *Changing European farming systems for a better future* (pp. 212-216). Wageningen: Wageningen Academic Publishers.

Brunori, G., Barjolle, D., Dockes, A.-C., Helmle, S., Ingram, J., Klerkx, L., … Tisenkopfs, T. (2013). CAP Reform and Innovation: The Role of Learning and Innovation Networks. *EuroChoices, 12*(2), 27-33. https://doi.org/10.1111/1746-692X.12025

Chambers, R., Pacey, A., & Thrupp, L. A. (Eds.). (1998). *Farmer First: Farmer Innovation and Agricultural Research.* London: Intermediate Technology Publications.

Chikozho, C. (2005). Policy and institutional dimensions of small-holder farmer innovations in the Thukela River Basin of South Africa and the Pangani River Basin of Tanzania: a comparative perspective. *Physics and Chemistry of the Earth, 30*, 913-924. http://dx.doi.org/10.1016/j.pce.2005.08.038

Cournut, S., & Dedieu, B. (2006). How do dairy farmers face work problems? Some elements about the adaptation of the technical systems in Central France. In H. Langeveld & N. Röling (Eds.), *Changing European farming systems for a better future. New visions of rural areas* (pp. 227-231). Wageningen: Wageningen Academic Publishers.

Critchley, W. R. S. (2000). Inquiry, Initiative and Inventiveness: Farmer Innovators in East Africa. *Physics and Chemistry of the Earth, 25*(3), 285-288. http://dx.doi.org/10.1016/S1464-1909(00)00016-2

Critchley, W. R. S., & Mutunga, K. (2003). Local innovation in a global context: Documenting farmer initiatives in land husbandry through WOCAT. *Land Degradation & Development, 14*(1), 143-162. http://dx.doi.org/10.1002/ldr.537

Darnhofer, I. (2005). Organic Farming and Rural Development: Some Evidence from Austria. *Sociologia Ruralis, 45*(4), 308-323. https://doi.org/10.1111/j.1467-9523.2005.00307.x

Darnhofer, I., Bellon, S., Dedieu, B., & Milestad, R. (2010). Adaptiveness to enhance the sustainability of farming systems. A review. *Agronomy for Sustainable Development, 30*(3), 545-555. http://dx.doi.org/10.1051/agro/2009053

European Commission [EC] (2012). *Communication from the Commission to the European Parliament and the Council on the European Innovation Partnership 'Agricultural Productivity and Sustainability'.* COM(2012) 79 final, Brussels.

Gerber, A., Hoffmann, V., & Kügler, M. (1996). Das Wissensystem im Ökologischen Landbau in Deutschland: Zur Entstehung und Weitergabe von Wissen im Diffusionsprozeß. *Berichte über Landwirtschaft, 74,* 591-627.

Haverkort, B., van der Kamp, J., & Waters-Bayer, A. (Eds.). (1991). *Joining farmers' experiments. Experiences in participatory technology development.* London: Intermediate Technology Publications.

Hermans, F., Klerkx, L., & Roep, D. (2015). Structural Conditions for Collaboration and Learning in Innovation Networks: Using an Innovation System Performance Lens to Analyse Agricultural Knowledge Systems. *The Journal of Agricultural Education and Extension, 21*(1), 35-54. http://dx.doi.org/10.1080/1389224x.2014.991113

Hoffmann, V., Probst, K., & Christinck, A. (2007). Farmers and researchers: how can collaborative advantages be created in participatory research and technology development? *Agriculture and Human Values, 24*(3), 355-368. http://dx.doi.org/10.1007/s10460-007-9072-2

Home, R., & Rump, N. (2015). Evaluation of a Multi-case Participatory Action Research Project: The Case of SOLINSA. *Journal of Agricultural Education and Extension, 21*(1), 73-89. http://dx.doi.org/10.1080/1389224X.2014.991112

Hubert, B., Ison, R. L., & Röling, N. (2000). The 'Problematique' with Respect to Industrialised-Country Agricultures. In M. Cerf, D. Gibbon, B. Hubert, R. Ison, J. Jiggins, M. Paine, J. Proost & N. Röling (Eds.), *Cow Up A Tree. Knowing and Learning for Change in Agriculture. Case Studies from Industrialised Countries* (pp. 13-29). Paris: INRA Editions.

Ingram, J. (2010). Technical and Social Dimensions of Farmer Learning: An Analysis of the Emergence of Reduced Tillage Systems in England. *Journal of Sustainable Agriculture, 34*(2), 183-201. http://dx.doi.org/10.1080/10440040903482589

Johnson, A. W. (1972). Individuality and experimentation in traditional agriculture. *Human Ecology, 1,* 149-159. https://doi.org/10.1007/BF01531352

Kandel, H. J., Porter, P. M., Carr, P. M., & Zwinger, S. F. (2008). Producer participatory spring wheat variety evaluation for organic systems in Minnesota and North Dakota. *Renewable Agriculture and Food Systems, 23*(3), 228-234. http://dx.doi.org/10.1017/S1742170508002263

Klerkx, L., van Mierlo, B., & Leeuwis, C. (2012). Evolution of systems approaches to agricultural innovation: concepts, analysis and intervention. In I. Darnhofer, D. Gibbon & B. Dedieu (Eds.), *Farming Systems Research into the 21st Century: The New Dynamic.* Dordrecht: Springer.

Kummer, S. (2011). *Organic farmers' experiments in Austria. Learning processes and resilience building in farmers' own experimentation activities.* (Doctoral thesis), University of Natural Resources and Life Sciences, Vienna. Retrieved from http://permalink.obvsg.at/bok/AC07810093

Kummer, S., Milestad, R., Leitgeb, F., & Vogl, C. R. (2012). Building Resilience through Farmers' Experiments in Organic Agriculture: Examples from Eastern Austria. *Sustainable Agriculture Research, 1*(2), 308-321. http://dx.doi.org/10.5539/sar.v1n2p308

Lawrence, D., Christodoulou, N., & Whish, J. (2007). Designing better on-farm research in Australia using a participatory workshop process. *Field Crops Research, 104*(1-3), 157-164. http://dx.doi.org/10.1016/j.fcr.2007.03.018

Leitgeb, F., Funes-Monzote, F. R., Kummer, S., & Vogl, C. R. (2011). Contribution of farmers' experiments and innovations to Cuba's agricultural innovation system. *Renewable Agriculture and Food Systems, 26*(04), 354-367. http://dx.doi.org/10.1017/s1742170511000251

Leitgeb, F., Kummer, S., Funes-Monzote, F. R., & Vogl, C. R. (2014). Farmers' experiments in Cuba. *Renewable*

Agriculture and Food Systems, 29(1), 48-64. https://doi.org/10.1017/S1742170512000336

Lyon, A., Bell, M. M., Gratton, C., & Jackson, R. (2011). Farming without a recipe: Wisconsin graziers and new directions for agricultural science. *Journal of Rural Studies, 27*(4), 384-393. http://dx.doi.org/10.1016/j.jrurstud.2011.04.002

Maat, H. (2011). The history and future of agricultural experiments. *NJAS - Wageningen Journal of Life Sciences, 57*(3-4), 187-195. http://dx.doi.org/10.1016/j.njas.2010.11.001

McKenzie, F. (2013). Farmer-driven Innovation in New South Wales, Australia. *Australian Geographer, 44*(1), 81-95. http://dx.doi.org/10.1080/00049182.2013.765349

Milestad, R., Dedieu, B., Darnhofer, I., & Bellon, S. (2012). Farms and farmers facing change: The adaptive approach. In I. Darnhofer, D. Gibbon & B. Dedieu (Eds.), *Farming Systems Research into the 21st Century: The New Dynamic*. Dordrecht: Springer.

Moeskops, B., Blake, F., Tort, M-C., & Torremocha, E. (Eds.) (2014). *Action Plan for Innovation and Learning*, Brussels, Belgium: TP Organics.

Morgan, S. L., Marsden, T., Miele, M., & Morley, A. (2010). Agricultural multifunctionality and farmers' entrepreneurial skills: A study of Tuscan and Welsh farmers. *Journal of Rural Studies, 26*(2), 116-129. http://dx.doi.org/10.1016/j.jrurstud.2009.09.002

Moschitz, H., Roep, D., Brunori, G., & Tisenkopfs, T. (2015). Learning and Innovation Networks for Sustainable Agriculture: Processes of Co-evolution, Joint Reflection and Facilitation. *The Journal of Agricultural Education and Extension, 21*(1), 1-11. http://dx.doi.org/10.1080/1389224X.2014.991111

Oreszczyn, S., Lane, A., & Carr, S. (2010). The role of networks of practice and webs of influencers on farmers' engagement with and learning about agricultural innovations. *Journal of Rural Studies, 26*, 404-417. http://dx.doi.org/10.1016/j.jrurstud.2010.03.003

Padel, S. (2001). Conversion to Organic Farming: A Typical Example of the Diffusion of an Innovation? *Sociologia Ruralis, 41*(1), 40-61. https://doi.org/10.1111/1467-9523.00169

Quiroz, C. (1999). Farmer experimentation in a Venezuelan Andean group. In G. Prain, S. Fujisaka & M. D. Warren (Eds.), *Biological and Cultural Diversity. The Role of Indigenous Agricultural Experimentation in Development* (pp. 113-124). London: Intermediate Technology Publications.

Reij, C., & Waters-Bayer, A. (2001a). An initial analysis of farmer innovators and their innovations. In C. Reij & A. Waters-Bayer (Eds.), *Farmer Innovation in Africa. A Source of Inspiration for Agricultural Development*. London, UK: Earthscan.

Reij, C., & Waters-Bayer, A. (Eds.). (2001b). *Farmer Innovation in Africa: A Source of Inspiration for Agricultural Development*. London: Earthscan.

Rivera, W. M., & Sulaiman, R. V. (2009). Extension: object of reform, engine for innovation. *Outlook on Agriculture, 38*(3), 267-273. https://doi.org/10.5367/000000009789396810

Röling, N., & Brouwers, J. (1999). Living local knowledge for sustainable development. In G. Prain, S. Fujisaka & M. D. Warren (Eds.), *Biological and Cultural Diversity. The Role of Indigenous Agricultural Experimentation in Development* (pp. 147-157). London: Intermediate Technology Publications.

Saad, N. (2002). Farmer processes of experimentation and innovation. A review of the literature.: CGIAR systemwide program on participatory research and gender analysis.

Schneider, F., Fry, P., Ledermann, T., & Rist, S. (2009). Social Learning Processes in Swiss Soil Protection - The 'From Farmer - To Farmer' Project. *Human Ecology, 37*, 475-489. http://dx.doi.org/10.1007/s10745-009-9262-1

Schut, M., Rodenburg, J., Klerkx, L., van Ast, A., & Bastiaans, L. (2014). Systems approaches to innovation in crop protection. A systematic literature review. *Crop Protection, 56*, 98-108. http://dx.doi.org/10.1016/j.cropro.2013.11.017

Sewell, A. M., Gray, D. I., Blair, H. T., Kemp, P. D., Kenyon, P. R., Morris, S. T., & Wood, B. A. (2014). Hatching new ideas about herb pastures: Learning together in a community of New Zealand farmers and agricultural scientists. *Agricultural Systems, 125*, 63-73. http://dx.doi.org/10.1016/j.agsy.2013.12.002

Stolzenbach, A. (1999). The indigenous concept of experimentation among Malian farmers. In G. Prain, S. Fujisaka & M. D. Warren (Eds.), *Biological and Cultural Diversity. The role of indigenous agricultural experimentation in development* (pp. 163-171). London, UK: Intermediate Technology Publications.

Stuiver, M., Leeuwis, C., & van der Ploeg, J. D. (2004). The Power of Experience: Farmers' Knowledge and Sustainable Innovations in Agriculture. In J. S. C. Wiskerke & J. D. van der Ploeg (Eds.), *Seeds of Transition. Essays on novelty production, niches and regimes in agriculture* (pp. 93-118). Assen, The Netherlands: Royal van Gorcum.

Sturdy, J. D., Jewitt, G. P. W., & Lorentz, S. A. (2008). Building an understanding of water use innovation adoption processes through farmer-driven experimentation. *Physics and Chemistry of the Earth, 33*, 859-873. http://dx.doi.org/10.1016/j.pce.2008.06.022

Sumberg, J., & Okali, C. (1997). *Farmers' Experiments: Creating Local Knowledge*. London: Lynne Rienner Publishers, Inc.

Sumberg, J., Okali, C., & Reece, D. (2003). Agricultural research in the face of diversity, local knowledge and the participation imperative: theoretical considerations. *Agricultural Systems, 76*(2), 739-753. https://doi.org/10.1016/S0308-521X(02)00153-1

Tambo, J. A., & Wünscher, T. (2014). Identification and prioritization of farmers' innovations in northern Ghana. *Renewable Agriculture and Food Systems, 30*(06), 537-549. http://dx.doi.org/10.1017/s1742170514000374

Vaarst, M., Nissen, T. B., Østergaard, S., Klaas, I. C., Bennedsgaard, T. W., & Christensen, J. (2007). Danish Stable Schools for Experiential Common Learning in Groups of Organic Dairy Farmers. *Journal of Dairy Science, 90*(5), 2543-2554. http://dx.doi.org/10.3168/jds.2006-607

Van Veldhuizen, L., Waters-Bayer, A., Ramírez, R., Johnson, D. A., & Thompson, J. (Eds.). (1997). *Farmers' Research in Practice. Lessons from the Field*. London, UK: Intermediate Technology Publications.

Vogl, C. R., Kummer, S., Leitgeb, F., Schunko, C., & Aigner, M. (2015). Keeping the Actors in the Organic System Learning: The Role of Organic Farmers' Experiments. *Sustainable Agriculture Research, 4*(3), 136-144. http://dx.doi.org/10.5539/sar.v4n3p140

Wilbois, K.-P., Schwab, A., Fischer, H., Bachinger, J., Palme, S., Peters, H., & Dongus, S. (Eds.). (2004). *Leitfaden Praxisversuche*. Frankfurt: Forschungsinstitut für biologischen Landbau [FiBL] Deutschland.

Wortmann, C. S., Christiansen, A. P., Glewen, K. L., Hejny, T. A., Mulliken, J., Peterson, J. M., ... Zoubek, G. L. (2005). Farmer research: Conventional experiences and guidelines for alternative agriculture and multi-functional agro-ecosystems. *Renewable Agriculture and Food Systems, 20*(4), 243-251. http://dx.doi.org/10.1079/raf2005110

10

Labor Use and Profitability Associated with Pasture Systems in Grass-Fed Beef Production

Basu Bhandari[1], Jeffrey Gillespie[2], & Guillermo Scaglia[3]

[1]Dept. of Agricultural Economics and Agribusiness, Louisiana State University Agricultural Center, Baton Rouge, LA, USA

[2]Animal Products and Cost of Production, Market and Trade Economics Division, USDA-Economic Research Service, Washington, DC, USA

[3]Iberia Research Station, Louisiana State University Agricultural Center, Jeanerette, LA, USA

Correspondence: Jeffrey Gillespie, Animal Products and Cost of Production, Market and Trade Economics Division, USDA-Economic Research Service, Washington, DC, USA. E-mail: Jeffrey.Gillespie@ers.usda.gov

Abstract

Three pasture systems for grass-fed beef that are representative of those used in the U.S. Gulf Coast region are compared by labor use and profitability. In addition to means comparisons, stochastic efficiency with respect to a function analysis allows us to incorporate the role of risk preference in determining the most preferred production system. Five years of experimental data from the Iberia Research Station in Louisiana are used to develop revenue, expense, and labor use estimates for the three systems. Results suggest that, with or without including charges for labor, the most profitable system is the least complex bermudagrass-ryegrass system. If labor is included, a medium-complexity forage system becomes preferred for more risk averse farmers. The most complex forage system might become competitive if a carbon market were developed and/or farmers were able to realize higher grass-fed beef prices on the basis of product quality.

Keywords: grass-fed beef, stochastic efficiency with respect to a function, labor requirements

1. Introduction

Labor is a major input in livestock production. Grass-fed beef (GFB) operations are particularly labor-intensive, with labor requirements differing by production system. The major work performed by labor in a GFB beef operation includes moving, checking, and working animals, as well as operating machinery and equipment in managing forage. According to USDA-NASS (2007), most U.S. beef operations are small relative to other agricultural operations; 50% of beef farms have fewer than 20 cows and operate on a fixed land area. The GFB segment of the U.S. beef industry is similarly characterized by relatively small operations, with the labor requirements of these farms being fulfilled mostly by landowners and their family members.

Our observations suggest that a wide range of pasture management systems are used for GFB production throughout the U.S. with considerable differences in labor and management complexity. Grass-fed beef producers are expected to be interested in pasture systems that simultaneously require less labor but yield greater profit. Though GFB production accounts for a very small share of the U.S. beef industry as a whole, i.e. less than 1% of the production (Pelletier, Pirog, & Rasmussen, 2010), it has gained interest over the last two decades due to human health, environmental and animal welfare concerns (Wright, 2005; Mills, 2003; McCluskey et al., 2005).

Several studies have examined farm labor differences by agricultural production system in the beef industry. Gillespie, Wyatt, Venuto, Blouin, and Boucher (2008) analyzed the roles of labor and profitability in choosing a grazing strategy for cow-calf production in the U.S. Gulf Coast Region. They found that the greater labor requirements associated with rotational grazing systems relative to continuous grazing systems reduced the profitability associated with rotational grazing. Wyatt, Gillespie, Blouin, Venuto, Boucher, and Qushim (2013) evaluated the effects of year-round stocking rate and stocking method on cow-calf production systems considering costs, returns and labor considerations. Neither of these studies, however, have focused on GFB production. In this paper, we estimate the relative profitability of three pasture systems for GFB production with and without considering the costs associated with labor. The specific objectives of this study are to: 1) determine the operating expenses, fixed expenses, revenue, and returns over expenses of three pasture systems under GFB production; 2) determine the involvement of labor in specific activities in the three pasture systems; and 3) determine the most profitable pasture system for GFB production in the U.S. Gulf Coast Region with and without considering labor.

2. Conceptual Model

The conceptual model for this research is represented by the following profit maximizing problem for the GFB producer:

$$max \; \pi(x) = \; \pi(X_i) = \{\{P_{slaugh} * f(X_i) + P_{hay} * g[f(X_i)] - \sum_{i=1}^{n} W_i * X_i\} \tag{1}$$

where $\pi(.)$ is farm profit; X_i is the amount of input i; P_{slaugh} is the price of a slaughter animal; $f(X_i)$ is the production function for a GFB slaughter animal; P_{hay} is the price of hay; $g[f(X_i)]$ is the production function for hay produced in pastures, which is a function of the production of slaughter animals; and W_i is the price of input i. Here, the production function for hay from pasture is a function of slaughter animal production because the primary purpose of growing and maintaining pasture for GFB production is to produce beef, not hay. Since the primary purpose of growing forage is for grazing animals, only the left-over or excess forage is generally used to produce hay, which is in turn generally fed during periods of low grazing potential. Hay remaining after feeding animals is sold.

By solving for the first order conditions, the optimum quantity of use of input j for profit maximization can be estimated as:

$$\{P_{slaugh} * \frac{\partial f(X_i)}{\partial X_j} + P_{hay} * \frac{\partial g[f(X_i)]}{\partial X_j}\} = W_j \tag{2}$$

where the left hand side value represents the marginal value product and the right hand side represents the marginal factor cost, showing that the profit-maximizing producer determines optimal input usage by considering the marginal physical productivity, output prices, and input prices. In the case of using multiple forage species for pasture and/or hay, additional labor costs will be incurred if the additional value of the product (finished animals and hay) is greater than the additional cost associated with the labor input.

If extensive data were available, solving the profit-maximizing problem using the production function could provide the optimum level of input usage. It is often impractical, however, to collect such extensive data via experimental research, precluding the estimation of a precise optimum input-output combination based on the above conceptual model. Optimal solutions can, however, be approximated at discrete points in the production function. In this study, comparisons were made between three different pasture combinations evaluating operating expenses including labor involvement for different activities, fixed expenses, steer revenue, hay revenue, and return over expenses with each system.

3. Data and Methods

Three treatments used in a field experiment at the LSU AgCenter Iberia Research Station (IRS) in Jeanerette, LA, from 2009-2010 to 2013-2014 represented pasture systems with different degrees of management complexity. The three forage systems were: (1) bermudagrass as summer pasture, annual ryegrass as winter pasture; (2) bermudagrass as summer pasture, annual ryegrass, rye, and clover mix (berseem, red, and white clovers) as winter pastures; and (3) bermudagrass, sorghum-sudan hybrid, and forage soybean as summer pastures, and annual ryegrass, rye, clover mix (berseem, red, and white clovers), and dallisgrass as winter pastures. These systems were chosen as representative of the types of systems currently being used by GFB producers in the U.S. Gulf Coast Region (Scaglia, Rodriguez, Gillespie, Bhandari, Wang, & McMillin, 2014). System 1 consists of only two forage types and is the least complex while System 3 consists of nine forage types and is the most complex among these systems.

System 1 consists of three sub-paddocks of bermudagrass, System 2 consists of two sub-paddocks of bermudagrass, and System 3 consists of only one bermudagrass paddock. Since Systems 2 and 3 included other forages, System 1 included the greatest number of bermudagrass sub-paddocks. These sub-paddocks were divided using temporary fencing as per the availability of green forage and appropriate grazing management.

Annually, 54 seven to eight month old Fall-born steers were assigned to one of the three pasture systems immediately after weaning and remained until time of harvest at age 17-19 months. The same pastures were used for each treatment each year. The experimental year began in May and ended by the end of April the following year. The three forage systems were managed in different sub-paddocks at the IRS, and animals were rotated among the sub-paddocks based on forage availability. The steers were blocked at weaning by weight into nine groups (six steers/group). Each group was randomly assigned to one of the three treatments, each of which was replicated three times. During the transition period when forage availability was low (mid-November to December), animals were fed hay produced in the paddocks allocated to the system/replication group. Constructed portable shades were made available for the animals in each group. They were moved along with the animals when rotated. Water and mineral mix were available at all times. The stocking rate was one hectare per animal for each entire system.

Detailed expense and input records were kept for each pasture by year. These records were used to develop detailed revenue and expense estimates for each treatment/replication. These estimates included revenue, operating expenses, fixed expenses, and land rent. Expenses for seed, fertilizer, pesticides, minerals, medication, twine, fuel, purchased weaned steers, repair and maintenance of machinery, and interest on operating capital were included in the operating expenses. Depreciation and interest on machinery (trucks, tractors, and other implements), permanent fencing, and temporary fencing were included in the fixed expenses. The fixed expenses associated with machinery and equipment were allocated according to use, assuming their useful life and performance rates as shown in Boucher and Gillespie (2009, 2010, 2011, 2012, and 2013). The opportunity cost of land rental was included. Similarly, labor used for each activity was kept by pasture system. A total of 45 revenue and expense estimates were made for the project: (3 treatments × 3 replications × 5 years).

Labor usage was categorized into four subgroups. *Moving Animals and Shades* involved measuring forage availability and moving the animals among paddocks accordingly. It also included the movement of shades and water troughs. The second category was *Checking Animals and Routine Tasks*, which included checking animals twice per day Monday-Friday and once per day during the weekend. On days when the animals were moved, the checking task was conducted at the same time. Therefore, no separate labor was required for this task on the animal moving day. A third labor category was *Vaccinating Animals*. This was done as per vaccination requirements and included the labor required for working animals. The final category was *Operator Labor*, which included the labor required for operating machinery as well as labor involved in machinery, equipment, building, and fencing repair and maintenance. Much of this category involved planting and harvesting forage. Previous work examining labor use by stocking strategy includes Gillespie et al. (2008) and Wyatt et al. (2013).

The fifth year data differed somewhat from that of previous years because berseem clover, which was used in Systems 2 and 3, was not available in the local market that year. Furthermore, sorghum-sudan (System 3) was not available, but was replaced with pearl millet in the 5th year. In addition, there was a labor shortage at the IRS, so application of fertilizer and moving of animals were conducted only two-thirds of the times of the earlier years, consistently reduced across the treatments. Thus, input use differed and was somewhat lower in the fifth year. We included fifth-year data, however, since those conditions sometimes prevail in actual farm situations. Thus, it can be argued that analysis including the fifth-year data reflects the reality of resource constraints of a commercial farm.

Annual input and output prices are presented in Table 1. With the exception of those listed in subsequent discussion, these prices were those used by Boucher and Gillespie (2009, 2010, 2011, 2012, and 2013) in revenue and expense estimates for Louisiana cattle and forage production. The prices of weaned calves were obtained from 2011 Louisiana Agricultural Statistics (LSU Agricultural Center; USDA-NASS, 2012) for Years 1-3. For Years 4-5, the weaned calf price was obtained from Boucher Gillespie (2012, 2013) revenue and expense estimates due to the unavailability of Louisiana Agricultural Statistics data for those years.

Hay was measured as a large bale of average weight 430 kg. We used the Weekly Texas Hay Report for hay prices (USDA-TX, 2010, 2011, 2012, 2013, and 2014). The hay price was at its peak ($82.50 per large round bale) in 2012 due to unfavorable weather and low hay production that year. The price was approximately double that of the other years. We used the USDA-ERS (2014) published prices for fed steers as a base, adjusted for the grass-fed steer price by adding $0.44/kg as suggested by the manager of one of the larger GFB production firms in the Gulf Coast Region.

Table 1. Prices of Inputs and Outputs for the Experimental Years

Inputs/Outputs	Unit	Price in $				
		2009	2010	2011	2012	2013
Urea	Kg	0.40	0.35	0.42	0.42	0.62
Gramoxone Max	Liter	10.50	11.54	11.54	11.54	11.54
Grazon P+D	Liter	8.47	10.44	8.18	8.79	8.94
Outrider	Liter	676.28	N/A	N/A	N/A	N/A
Roundup Original Max	Liter	13.86	15.32	15.22	12.85	12.68
Malathion	Liter	N/A	8.98	8.94	8.94	N/A
Sevin 80% WP	Kg	13.51	15.01	16.20	16.20	16.20
Bovishield	Dose	2.50	2.50	2.50	2.50	2.50
One Shot	Dose	2.50	2.50	2.50	2.50	2.50
Sweetlix	Block	18.00	18.00	18.00	18.00	18.00
Ultrabac 8	Dose	0.40	0.40	0.40	0.40	0.40
Vigortone 3V2	Bag	26.20	26.20	26.20	26.20	26.20
Vigortone 3V5	Bag	17.13	17.13	17.13	17.13	17.13
Weaning Calf	CWT	98.30	114.00	114.00	125.00	150.00
Twine	Ton	0.75	0.75	0.75	0.75	0.75
Berseem Clover Seed	Kg	4.72	4.74	7.72	7.72	N/A
Red Clover Seed	Kg	5.51	6.61	2.65	2.65	3.96
White Clover Seed	Kg	5.51	7.05	6.83	6.61	6.61
Rye Seed	Kg	0.49	0.97	0.97	0.99	1.10
Ryegrass Seed	Kg	1.34	1.54	1.10	1.06	1.10
Cowpea Seed	Kg	N/A	N/A	N/A	2.20	2.20
Soybean Seed	Kg	1.23	1.17	1.32	1.32	1.32
Sorghum Sudan Seed	Kg	1.04	1.76	1.76	1.85	N/A
Pearl Millet	Kg	N/A	N/A	N/A	N/A	3.08
Hay*	Bale	45.00	40.00	82.50	37.50	40.00
Steers at Harvest*	CWT	116.00	133.00	141.00	147.00	168.00
Diesel Fuel	Liter	0.58	0.61	0.73	0.93	0.87

*Although the prices of hay and steer at harvest were tabulated as 2009, 2010, 2011, 2012, and 2013, those were based on USDA prices in the following years (2010, 2011, 2012, 2013, and 2014) since the harvesting and selling of hay and steers was in the second calendar year of the experiment.

Note: N/A indicates data not available.

Annual fixed expenses and the repair and maintenance expenses for fixed inputs are presented in Table 2. Fixed expenses of machinery and equipment were determined using the capital recovery method (Boehlje & Eidman, 1984). Annual capital recovery was calculated as:

Annual Capital Recovery Charge = {(Purchase Price - Salvage Value) * Capital Recovery Factor} + (Salvage Value * Interest Rate)

Table 2. Prices of Fixed Inputs, Machinery, and Equipment

Fixed Input Annual Costs in US$			
Input Structure	Units	Repair and Maintenance	Fixed Costs
Fence Electric	Km	23.61	156.19
Fence 5 wire	Km	130.49	302.30
hay rack	Each	9.04	26.27
Shade structure	Each	3.48	72.65
Shade cloth	Each	5.30	64.25
Water tank and pump	Each	40.00	132.50
Machinery and Equipment Costs in US$			
Machinery/Equipment	Direct Costs / Hour		Fixed Costs / Hour
Mower Conditioner	10.79		12.89
Hay Rake	2.43		3.16
Hay Tedder	2.45		3.67
Hay Fork	0.09		0.22
Baler Round	13.98		18.56
Mower Drum	4.68		5.59
Boom Sprayer	2.35		3.12
Tractor (40-59hp)	6.48		4.42
Tractor (60-89hp)	10.05		7.81
Tractor (90-115hp)	14.31		12.52

The capital recovery factor is the tabulated value based on the useful life of equipment in years and the interest rate. The fixed expense per hour was calculated by dividing the annual capital recovery charge by annual hours of use. Similarly, the operating expense per hour was estimated by computing the total repair and maintenance costs over the life of the machinery and dividing by total hours of use of the machinery.

Total Revenue and its components, Operating Expenses and its components, Return over Operating Expenses, Fixed Expenses, Total Specified Expenses, Return over Specified Expenses, and Residual Income were estimated. Residual Income is Return over Specified Expenses less land rent. Similarly, differences in the labor involved in each of the four labor categories were estimated. Differences were determined using the Kenward-Roger Degrees of Freedom method (Kenward & Roger, 1997).

Since this research is based on only 5 years of data, i.e. 45 observations, simulation and dominance techniques were used to strengthen the analysis. Simetar, a commercial mathematical simulation software developed by Richardson, Schumann, and Feldman (2008), was used to develop 1,000 randomly simulated input (fertilizer, fuel, and calves) and output (steers, hay) prices based on historical data (13 years; 2001-2013). Hay yield was estimated based on 13 years of historical rainfall data at the IRS and 1,000 randomly simulated values were developed using the same software. We did not observe significant variation in the other input variables and prices and quantities of steers, so these were taken as constant for the analysis. Based on these simulated values, 1,000 return over expense measures for each of the systems were developed.

Certainty equivalents (CE) were estimated assuming different risk preferences using the 1,000 simulated return over expenses for each system as per the relationship outlined by Hardakar, Richardson, Gudbrand, and Schuman (2004). The CE is defined as the return over expenses held with certainty at which the decision maker would be indifferent to a risky distribution of return over expense values. The utility function of the decision maker is used to estimate the CE. The relationship between the utility function $U(w)$ and the Arrow-Pratt coefficient of absolute risk aversion $r_a(w)$ (Arrow, 1971; Pratt, 1964) is shown in equation (3):

$$U(w) = -exp(-r_a(w)) \tag{3}$$

where w is return over expenses. Equation (4) defines the absolute risk aversion coefficient as the negative ratio of the second and first derivatives of the utility function:

$$r_a(w) = -\frac{u''(w)}{u'(w)} \tag{4}$$

A higher $r_a(w)$ indicates the producer is willing to accept less risk associated with return over expense, or variability of return over expense. Such an individual is more risk averse. The relationship between the coefficient of absolute risk aversion and the relative risk aversion coefficient, $r_r(w)$, is expressed as:

$$r_a(w) = r_r(w)/w \tag{5}$$

The CE for a random sample of size n from risky alternatives w is estimated as follows, as shown by Hardaker et al. (2004):

$$CE(w, r_a(w)) = ln\left\{\left(\frac{1}{n}\sum_i^n exp(-r_a(w)w_i)\right)^{-1/r_a(w)}\right\}. \tag{6}$$

A general classification of relative risk aversion coefficients falling in the range of 0 for risk neutral to 4 for highly risk averse was proposed by Anderson and Dillon (1992). Coefficients of absolute risk aversion were obtained by dividing a range of relative risk aversion coefficients (0 to 4) by the estimated mean return over expense. This yields a maximum coefficient of absolute risk aversion of 0.0024, which is used in a stochastic efficiency with respect to function analysis. Stochastic efficiency with respect to a function provides a means for evaluating the risky alternatives in terms of CEs for a specified range of coefficients of absolute risk aversion. The result is graphed to analyze which system would be preferred by GFB producers from the perspective of return over expenses and risk preference.

4. Results and Discussion

Revenue and expenses per steer excluding labor are presented in Table 3. Mean revenues from steers were $1,434, $1,446, and $1,441 for Systems 1, 2, and 3, respectively, which did not differ significantly at p ≤ 0.10 among the systems. Mean weights per steer per year were 462 kg, 461 kg, and 464 kg, respectively, for Systems 1, 2, and 3 (Table 4).

Table 3. Revenue, Expenses, and Return over Expenses (Without Labor Included), Dollars per Animal

Revenue / Expenses	System 1	System 2	System 3
Revenue			
Steers	1,434.42	1,445.68	1,440.78
Hay	667.5[bc]	527.24[ca]	350.91[ab]
Total Revenue	2,109.94[c]	1,972.93[c]	1,791.70[ab]
Operating Expenses			
Fertilizer	293.48[bc]	230.44[ca]	195.73[ab]
Pesticides	39.47[c]	37.67[c]	48.02[ab]
Livestock	690.77	690.54	692.61
Seed	55.08[bc]	134.46[ca]	188.34[ab]
Twine	3.44[bc]	2.52[ca]	2.01[ab]
Medication, Minerals	22.17	22.67	22.67
Diesel	68.22[bc]	55.14[ca]	43.27[ab]
Repair and Maintenance Expense	59.72[bc]	48.54[ca]	41.10[ab]
Interest on Operating Capital	42.72	46.68	41.82
Total Operating Expenses	1,275.68	1,264.27	1,279.27
Return over Operating Expenses	826.19[c]	708.60[c]	512.34[ab]
Fixed Expenses	198.03[bc]	158.82[ac]	135.03[ab]
Total Specified Expenses	1,473.73[c]	1,423.20	1,414.42[a]
Return over Specified Expenses	628.08[c]	549.68[c]	377.18[ab]
Residual Income	545.70[c]	457.58[c]	305.09[ab]

Note: Superscript a means differ significantly from System 1, superscript b means differ significantly from System 2, and superscript c means differ significantly from System 3 within rows at p<0.10.

Table 4. Steer Initial and Final Body Weights and Numbers of Hay Bales Produced and Fed Each Year

System	Average Weight per Steer in Kg		Number of Hay Bales	
	Initial	Final	Produced	Fed
System 1 Average	260	462	87	5
2009	255	461	54	7
2010	247	459	148	4
2011	273	466	86	6
2012	260	472	89	4
2013	266	451	59	4
System 2 Average	260	461	70	5
2009	258	445	81	7
2010	246	469	101	3
2011	275	459	58	4
2012	260	474	68	4
2013	263	460	42	5
System 3 Average	261	464	49	6
2009	256	440	64	6
2010	247	463	73	5
2011	275	474	40	5
2012	259	482	37	6
2013	266	461	29	8

Hay revenues were $668, $527, and $351 for Systems 1, 2, and 3, respectively, which differed among these systems. Hay was made from surplus green forage after grazing the animals. Of the hay produced, part of it was fed to the steers of the respective systems during the lean season of the fall when green forages were not available. Left-over hay was sold, constituting hay revenue. System 1 yielded the highest hay revenue while System 3 yielded the lowest, as more hay was harvested in System 1 than System 2 and more harvested in System 2 than in System 3. Hay produced and consumed within systems is shown in Table 4. Average hay amounts produced per system per year were 87, 70, and 49 bales in Systems 1, 2, and 3, respectively. Average hay consumption per group of 6 steers was 5, 5, and 6 bales for Systems 1, 2, and 3, respectively. Total revenues per steer per year were estimated to be $2,110, $1,973 and $1,792 for Systems 1, 2, and 3, respectively. Systems 1 and 2 had higher total revenue than System 3. The major determinant of differences in revenue by system was hay production.

Operating expenses included seed, fertilizer, pesticides, weanling animals, minerals, vaccinations, diesel, repair and maintenance, and interest on operating capital. Fertilizer expense differed among systems with the highest in System 1 and the lowest in System 3. This was due to the inclusion of leguminous nitrogen-fixing forages in Systems 2 and 3. System 3 included more leguminous forages than System 2; therefore, System 3 required less fertilizer expense than System 2. Seed expenses were greatest in System 3, which included more forage types than the other systems. Seed expense in System 1 was lowest because it included only bermudagrass and ryegrass. Similarly, diesel and repair and maintenance expenses differed among the systems because of different levels of machinery and equipment use for harvesting hay by system. Since System 1 produced more hay, machinery usage was greatest in System 1, thus greater machinery expense in System 1 than in Systems 2 and 3.

  Overall, total operating expenses excluding labor were \$1,276, \$1,264, and \$1,279 for Systems 1, 2, and 3, respectively, which did not differ statistically at $P \leq 0.10$. Return over operating expenses is total revenue less total operating costs. System 3 yielded lower return over operating expenses than Systems 1 and 2.

Fixed expenses differed among the systems due mostly to differences in the use of machinery and equipment for cutting and baling hay. Total specified expenses include both operating and fixed expenses. Return over specified expenses is estimated by subtracting total specified expenses from total revenue. System 3 yielded lower return over total specified expenses than Systems 1 and 2. Residual income was estimated after subtracting total specified expenses and an opportunity cost of land from total revenue. Residual incomes were \$546, \$458, and \$305, respectively, for Systems 1, 2, and 3 with Systems 1 and 2 having greater residual income than System 3.

Labor involvement in the 3 systems is presented in Table 5. In total, 17, 15, and 13 hours of labor per animal were involved annually in Systems 1, 2, and 3, respectively. Greater labor involvement in System 1 was due to the greater use of machinery for harvesting and making hay, thus greater *Operator Labor*. Similarly, the movement of animals was greatest in System 1 and least in System 3, which was due to greater movement among the bermudagrass sub-paddocks (System 1) than the movement of animals between paddocks of different forage types (Systems 2 and 3). The labor involved in *Vaccinating Animals* did not differ as all systems were treated the same in this regard. Although labor involved in *Checking Animals and Routine Tasks* should generally be the same across the different systems, it differed among the systems because checking animals was conducted at the same time as moving animals on the days animals were moved. More than 50% of the total labor involved was *Operator Labor*. *Moving Animals and Shades* was the second-most labor-consuming activity, while *Vaccinating Animals* was the least labor-consuming activity.

The results of the revenue and expense analysis including labor expenses are presented in Table 6. Labor expenses are divided into operator and other labor expenses. Total labor expenses were \$160, \$138, and \$123 for Systems 1, 2, and 3, respectively, which differed among the systems. Operator labor expenses were greatest in System 1 due to the greater use of machinery and equipment for harvesting and baling hay. Other labor expenses were also greatest in System 1 and least in System 3 due to greater movement of animals in System 1 and the least in System 3. Returns over operating expenses were \$826, \$709, and \$512 for Systems 1, 2, and 3, respectively, without accounting for the labor costs. System 3 had lower return over operating expenses than Systems 1 and 2. The returns over operating expenses when including labor costs were reduced to \$661, \$564, and \$385 for Systems 1, 2, and 3, respectively. Again, System 3 had lower return over operating expenses than the other systems, as shown in Table 6. Though labor used in System 1 was greater than that for the other systems, System 1 remained the most profitable of the systems.

Table 5. Annual Labor Usage Hours in the Different Systems, Hours per Animal

Labor Category	System 1	System 2	System 3
Moving Animals and Shades	4.26[bc]	3.87[ca]	3.42[ab]
Checking and Routine Tasks	2.93[c]	2.97[c]	3.02[ab]
Vaccinating Animals	0.37	0.37	0.37
Operator Labor	9.33[bc]	7.35[ac]	6.34[ab]
Total Labor	16.89[bc]	14.55[ac]	13.15[ab]

Note: Superscript a means differ significantly from System 1, superscript b means differ significantly from System 2, and superscript c means differ significantly from System 3 within rows at *p<0.10*

System 1 had greater total specified expenses than Systems 2 and 3. Return over total specified expenses was lowest in System 3 while Systems 1 and 2 did not differ statistically from each other. After accounting for labor, the residual incomes were \$381, \$331, and \$178 for Systems 1, 2, and 3, respectively. Similar to the results without including labor expenses, the residual incomes of Systems 1 and 2 were greater than with System 3. There was no statistical difference in the residual income between Systems 1 and 2 although System 1 yielded numerically greater income than System 2.

Table 6. Revenue, Expenses, and Return over Expenses (Labor $9.60/hr. Included), Dollars per Animal

Revenue / Expenses	System 1	System 2	System 3
Revenue			
Steers	1,434.42	1,445.68	1,440.78
Hay	667.5[bc]	527.24[ca]	350.91[ab]
Total Revenue	2,109.94[c]	1,972.93[c]	1,791.70[ab]
Operating Expenses			
Fertilizer	293.48[bc]	230.44[ca]	195.73[ab]
Pesticides	39.47[c]	37.67[c]	48.02[ab]
Livestock	690.77	690.54	692.61
Seed	55.08[bc]	134.46[ca]	188.34[ab]
Twine	3.44[bc]	2.52[ca]	2.01[ab]
Medication, Minerals	22.17	22.67	22.67
Other Labor	70.57[bc]	67.21[ca]	62.29[ab]
Operator Labor	89.60[bc]	70.52[ca]	60.91[ab]
Diesel	68.22[bc]	55.14[ca]	43.27[ab]
Repair and Maintenance	59.72[bc]	48.54[ca]	41.10[ab]
Interest on Operating Capital	48.40	49.82	46.26
Total Operating Expenses	1,442.06	1,408.67	1,406.52
Return over Operating Expenses	660.51[c]	564.19[c]	385.08[ab]
Fixed Expenses	198.03[bc]	158.82[ac]	135.03[ab]
Total Specified Expenses	1,639.17[bc]	1,567.59[a]	1,541.79[a]
Return over Specified Expenses	463.04[c]	405.27[c]	249.87[ab]
Residual Income	380.70[c]	331.02[c]	177.72[ab]

Note: Superscript a means differ significantly from System 1, superscript b means differ significantly from System 2, and superscript c means differ significantly from System 3 within rows at p<0.10

Residual Return = Total Income - Direct Expense - Fixed Expense - Land Rent

Sensitivity analysis showed that if the wage rate for the labor were more than $32 per hour, System 2 would become numerically more profitable than System 1. In all cases, Systems 1 and 2 dominated System 3. Results of the simulation and stochastic efficiency analysis are presented in Figures 1 and 2. Figure 1 shows the stochastic efficiency with respect to a function results without including labor. It clearly shows that System 1 dominates Systems 2 and 3 at all levels of risk aversion, though the margin of dominance narrows as the coefficient of absolute risk aversion becomes larger (risk aversion increases). Revenue and expense analysis did not show that Systems 1 and 2 differed statistically; however, results of the simulation and dominance analysis clearly show that System 1 dominates both Systems 2 and 3. Furthermore, both Systems 1 and 2 dominate System 3.

The situation changes when labor is included in the profitability estimates (Figure 2). In all cases, Systems 1 and 2 dominate System 3. With coefficients of absolute risk aversion of <0.0008, System 1 dominates System 2, but for more risk averse producers where coefficients of absolute risk aversion are >0.0008, System 2 dominates System 1. Thus, the producer would make his or her decision among Systems 1 and 2 based on his/ her risk preference. There was relatively greater variability of hay production in System 1 than in System 2, thus its greater level of production risk. Since the difference in residual income without accounting for labor was wider, System 1 dominated System 2 in the former case.

5. Conclusions

Without accounting for labor, Systems 1 and 2 were more profitable than System 3. Under this condition, there is no conclusive evidence that the least complex burmudagrass and ryegrass system differs in profitability from the more complex (but not most complex) bermudagrass, ryegrass, rye, dallisgrass and clover mix (berseem, red, and white clover) system. When accounting for labor, Systems 1 and 2 were again more profitable than System 3, with no significant difference between Systems 1 and 2. Though many farm operations are run by household members, accounting for the value of labor has a significant impact on return over expenses.

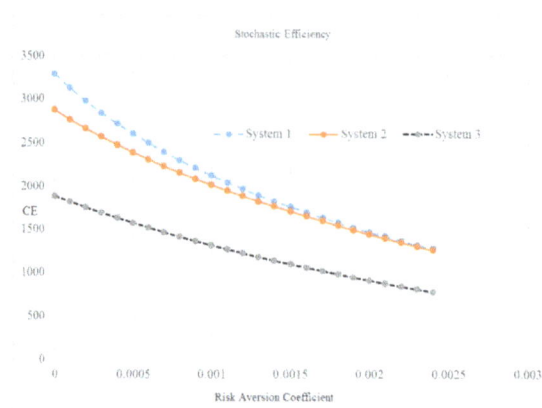

Figure 1. Stochastic Efficiency without Labor, Per Treatment (6 Animals)

System 1 was more profitable and more labor-consuming because of the greater use of machinery for hay making and harvesting. Therefore, there was less difference in the residual income among the various systems after accounting for labor. Since System 1 consists of only bermudagrass and ryegrass, it is the simplest system in the context of management complexity.

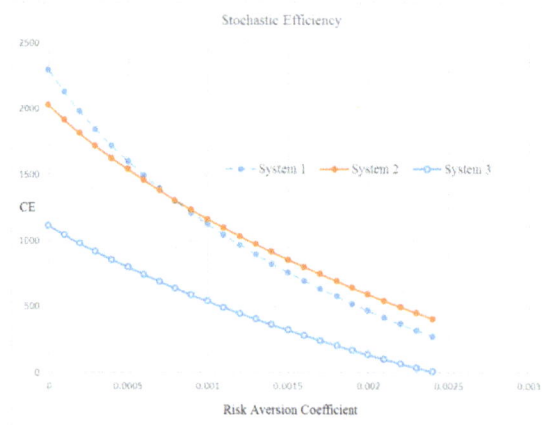

Figure 2. Stochastic Efficiency with Labor, Per Treatment (6 Animals)

On the one hand, results of simulation and stochastic efficiency analysis further confirm the results of the revenue and expense analysis. In both cases, with or without including labor inputs, Systems 1 and 2 dominate System 3. However, due to the narrower numeric difference in profitability after accounting labor, the choice between Systems 1 and 2 changes based on the risk aversion of the decision makers. The price of labor would have to be $32 or more before System 2 would become numerically more profitable holding all else equal.

If we were to consider, however, the carbon dioxide equivalent emissions from these systems, System 2 emits less than System 1 (Bhandari, Gillespie, Scaglia, Wang, & Salassi 2015). System 3 had the lowest carbon dioxide equivalent emissions. Furthermore, Torrico et al. (2014) analyzed sensory scores for the meats from the three systems and found greater sensory scores for System 3 by some groups. These results raise further concerns in determining the profitability of different systems. Further investigation on carbon emissions and the value of carbon reduction as well as the premium that could be expected for superior meat products would be needed to develop a more holistic evaluation of the economics of those systems. If reduction in carbon emissions has significant monetary value and/or farmers can receive premiums for meat that has more favorable sensory characteristics, then the more complex System 3 becomes more economically viable.

The findings of this study are useful in the context of developing a profitable GFB production program in the Southeastern U.S. Since the results are based on experimental data from a research station where conditions are more heavily controlled, there might be some variation in their wider application. Similar research can be replicated in other regions of the country to determine the appropriate pasture system by region.

Funding

This research has been funded in part by Southern SARE (LS09-211), NIFA/AFRI (2011-67023-30098), and HATCH funds (LAB 4178).

Statement

The views expressed here are not necessarily those of Economic Research Service or the U.S. Department of Agriculture.

References

Aberle, E. D., Reeves, E. S., Judge, M. D., Hunsley, R. E., & Perry, T. W. (1981), Palatability and muscle characteristics of cattle with controlled weight gain: time on a high energy diet. *Journal of Animal Science, 52*, 757-563. https:/doi.org/10.2527/jas1981.524757x

Arrow, K. J. (1971). *Essays in the theory of risk bearing*. Chicago: Markham.

Bhandari, B. D., Gillespie, J. M., Scaglia, G., Wang, J., & Salassi, M. (2015). Analysis of pasture systems to maximize the profitability and sustainability of grass-fed beef production. *Journal of Agricultural and Applied Economics, 47*, 193-212. http://search.proquest.com/docview/1718180676?pq-origsite=gscholar

Boehlje, M. D., & Eidman, V. R. (1984). *Farm Management*. New York: John Wiley and Sons.

Boucher, R. W., & Gillespie, J. M. (2009). *Projected costs and returns for beef cattle, dairy and forage crop production in Louisiana 2009*. Agricultural Economics and Agribusiness Information Series No. 263, Louisiana State University Agricultural Center, Baton Rouge, Louisiana.

Boucher, R. W., & Gillespie, J. M. (2010). *Projected costs and returns for beef cattle, dairy and forage crop production in Louisiana 2010*. Agricultural Economics and Agribusiness Information Series No. 268, Louisiana State University Agricultural Center, Baton Rouge, Louisiana.

Boucher, R. W., & Gillespie, J. M. (2011). *Projected costs and returns for beef cattle, dairy and forage crop production in Louisiana 2011*. Agricultural Economics and Agribusiness Information Series No. 274, Louisiana State University Agricultural Center, Baton Rouge, Louisiana.

Boucher R. W., & Gillespie, J. M. (2012). *Projected costs and returns for beef cattle, dairy and forage crop production in Louisiana 2012*. Agricultural Economics and Agribusiness Information Series No. 284, Louisiana State University Agricultural Center, Baton Rouge, Louisiana.

Cox, R. B., Kerth, C. R., Gentry, J. G., Prevatt, J. W., Braden, K. W., & Jones, W. R. (2006). Determining acceptance of domestic forage- or grain-finished beef by consumers from three Southeastern U.S. states. *Journal of Food Science, 71*, 2762-2771. https:/doi.org/10.1111/j.1750-3841.2006.00124.x

Fishell, V. K., Aberle, E. D., Judge, M. D., & Perry, T. W. (1985). Palatability and muscle properties of beef as influenced by pre-slaughter growth rate. *Journal of Animal Science, 61*, 151-157. https:/doi.org/10.2527/jas1985.611151x

Gerrish, J. (2006). Is pasture-finishing for you? *Beef Magazine*, 2 Oct. Available via: http://beefmagazine.com/americancowman/pasture-and-range/pasture_finished_beef/. Accessed 6 June 2011.

Gillespie, J. M., Wyatt, W., Venuto, B., Blouin, D., & Boucher, R. (2008). The role of labor and profitability in choosing a grazing strategy for beef production in the U.S. Gulf Coast Region. *Journal of Agricultural and Applied Economics, 40*, 301-313. https:/doi.org/10.1017/S1074070800028121

Gwin, L. (2009). Scaling-up sustainable livestock production: Innovation and challenges for grass-fed beef in the U.S. *Journal of Sustainable Agriculture, 33*, 189-209. https:/doi.org/10.1080/10440040802660095

Hardaker, J. B., Richardson, J. W., Gudbrand, G. L., & Schuman, K. D. (2004). Stochastic efficiency analysis with risk aversion bounds: a simplified approach. *Australian Journal of Agricultural and Resource Economics, 48*, 253-270. https:/doi.org/10.1111/j.1467-8489.2004.00239.x

Kenward, M. G., & Roger, J. H. (1997). Small sample inference for fixed effects from restricted maximum likelihood. *Biometrics, 53*, 983-997. https:/doi.org/10.2307/2533558

Louisiana State University Agricultural Center; U.S. Department of Agriculture, National Agricultural Statistics Service, Louisiana Field Office (2012). *2011 Louisiana Agricultural Statistics*. A.E.A. Information Series No. 287, Department of Agricultural Economics and Agribusiness, Louisiana State University Agricultural Center, Baton Rouge, Louisiana, October.

Oltjen, R. R., Rumsey, T. S., Puttnam, P. A. (1971). All-forage diets for finishing beef cattle. *Journal of Animal Science, 32,* 327-333. https:/doi.org/10.2527/jas1971.322327x

Pelletier, N., Pirog, R., & Rasmussen, R. (2010). Comparative 413 life cycle environmental impacts of three beef production strategies in the Upper Midwestern United States. *Agricultural Systems, 13,* 380-389. https:/doi.org/10.1016/j.agsy.2010.03.009

Pratt, J. (1964). Risk aversion in the small and large. *Econometrica, 32,* 122-136. https:/doi.org/10.2307/1913738

Prevatt, J. W., Kerth, C. R., & Fields, D. (2006). *Marketing Alabama-grown, Forage-fed Beef.* Alabama Agricultural Experiment Station Serial Report No. 4, Auburn, Alabama.

Richardson, J. W., Schumann, K. D., & Feldman, P. A. (2008) *Simetar: Simulation and Econometrics to Analyze Risk.* Simetar, Inc., College Station, Texas.

Scaglia, G., Rodriquez, J., Gillespie, J., Bhandari, B., Wang, J. J., & McMillin, K. W. (2014). Year-round forage systems for forage-fed beef production in the Gulf Coast. *Journal of Animal Science, 92,* 5704-5715. https:/doi.org/10.2527/jas.2014-7838

Steinberg, E. L., & Comerford, J. W. (2009). Case study: a survey of pasture-finished beef producers in the Northern United States. *The Professional Animal Scientist, 25,* 104-108. https:/doi.org/10.15232/S1080-7446(15)30682-3

Torrico, D. D., Jirangrat, W., Scaglia, G., Malekian, F., Janes, M. E., McMillin, K. W., & Prinyawiwatkul, W. (2014). Proximate and fatty acid compositions and sensory acceptability of Hispanic consumers towards rib-eye steaks from forage-finished steers. *Journal of Food Science and Technology, 49,* 1788-1798. https:/doi.org/10.1111/ijfs.12552

Umberger, W. J., Feuz, D. M., Calkins, C. R., & Killinger, K. (2002). U.S. consumer preference and willingness-to-pay for domestic corn-fed beef versus international grass-fed beef measured through an experimental auction. *Agribusiness, 18,* 491-504. https:/doi.org/10.1002/agr.10034

USDA Economic Research Service (2014). US red meat and poultry forecasts. *Livestock, Dairy, and Poultry Outlook*/LDP-M-215/ 16 May.

USDA National Agricultural Statistical Service (2007). *Census of Agriculture: United States Summary and State Data.* Available at: http://www.agcensus.usda.gov/Publications/2007/Full_Report/ usv1.pdf. Accessed 1 Oct 2014.

USDA-Texas (2010). *Weekly Texas Hay Report,* April 9, 2010. USDA-Texas Department of Agricultural Marketing News, Amarillo, Texas.

USDA-Texas (2011). *Weekly Texas Hay Report,* April 8, 2011. USDA-Texas Department of Agricultural Marketing News, Amarillo, Texas.

USDA-Texas (2012). *Weekly Texas Hay Report,* April 6, 2012. USDA-Texas Department of Agricultural Marketing News, Amarillo, Texas.

USDA-Texas (2013). *Weekly Texas Hay Report,* April 5, 2013. USDA-Texas Department of Agricultural Marketing News, Amarillo, Texas.

USDA-Texas (2014). *Weekly Texas Hay Report,* April 4, 2014. USDA-Texas Department of Agricultural Marketing News, Amarillo, Texas.

Wyatt, W. E., Gillespie, J. M., Blouin, D. C., Venuto, B. C., Boucher, R. W., & Qushim, B. (2013). Effects of year-round stocking rate and stocking method systems on cow-calf production in the Gulf Coast Region of the U.S.: costs, returns, and labor considerations. *The Professional Animal Scientist, 29,* 10-15. https:/doi.org/10.15232/S1080-7446(15)30189-3

Young, A. W., & Kauffman, R. G. (1978). Evaluation of beef from steers fed grain, corn silage or haylage-corn silage diets. *Journal of Animal Science, 46,* 41-47. https:/doi.org/10.2527/jas1978.46141x

11

Evaluation of Salt Affected Soils for Rice (*Oryza Sativa*) Production in Ndungu Irrigation Scheme Same District, Tanzania

Joel L. Meliyo[1], Sophia Kashenge-Killenga[2], Kongo M. Victor[3], Benjamin Mfupe[4], Samwel Hiza[5], Ashura L.Kihupi[6], Brian J. Boman[7], & Warren Dick[8]

[1]Selian Agricultural Research Institute, P.O Box 60248, Arusha, Tanzania

[2]Chollima Agro-scientific Research Institute, P.O Box 1892, Dakawa, Morogoro, Tanzania

[3]University of Dar-es-salaam P.O Box 35131, Dar es Salaam, Tanzania

[4]KATRIN Agriculture research Institute, Private bag, Ifakara, Morogoro, Tanzania

[5]Mlingano Agricultural Research Institute, P.O Box 5088, Muheza, Tanga, Tanzania

[6]Sokoine University of Agriculture. (SUA), Morogoro, Tanzania

[7]University of Florida, 2199 South Rock Road Fort Pierce, FL 34945-3138, USA

[8]The Ohio State University, School of Environment and Natural Resources, 1680 Madison Avenue; Wooster

Correspondence: Joel L. Meliyo, Selian Agricultural Research Institute, P.O Box 6024, Arusha, Tanzania. Email: joelmeliyo15@gmail.com

Abstract

A study was carried out to examine distribution of salt affected soils by types and extent in the Ndungu Agricultural Development Project (NADP) area of Tanzania. The objective was to generate information to guide salt-affected soil management for sustainable rice production. Conventional methods including use of mini-pits and profile pits, coupled with farmers' experiences were used to characterise soil. A total of seven randomly selected soil profile pits located in major soils were dug and described. Soil was sampled from natural horizons for laboratory analysis. In addition a total of 158 topsoil (0 – 20 cm depth) composites soil samples were randomly collected from 90 sites of NADP project area for laboratory analysis. Results showed that a few blocks (block is a piece of farm of 6 to 12 acres) had high exchangeable sodium percentage and high levels of bicarbonates, indicating salt-affected soils. Soil pH, exchangeable sodium percentage (ESP), and electrical conductivity of soil paste extract (ECe) values as high as 9.06, 28.7 $cmol_{(+)}Na\ kg^{-1}$, and 14d Sm^{-1} were measured. Out of 90 blocks, 10 blocks (11%) showed slight to strong salt effects. Two blocks (2%) have been abandoned, and in some cultivated blocks zero yields were recorded due to salt content. The different levels of salinity development in the project area suggest site-specific remediation and appropriate management options be developed to improve crop production. These include rehabilitation of the irrigation infrastructure, use of farmyard manure as a soil amendment and growing salt-tolerant rice varieties. Furthermore, it is important to create awareness among farmers of the problem of salt-affected soil on rice productivity.

Key words: saline-sodic, high pH soils, salinity, sodicity, soil degradation

1. Introduction

Salt-affected soils, where salts concentrate on the soil surface causing severe decline of crop yields, are a worldwide problem (Metternicht & Zunk, 2003; Yadava et al., 2011; Shahid & Al-Shankiti, 2013). Such soils are found in diverse climates but they are dominant in arid and semi-arid climates (FAO, 2000; Graaff & Patterson, 2001; Robert & Ulery, 2011; Qadir et al., 2015). Global projections show that salt-affected soils are increasing. The extent of salt affected soils particularly in irrigated areas has increased in the last two decades from 45 million hectares to 62 million hectares between 1990 through 2013 (Ghassemi et al., 1995; Metternicht & Zinck, 2003; Qadir et al., 2014). These figures suggest that at global scale every day an area of about 2,000 ha of irrigated cropland is affected by varying levels of salinity (Qadir et al., 2014). The consequences of salinity and sodicity are harmful effects on plant growth and food production, reduction of water quality for uses other than agriculture, sedimentation and soil erosion. The consequences impact severely developing countries. According to Keshavarzi & Sarmadian (2012) soluble salts affect the productivity of soils by changing the osmotic potential of soil solution and by increasing the content of exchangeable sodium.

Salt-affected soils are categorized into three groups which are (i) saline soils characterized by high electrical conductivity of paste extract (ECe > 4 dSm^{-1}); (ii) sodic soils with higher exchangeable sodium percentage (ESP > 15%) but low ECe; and (iii) saline-sodic soils are a combination of saline and sodic soils characterized by high electrical conductivity (ECe) which is > 4 dSm^{-1}and ESP is > 15 % (Eynard et al., 2005). Criteria for assessment and categorization of salt affected soils into saline, sodic, and saline-sodic is usually done using the sodium adsorption ratio (SAR) and the exchangeable sodium percentage (ESP) (US Salinity Laboratory Staff, 1954; Charman & Murphy, 2007; Brady & Weil, 2008; Seilsepour et al., 2009) which are defined by equations 1 and 2:

$$S.A.R. = \frac{Na^+}{\sqrt{\frac{1}{2}(Ca^{2+} + Mg^{2+})}}$$

(1)

where:

S.A.R. = sodium adsorption ratio,

Na$^+$, Ca^{2+} and Mg^{2+} are concentrations of soluble cations meqL^{-1}

$$(Exchangeable\ Na^+/CEC) \times 100$$

(2)

where:

ESP = exchangeable sodium percent, %.

Na$^+$ = measured exchangeable Na$^+$, meq/100g.

CEC = cation exchange capacity, meq/100g.

Literature shows that of the three salt affected soil categories, salinity is dominant (Rengasamy, 2006; Qureshi, et al., 2007; Thomas, 2010).

In Tanzania, salt-affected soils are a major constraint to production that contributes to low yields in most rice producing irrigation schemes (Kashenge-Killenga et al., 2012a; Makoi & Ndakidemi, 2007). Salt-affected soils are found in most semi-arid irrigated and non-irrigated, and even in lowland areas with high rainfall, which are characterised by high water tables (Kanyeka et al., 1995; FAO, 2001; Makoi & Ndakidemi, 2007).

Although there are a few scattered research reports on salt-affected soils in Tanzania (Mnkeni, 1996; Makoi & Ndakidemi, 2007; Kashenge_Killenga et al., 2012a), the extent of salt-affected soils in Tanzania is not well established. Mnkeni (1996) estimated that there were more than 2.9 million ha affected by salinity while 700 000 ha had high sodicity. On the other hand FAO (2000) estimated over 1.7 million ha are saline and 300,000 ha are sodic. Furthermore, FAO (2003) increased the estimated extent of salt affected soils to be over 3.5 million ha with proportions of 16% and 84% for sodicity and salinity, respectively. The disparity of these figures by FAO (2000, 2003) and Mnkeni (1996) suggest that the extent of salt-affected soils in the country is not well known. As a consequence, salt affected soils are not mentioned in research agendas and in policy formulation addressing agricultural problems in the nation. In reality, it is a serious problem that is turning agricultural land to barren, salty lands. In addition, several small rice irrigation schemes (classified as traditional irrigation schemes that are community managed) are experiencing reduced yields due to salt (salinity and sodicity) problems (Kashenge-Killenga et al., 2012b). Apart from the fact that the problem of salt-affected soils in rice production systems is increasing, information of type of salts and their extent is scanty or lacking (Kashenge-Killenga et al., 2012a).

Effective management of salt-affected soils requires correct diagnosis of the problem (Robbins & Gavalak, 1989; Eynard et al., 2005). Understanding the salinization processes and the limitations they cause to crop production is of significant importance for improvement of crop productivity with regards to constrains attributed to salinity and sodicity (Robbins & Gavalak, 1989). In the Ndungu Agricultural Development Project (NADP) area there is inadequate understanding of the prevailing salinization processes and the aerial extent affected and environmental factors attributed to salts build up the fact which limits the basis for recommendations for sustainable management options needed to mitigate the problem. This is the case also in most of the other existing irrigation schemes in Tanzania (Kashenge_Killenga et al., 2012a).

The NADP area that is located in Same District in the Kilimanjaro Region is among several projects launched by the Tanzanian Government, geared towards improving agricultural sector, particularly irrigated rice (*Oryza sativa*) production, which is the second staple after maize (*Zea mays*) to address food self-sufficiency and poverty reduction. The 680 ha NADP irrigation area was initiated in 1988, and it has developed into an important rice production area for the Kilimanjaro Region. However, yield reductions due to salt-affected soil is rapidly increasing to the extent that some farmers have had to quit severely affected soils because they were no longer productive (Kiangi, 2005). The genesis and fast spread of the sodicity and salinity problems may be accelerated by the prevailing climatic condition and geographical setting of the area which is characterised by high evapotranspiration that exceeds precipitation. However, information on distribution and identification of soil salt types and their extent in the NADP area is inadequate.

The objective of the study was to examine distribution and extent of salt-affected soils by their types in the NADP area so as to provide information that will guide development of salt affected soil management options. We hypothesize that a clear understanding of the distribution of salt affected soils by their types will enable the development of best management packages to help overcome the salinity and sodicity problems. Adoption of these practices should not only limit further development of salt problems, but also reverse the detrimental effects of salinity and result in sustained rice production in the NADP area.

2. Materials and Methods

2.1 Description of Survey Areas

NADP is located in the Same District, Kilimanjaro Region in the Pangani water catchment at the foot slope of South Pare Mountains. The alluvial/colluvial plain extends into Mkomazi Plain where the Mkomazi National Park is located. The geographic location is latitude -4^{0}22'0.34" S and longitude 38^{0}4'28.06" E, at an elevation of 510 m above seas level, and with a slope gradient ranging from 0.5 to 1%. Soil parent materials are mainly the neogene alluvium of diverse origin, forming mainly heavy clay soils that during dry season crack like Vertisols. Soils near river channels are Fluvisols. The Yongoma River that traverses the project area has been harnessed through a diversion and canal infrastructure for gravity irrigation.

Temperatures in the area are generally high during December-January with daily average temperatures over 30^{0}C, and lowest in July-August at 22^{0}C.Mean annual rainfall is 660 mm, most it falling during the rainy season. There are two rainfall seasons which are short and long rain-seasons which start on March through April and late October through December, respectively (Figure 1). Rainfall fluctuation and evapotranspiration influence the extent of the seasonal growing period (Figure 2). The unreliable, short rainfall locally known as 'Vuli' is used for crop cultivation. *Masika* (local name) for a long rainy season often is of short duration lasting for one month between March and April. This is a shorter period for most traditional crops grown for example a crop like maize (*Zea mays*) which stay in the field for 4 to 6 months. The shorter growing periods experienced in the NADP area has been a cause to frequent crop failures. The NADP area experiences long dry season, which starts from June through October.

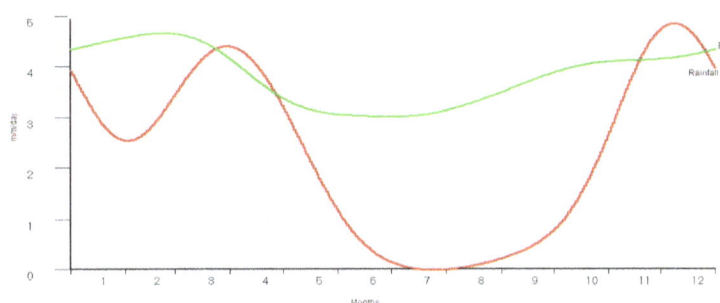

Figure1. Annual rainfall distribution in the NADP area, Same District, Tanzania

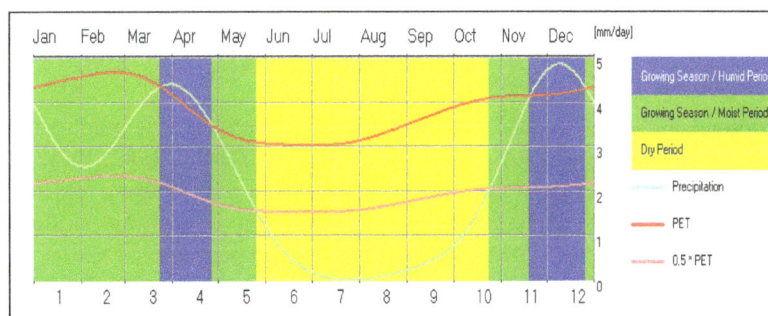

Figure 2. Dependable growing period for the NADP area, Same District, Tanzania

2.2 Soil Sampling

Fieldwork started by doing reconnaissance in the NADP area for familiarisation. Differences between soils were examined and delineated based upon digging of mini-pits (50 cm deep pits), profile pits, and farmers' experience. Auguring was not done because of soil hardness, attributed to both farm operations and soil properties (heavy clays). A total of seven soil profile pits randomly located in representative major soils were dug, described, and soil was sampled from natural horizons for laboratory analysis according to Soil Profile Description Guidelines by FAO (2006a). The soil samples were collected from natural soil profile horizons for laboratory analysis. Soil colours were named according to the Munsell Colour Notation (Munsell Colour Charts Inc., 1992). The scheme has 90 blocks of varied sizes in acres. Topsoil samples were collected from all 90 blocks, at a depth 0-20cm at randomly selected sites by a field team that included researchers, farmers, key informants, and NADP farm staff. Typically, soil composites was done by collecting 8 sub-samples randomly from three plots within a single block, mixed well, and by quarterly, it was reduced to 1 kg that formed a composite representing a block for laboratory analysis. A total of 158 composite soil samples were collected for laboratory analysis.

2.3 Laboratory Analysis

Soil pH was measured potentiometrically in water and in 1M KCl at the ratio of 1:2.5 soil-water and soil-KCl, respectively. ECe was determined using a paste extract where the leachate was used to take readings with an EC meter (Moberg, 2000). Organic carbon (SOC) was determined by the wet digestion method of Walkley and Black (Nelson & Sommers, 1982). Total nitrogen was determined by the Kjeldahl method (Bremner & Mulvaney, 1982). Phosphorus was extracted by the Bray and Kurtz-1 method and determined spectrophotometrically (Olsen & Sommers, 1982). The cations exchange capacity (CEC) and exchangeable bases were extracted by saturating soils with neutral 1M NH4OAc (Thomas, 1982) and the adsorbed $NH4^+$ was displaced by K^+ using 1M KCl and then determined by Kjeldahl distillation method for the estimation of CEC of soil. The bases Ca^{2+}, Mg^{2+}, K^+ and Na^+ displaced by $NH4^+$ were measured by atomic absorption spectrophotometry. Soil texture was determined by the pipette method after dispersing soil with sodium hexametaphosphate (Klute, 1986).

2.4 Data Analysis and Soil Classification

Assessment of soil soluble salts was done by measuring electrical conductivity and by calculating the sodium adsorption ratio (SAR) and the exchangeable sodium percentage (ESP) (Graaff & Patterson, 2001; Brady & Weil, 2008; Seilsepour et al., 2009). These were rated using established thresholds (FAO, 1999a&b). Descriptive statistics were used to derive tables and graphs depicting patterns of studied parameters and the results were used to compare severity of soil salt between different locations (soil profiles) in different blocks in the NADP area. Using both field and laboratory data the identified soil types were classified up to level-2 soil unit names according to the World References Base for soil resources 2014 (FAO, 2015).

3. Results and Discussion

3.1 Soil Texture

Figure 3 and Table 1 present particle size distribution of soils collected from the NADP area, indicating that texture is dominated by clay. Figure 3 shows individual textural separates with depth whereby sand, clay and silt particles are significantly different ($P \leq 0.05$) in the upper part of the profile, but not significantly different for sand and clay in the subsoil (>100 cm from the surface). The subsoils are sandier that might negatively influence both the amount of water that can be stored and the crop vigour that is typically poor in sandy soils. This situation was observed randomly distributed in the NADP area, but generally the detrimental effects were not evident because sands are buried in the sub-surface layers (Figure 3). Some of these sandier spots noted to have poor crop stands were in the old riverbed, which had changed its course in past history. Other areas were affected by land levelling during the establishment of the scheme. Table 1 shows a tendency of clay declining with depth in some blocks, as indicated with a minimum value of 36% in topsoils but declines to 6% and 18% respectively, in subsoils (20-60cm and 60-200cm).

Soil texture is an important soil characteristic, particularly for irrigation and rice production. It influences water holding capacity, rootability, and retention of plant nutrients in soils (Hillel, 1982). In the NADP area, texture variability actually influences land use and field operations like tillage, and watering regime. The sites within the NADP area with sandy clay and sandy clay loam soils are used for maize growing during the dry season while areas with heavy clay soils are rarely used due to poor workability and poor drainage.

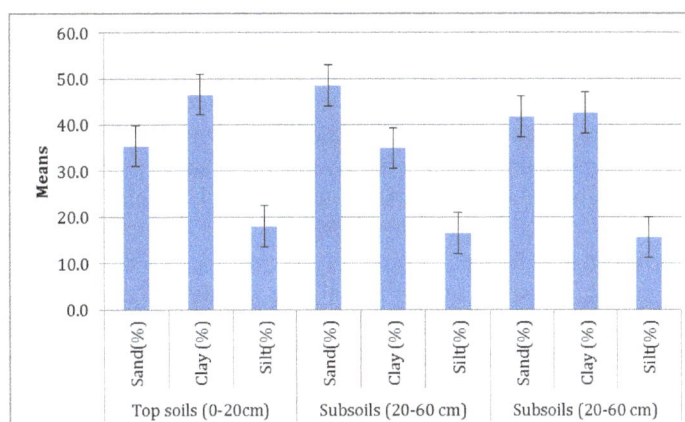

Figure 3. Distribution of soil particle size with depth, in soils from farms in the NADP area, in Same district, Tanzania

Table 1. Soil particle size distribution of the NADP area, Same District, Tanzania

Statistic	Top soils (0-20cm)			Subsoils (20-60 cm)			Subsoils (20-60 cm)		
	Sand (%)	Clay (%)	Silt (%)	Sand (%)	Clay (%)	Silt (%)	Sand (%)	Clay (%)	Silt (%)
Mean	35.4	46.6	18.0	48.5	35.0	16.5	41.8	42.6	15.6
SE	4.1	4.8	2.0	10.2	11.8	3.3	5.1	4.8	1.7
Median	40.0	40.0	14.0	47.0	35.0	16.0	39.0	41.0	16.0
SD	10.8	12.7	5.3	20.5	23.7	6.6	16.1	15.1	5.4
Range	28.0	30.0	12.0	48.0	58.0	14.0	56.0	50.0	18.0
Minimum	20.0	36.0	14.0	26.0	6.0	10.0	20.0	18.0	6.0
Maximum	48.0	66.0	26.0	74.0	64.0	24.0	76.0	68.0	24.0
n	158	158	158	158	158	158	158	158	158

3.2 Chemical Properties

3.2.1 Soil Reaction

Table 2 and Figure 4 present a summary of chemical properties of soils collected from sites within the NADP area. Table 2 shows variations of parameters between blocks and within soils. Figure 4 shows results of soil reactions, a measure of hydrogen ion concentration in the soil suspension as the pH values (Moberg, 2000). The results of soil pH indicate that it is predominantly alkaline, with pH values ranging from mildly alkaline (pH value 7.5) to strongly alkaline (pH value 8.5). Results show that the median pH value is 7.45, which implies that over 50% of soils are mildly alkaline. However, the minimum and maximum pH values range from moderately acid to strongly alkaline, with pH values of 5.7 and 9.06, respectively (Hazelton & Murphy, 2007).

Some profiles show pH declining with depth, while others have a pH increase with depth (Figure 4). The observed results imply that pH values at sites within the NADP area varies between and within soils. This could be attributed to the high concentration of exchangeable sodium as well as variations in calcium and magnesium concentrations (Table 3). The high pH values may also be attributed to the presence of bicarbonates. Most plants, including rice, grow well in soils of pH between 6.5 and 7.5 (Baize, 1993; Brady & Weil, 2008). This suggests that some of the NADP area soils are slightly to severely limiting to crop nutrients, particularly phosphorus and most micronutrients.

As the soil pH increases, the solubility of many nutrients is reduced because the nutrients are precipitated to solid materials that plants cannot use (Fernandez & Hoeft, 2009). It has been observed, for example, that the solubility of iron at pH 4.0 is 100 mgFe/kg soil but when the pH is increased to 6.0, the solubility drops to 0.01 mgFe/kg soil. At pH values above 7.5, the amount of iron in solution is often too low to sustain healthy plant growth. Furthermore, at high soil pH values, phosphorus, manganese, zinc, copper and boron also become minimally soluble, hence deficient. Many alkaline soils also contain low amounts of magnesium (Marschner, 2011; Fernandez & Hoeft, 2009). Thus, close to 50% of Ndungu soils in the NADP area may be experiencing macro and micronutrient deficiencies as influenced by high soil pH.

3.2.2 Soil Organic and Total Nitrogen

Table 3 and Figure 5 show results of both soil organic carbon (SOC) and total nitrogen percent in NADP soils.

According to Landon (1991) and Baize (1993), the mean and median values for organic carbon are rated as medium (> 1.26 %). More than 50% of the NADP soils had organic carbon levels in the medium level (Table 2), that is an amount common for tropical soils (Landon, 1991). However, exceptionally low soil organic carbon levels were recorded for sites LTB2 and RTB1. This means that some NADP soils had soil organic carbon concentrations that would be considered minimum for sustainable crop production. Some research works have indicated that most of the productive agricultural soils have organic matter levels between 3% and 6% (Fenton, Alber, & Ketterings, 2008). The variability of soil organic carbon levels in the NADP area is attributed to farming practices which in the NADP area involves grazing and residual removal by burning. Turnover of soil organic carbon which is converted to soil organic matter by multiplying by a factor 1.72 depends on composition of the organic matter and cropping system, its use and tillage methods (Szajdak, Jezierski & Cabrera, 2003; Helfrich, Ludwig, Buurman & Flessa, 2006). The organic carbon serves as a reserve for many essential nutrients, especially nitrogen. During the growing season, part of the SOC reserve is turned into mineral nitrogen through bacterial activity and made available to plants (Korsaeth, Henriksen, & Bakken, 2002).

Table 3 also presents result for total nitrogen describing, the mean, median, minimum and maximum values. The minimum value is rated to be very low (<0.1%), while the mean and maximum values are rated as low to medium (Landon, 1991; Baize, 1993). The low levels of total nitrogen observed in Table 3 could be attributed to the inherent soil properties and management practices prevailing in the irrigation scheme. Total nitrogen and available nitrogen in the soil are affected by such factor as soil texture, drainage, slope steepness, rainfall, temperature, soil aeration and salt content (electrical conductivity/EC) that in turn, affect the rate of N mineralization from organic matter decomposition, nitrogen cycling, and nitrogen losses through leaching, runoff, or denitrification (Korsaeth, Henriksen, & Bakken, 2002; Meysner, Szajdak & Ku, 2006). Although nitrogen is required N in largest amounts compared to all other macronutrients, it is also the most often deficient because of the dynamic nature of its cycle in the soil and many pathways of loss (Marschner, 2011; Brady & Weil, 2008). Nitrogen is an essential component of proteins such as chlorophyll, (the green pigment in leaves) responsible for the dark green colour of leaves, stem and is an essential constituent of all proteins hence responsible for vigorous growth, branching/tillering, leaf production, size enlargement and yield formation of plants (Roy, Fink, Blair & Tandon, 2006; Brady & Weil, 2008).

Table 2. Soil chemical properties of composites samples from 90 rice fields of NADP farm, Same District, Tanzania

Statistic	pH (extract)	ECe(dSm⁻¹)	TN (%)	OC(%)	P(mg/ kg soil)	Na⁺(cmo¹l₍₊₎/ kg soil)	K⁺(cmol₍₊₎/ kg soil)	Mg²⁺ (cmol₍₊₎/ kg soil)	Ca²⁺ (cmol₍₊₎/kg soil)	CEC(cmol₍₊₎/ kg soil)	BS(%)
Mean	7.45	1.04	0.14	1.68	5.09	1.42	0.66	5.85	11.46	26.77	70.85
SE	0.06	0.18	0.01	0.08	1.25	0.44	0.15	0.48	0.98	2.20	0.90
Median	7.60	0.60	0.13	1.59	2.01	0.63	0.29	4.82	8.57	20.32	71.82
St Dev	0.52	1.52	0.05	0.64	10.37	3.66	1.24	3.98	8.18	18.24	7.46
Variance	0.27	2.32	0.00	0.42	107.56	13.40	1.53	15.85	66.87	332.65	55.64
Kurtosis	1.65	17.12	1.79	-0.39	9.57	47.35	16.50	3.69	0.46	4.48	6.33
Skewness	-1.33	4.06	0.96	0.24	3.35	6.63	4.09	1.82	1.21	1.94	-1.51
Range	2.40	8.38	0.27	3.04	41.72	28.41	6.29	18.90	33.62	94.40	49.65
Minimum	5.70	0.25	0.03	0.27	1.56	0.25	0.08	1.41	1.78	7.52	35.22
Maximum	9.06	8.63	0.30	3.31	43.29	28.66	6.38	20.31	35.40	101.92	100.0
n	158	158	158	158	158	158	158	158	158	158	158
CI (95.0%)	0.13	0.37	0.01	0.15	2.49	0.88	0.30	0.96	1.96	4.38	1.79

---

[1] TN = Total nitrogen; OC = soil organic carbon; cmol₍₊₎ = centimol-charge; dSm⁻¹= decisiemen per metre; SE = Standard error; St Dev = Standard deviation; CI = Confidence interval

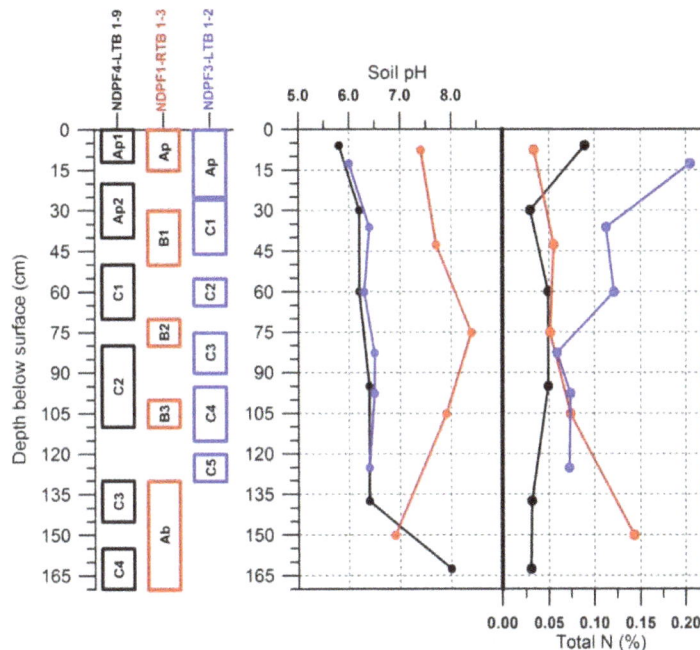

Key: Ap, B1, B2... and Ap1, Ap2...C4 are different profile horizons

Figure 4. Distribution and variation of soil pH and total nitrogen (%) between and within profiles in NADP farm, in Same District, Tanzania

Soils represented by the two of the selected profiles indicate that total nitrogen declines with soil depth (NDPF3 LTB1-2 and NDPF1-LTB 1-9) while in NDPF1 RTB1-3, soil nitrogen increases with soil depth (Figure. 4). Heavy clay cracking soils that is commonly observed in Vertisols soils, and that is represented by profile NDPF1 RTB1-3, had higher total nitrogen levels in subsoils as compared to the light sandy clay loam soil (Cambisols) represented by profile NDPF3 LTB1-2. The observed nitrogen levels could be explained that in heavy clay soils that crack, materials from topsoils composed of organic matter fall in wide cracks and get trapped. This material then decomposes and through microbial decomposition adds to the soil nitrogen in subsoils. However, soil nitrogen according to Brady & Weil (2008) is a dynamic plant nutrient, which in most cases needs replenishment, either as organic manure or as mineral fertilizer. In the NADP area, some sites had total nitrogen values that were generally considered very low, i.e. total soil nitrogen values ranged between 0.1% and 0.2% (Landon, 1991; Baize, 1993). Generally, the results imply that NADP soils require nitrogen fertilisers for sustained optimal rice production.

3.2.3 Available Phosphorus

The median value of 2.0 mg P/kg soil (Table 2) indicates that over 50% of NADP samples had very low levels of available phosphorus, well below the critical minimum value of <7 mg P/kg soil (Baize, 1993) where deficiency is usually observed. Phosphorus is important in plants because it plays a critical role in physiological and biochemical processes such as photosynthesis, respiration, N fixation, root development, maturation, flowering, fruiting, and seed production (Johnston & Steen, 2000). Phosphorus deficiency can restrain plant growth, delay maturity, and reduce crop yield. The P deficiency symptoms are expressed in the older leaves because of its poor immobility in the plants (Ketterings, Czymmek, Albrecht, & Barney, 2014). In the NADP area, exceptionally low P-values were observed throughout. The low available P levels can be attributed to high soil pH and presence of carbonates that precipitates P and render it unavailable for plant uptake (Johnston & Steen, 2000; Fernandez & Hoeft, 2009). These results imply that NADP area soils are mainly deficient in phosphorus, suggesting that application of P fertilisers is essential for sustained and optimal rice yield production.

3.2.4 Cations Exchange Capacity (CEC)

The CEC represents the total amount of negative charges available to attract positively charged ions (cations) in soil solution. Results (Table 2) show that CEC levels ranged from low (minimum value of 7.5 cmol$_{(+)}$/kg soil) to very high (maximum value of 101.9 cmol$_{(+)}$/kg soil). Over 50% of samples had CEC levels in the acceptable medium rate (Baize, 1993; Landon, 1991). These results also show that there is variation of CEC between soils. CEC was also studied in the profile pits from sites within the NADP area to establish CEC status with soil depth.

Profile NDPF1-RTB1-3 had very high CEC levels in the topsoil, particularly the first two natural horizons with clay contents of 66% and 64%, respectively (Figure 5). There was a clear relation between high CEC and clay content. As clay content declined with soil depth to 48%, there was a corresponding low CEC, regardless of an increase of soil organic carbon. Profile NDPF1-RTB1-3 was from soil disturbed during levelling and the original topsoil was buried. This could be the reason that the typical decline of soil organic carbon with depth was not observed. The CEC at many of the NADP area sites is mainly determined by the clay content of the soil. Although some profiles indicate irregular increases of organic carbon with depth, this also could be attributed to changes in soil characteristics attributed to levelling. The soil represented by Profile NDPF1_RTB1-3 exhibit the vertic characteristics because of high clay contents and they have wide and deep cracks whereby topsoil falls into cracks, resulting in a mixing of top and subsoils.

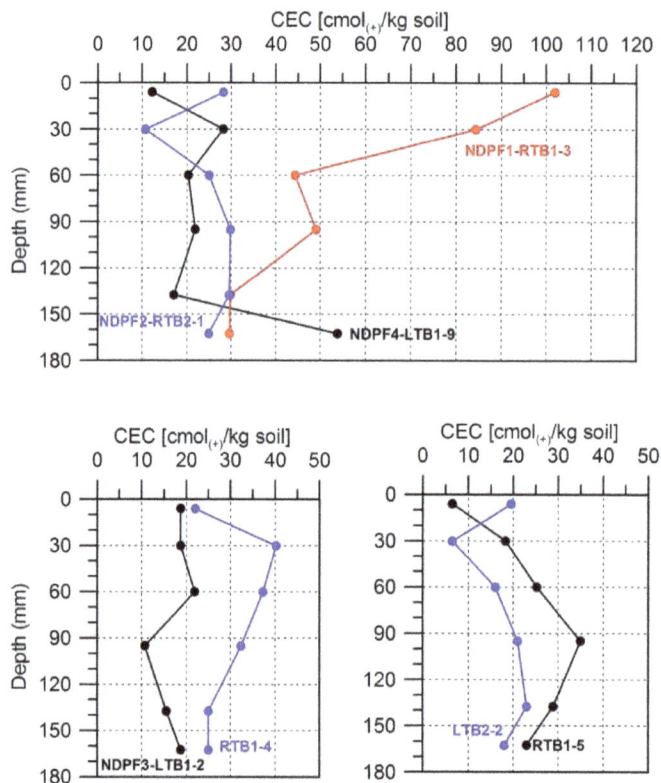

Figure 5. Variation of CEC between representative soil profiles and with depth at NADP area farms, Same District, Tanzania. The observed difference between NDPF1 RTB1-3 and the other two profiles may be attributed to levelling and soil texture

The CEC of NADP soils is dependent upon the amounts and types of clay minerals and organic matter present, as observed in NDPF1 RTB 1-3. Soils withhigh CEC were mainly characterised by high levels of clay content and organic carbon percentage. It is important to note that, although high CEC soils can hold more nutrients, a good soil management program is still required if these soils are to be sustainably productive over a long term period.

3.2.5 Exchangeable Bases (Ca^{2+}, Mg^{2+}, K^+ And Na^+)

Descriptive statistical results show wide variations between soils that can only be explained by different management of the various sites and the effects of land levelling when the project was being constructed. Exchangeable calcium levels in the NADP area soils are variable, ranging from medium (5-10 $cmol_{(+)}$/kg soil) to very high, with levels above 20 $cmol_{(+)}$Ca/kg soil (Table 2). Fifty of 90 studied sites had very high calcium levels according to the rating by Baize (1993), Landon (1991), and EUROCONSULT (1989) for tropical clayey soils. In addition, exchangeable calcium of < 2.0 $cmol_{(+)}$/kg soil was observed in some NADP sites.

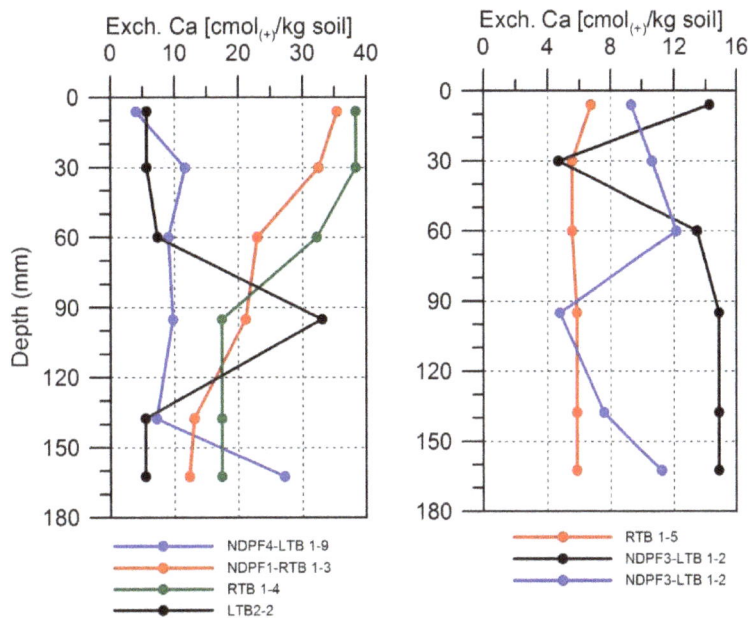

Figure 6. Distribution of calcium levels among selected soil profiles and with soil depth in studied NADP area farms, Same District, Tanzania

Profiles NDPF1 RTB-3 and RTB1-4 have excessively high levels of exchangeable calcium in topsoils but decline with soil depth (Figure 6). Profiles NDPF4-LTB1-9 and LTB2-2 show high calcium levels in the subsoils. The high levels are indicators of salt development. Magnesium in NADP soils had a median value of 4.8 $cmol_{(+)}Mg/kg$ soil and over 50% of soils have high magnesium levels (Landon, 1991). There were variations, as indicated by minimum and maximum values of 1.41 and 20.3 $cmol_{(+)}Mg/kg$ soil, rated to be medium to extremely high (Table 2) (Baize, 1993; Landon, 1991; EUROCONSULT, 1989) for tropical clayey soils. The high levels were localised in sub-soils at few sites represented by profile NDPF4 LTB1-9. These high Mg levels suggest the presence of localised salt-affected soils, which are present in some sites within the NADP area. In general the distribution of exchangeable magnesium between soils and with soil depth resembles that of calcium.

The minimum and median potassium levels in the composite soil samples were 0.08 and 0.29 $cmol_{(+)}/kg$ soil, respectively. These levels are considered to be very low to low (Baize, 1993). However, there ware spots with extremely high values (6.4 $cmol_{(+)}/kg$ soil)which are located in salt-affected sites. Representative profiles show large variations among soils (Figure 7).

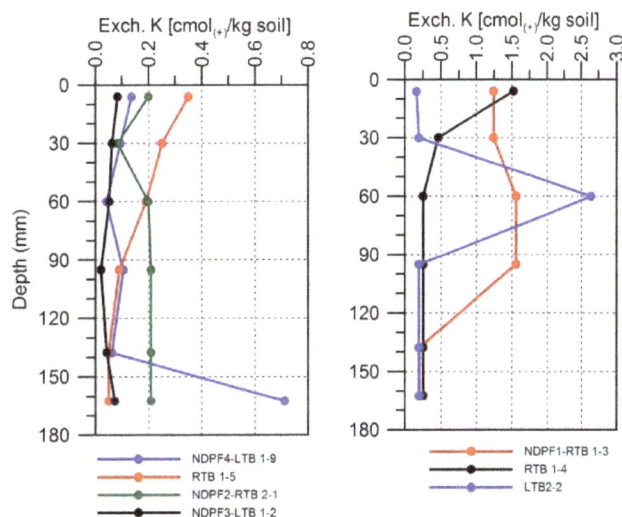

Figure 7. Distribution of exchangeable potassium among selected soil profiles and with soil depth in NADP farm, Same District, Tanzania

Composites soils samples show exchangeable sodium levels in NADP soils range from low to very high (0.25 to above 2 cmol$_{(+)}$Na/kg soil) (Table 2). The median exchangeable sodium value was 0.63 cmol(+)/kg soil, which is rated as medium (Landon, 1991; Baize, 1993). However, the maximum value of 28.7 cmol$_{(+)}$Na/kg soil was extremely high (Landon, 1991; Baize, 1993). High exchangeable Na levels were also observed in profilesNDPF4 LTB1-9 and RTB1-4,and these sodium levels were significantly higher in the subsoils. In contrast, NDPF1 RTB1-3 had high sodium levels in topsoils. These and other are salt affected spots in the NADP area need to be identified and managed properly before the fields become unproductive. These results on exchangeable bases at sites with high levels of Na, Mg, K and Ca coincide with areas where complaints of salt problems were put forth by farmers. These soluble salts accumulate and become concentrated to high levels, thus leading to salinity and/or sodicity problems

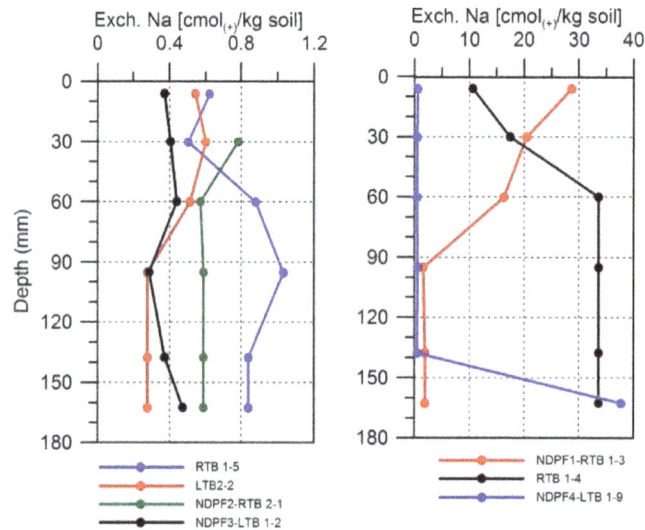

Figure 8. Distribution of exchangeable sodium between soil profiles and with soil depth in NADP farm, Same District, Tanzania

3.2.6 Base Saturation

The percent base saturation (BS) levels ranges from low with values of 20-30% to very high with values >80% (Table 2) (Hazelton & Murphy, 2007). The dominant soils in the NADP area farm have high BS as indicated by the median BS value of 71.8%. The percent BS refers to the proportion of the soil CEC occupied by the basic cations (Ca^{2+}, Mg^{2+}, K$^+$, and Na$^+$). It is important to note that BS is influenced by the relative proportion of acid and basic cations. The acidic cations are H$^+$ and Al^{3+}. The two types, basic and acidc derive their names from the way they influence the soils' pH. As the concentration of Ca^{2+} and Mg^{2+}ions increase, as is the case for most of the NADP area soils,and the number of H$^+$ and Al^{3+}ions decrease. Base saturation is relatively high in moderately weathered soils formed from basic igneous rocks, such as the basalts. The pH of soil increases as BS increases, while in contrast, highly weathered and/or acidic soils tend to have low BS (Brady & Weil, 2008).The higher BS in NADP area farms may be reflecting high concentrations of bicarbonates and basic cations to hazardous levels, particularly sodium.

3.3 Salt Indicators in NADP Farm

Several sites (approximately10 representing 11% of affected areas), have high levels of exchangeable sodium, high exchangeable sodium percent (ESP), and high levels of bicarbonates (Tabl 3 and Figure 9). These are indicators of salt-affected soils, mostly saline-sodic, sodic, and saline in few blocks (Table 3). The cause of salts build-up in soils is due to changes in the local water balance, commonly brought about by mismanagement of irrigation and the lack of adequate drainage. The farm has irrigation supply canals but few drainage ditches. Therefore, irrigation water (which contains salts flushed from the soil) is not removed from the system. It may be re-used at downstream sites or may be allowed to evaporate in the field. This poor irrigation practice adds salts to the water table, which further exacerbates the problem by increasing subsurface salts. Rice (*Oryza sativa*) typically can only grow in soils with ECe of 3dSm^{-1} or less (Kashenge-Killenga et al., 2012b). It cannot grow well at the RTB1-3 and RTB1-4 sites, which have high ECe unless a salt-tolerant variety is used which a threshold of ECe not more than 9 dSm^{-1}. When the ECe value is above 3 dSm^{-1}, the plant fail to set seeds. If proper reclamation measures and good management practices are not put in place, rice growing will also fail in the LTB2-2 and LTB1-9 sites, which have very high ECe in subsoils (Table 3 and Figure 9).

Table 3.Soil salt indicators in the selected sites within the NADP area farm, Same District, Tanzania

Block	pH (H$_2$O)	ECe dSm^{-1}	Exch. Na$^+$	SAR	ESP	HCO$_3^-$ MgL^{-1}	Salt status
LTB2-1	8.03	1.71	4.85	1.43	29.22	54.9	Slightly saline
LTB2-2	5.68	0.51	0.83	0.26	7.36	-	Normal
LTB1-5	6.06	0.34	1.10	0.28	6.38	-	Normal
RTB2-1	6.85	0.12	0.48	0.12	2.77	79.3	Normal
RTB1-1	7.63	0.11	0.95	0.16	2.45	140.0	Normal
LTB 2-2	9.06	0.96	16.88	2.94	33.43	115.9	Sodic
RTB1-4	8.02	13.08	16.96	2.41	23.85	78.2	Saline-Sodic
RTB1-4	8.11	8.31	13.87	3.1	48.25	73.2	Saline-Sodic
RTB1-3	7.41	28.66	28.66	7.69	28.12	30.5	Saline-Sodic
RTB 1-5	5.8	0.14	0.62	0.40	9.57	-	Normal

Slightly = just at the start /beginning of salinity

Also, when exchangeable Na levels exceed 15% of the CEC, physical and chemical soil properties are negatively affected. Under these conditions, water and air infiltration into the soil may be reduced and poor growing conditions may result (Ondrasek, Rengel, & Veres, 2011). To overcome this problem, Ca^{2+} is added to replace the Na$^+$on the CEC sites (Horneck, Ellsworth, Hopkins, Sullivan & Stevens, 2007; Syers, Johnston & Curtin, 2008). Sodium in the soil water is then leached out of the rootzone by excess irrigation water or rainfall. The amount of Ca^{2+} needed to replace the Na$^+$ is based on the amount of exchangeable Na$^+$ as well as Na$^+$ saturation.

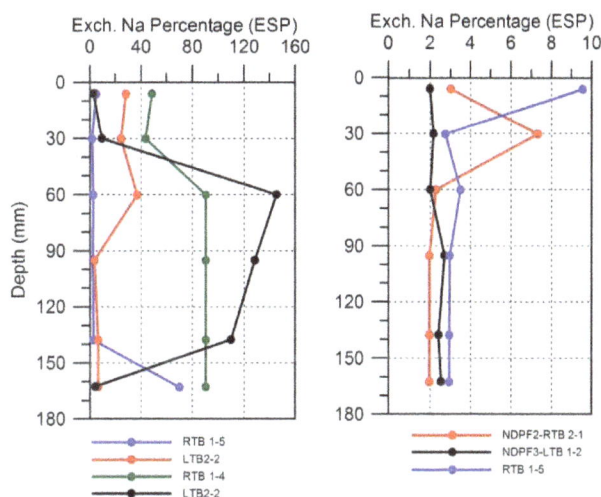

Figure 9. Distribution of exchangeable sodium percentage (ESP) among selected soil profiles in NADP area farms, Same District, Tanzania

The results of this work show that the NADP area definitely has a salinity problem that needs to be addressed before additional areas become unproductive and abandoned. Out of over 90 sites within the NADP area, 10 show slight to strong salt effects, representing 11% of the whole irrigation scheme. Similar effects have been reported on fields located outside the NADP area. Farmers have already abandoned the RTB 1-3 and RTB 1-4 sites. According to farmers, the value and yield of soils with high contents of salts are reduced to zero. This has caused severe socioeconomic problems, one being exodus of men to town and abandoning families. The low soil productivity has also prompted farmers to resort to clearing natural forests by illegal timbering and charcoal burning in order to support their families.

The symptoms of soil erosion are everywhere and some houses were being threatened by gully erosion. This phenomenon has been observed not only in Ndungu but also in the nearby villages. The consequences of low production and abandoned farmland in these areas are enormous, since farmers depend on the rice crop for both food and cash for their families. Salts accumulation in farms have been attributed primarily to poor irrigation water management. Other contributing factors include lack of maintenance and repairs of irrigation infrastructure, inadequate knowledge about salinity and its effects by farmers, water managers, government leaders, and inadequate resources to address the problem. The problem encountered in the NADP area has been reported as a serious problem worldwide. For example, it has been reported that the global cost of irrigation-induced salinity/sodicity is equivalent to an estimated US$11 billion per year (FAO, 2006b).

4. Conclusions and Recommendations

A study was carried out in the Ndungu Agricultural Development Project region to determine the causes of farmer's complaints concerning reduced rice yields, which in some locations were to the extent that some fields were abandoned. Yield losses will continue to occur if management practices and remediation measures are not taken to address the problem. Based on our study results, soils were classified as salt-affected when the pH values was > 7.5; the electrical conductivity (ECe) was > 3 dSm^{-1}, and exchangeable sodium percentage (ESP) was > 15%. The salt affected soils are categorised into three types: saline, sodic, and saline-sodic. The soils of the NADP region are mainly saline-sodic and, to a lesser extent, sodic. Salinity was also noted in few spots.

Social and ecological factors have contributed to the development of the salt affected soils. The parent material from which the soil in the area developed is rich with bases. This, combined with inadequate precipitation, resulted in limited leaching of bases during soil formation. The problem has been intensified through poor irrigation practices, a broken irrigation system (which increases seepage), as well the as lack of knowledge of the impact of salt-affected soils and their management on rice yields. In addition, soil fertility is quite variable between and within soils. Some of this can be attributed to levelling done during establishment of the project area, where cutting of topsoils in some places and filling in others was done. Other on-going soils and farm management practices such as grazing of animals, burning, and collection of grasses have contributed to soil organic matter decline and general soil degradation. As a result, the major nutrients are at very low to deficient levels in many of the salt-affected soils.

To minimize productivity losses due to salt accumulation, the use of soil and water best management practices are strongly recommended as follows:

1. Renovate and close supervision of water inlet and outlet canals to prevent leakage of water between blocks.
2. Affected fields require reclamation by the use of gypsum, farm-yard manure, good drainage and flushing using good quality water. It is essential that drainage water, with its salt load from the affected soil, be removed from scheme.
3. The use of farmyard manure at a rate of 3 to 8 tons/ha is recommended to boost soil organic matter and improved soil infiltration to facilitate washing of salts to minimize the effect of salts. The farmyard manure will also supply some nutrients (N, P and K) that were limiting in some soils and therefore need to be added to improve rice productivity.
4. Soil reclamation will probably need to be a continuing practice in the most affected areas. The most economical solution to maintain productivity will probably be a combination of salt tolerant variety selection and soil amelioration practices.

Acknowledgment

There are not enough good words that can express our gratitude to the "Innovative Agricultural Research Initiative (iAGRI) for giving us funds to carry out this work. We are also indebted to our collaborators from USA Prof. Brian Boman and Prof. Warren Dick who were very instrumental making the work successful. We very much appreciate the efforts and devotion of our main collaborators the farmers in NADP and the Same District management for providing us enabling environment and their participation.

References

Baize, D. (1993). *Soil Science Analyses. A Guide to Current Use*. John Wiley & Sons Ltd. West Sussex.

Brady, N. C., & Weil, R. R. (2008). *The Nature and Properties of Soils*. (14th.ed.). Upper Saddle River: Prentice Hall.

Bremner, J. M., & Mulvaney, C. S. (1982). Total nitrogen. In: A.L. Page, R.H. Miller & D.R. Keeney (Eds.). *Methods of Soil Analysis, Part 2. (2^{nd}ed.)* (pp. 595-624). ASA, SSSA Monograph no. 9, Madison, Wisconsin.

Charman, P. E. V., & Murphy, B. W. (eds.). (2007). *Soils-Their Properties and Management*, (3rd ed.). Oxford University Press in Association with Department of Natural Resources.

EUROCONSULT. (1989). *Agricultural Compendium for Rural Development in the Tropics and Subtropics*. Elsevier Science Publishers, Amsterdam.

Eynard, A., Lal, R., & Wiebe, K. (2005). Crop response in salt affected soils. Research reviews, practices policy and technology. *Journal of Sustainable Agriculture*. Jan 2005. Doi: 10.1300/J064V27no1_03.source OA1. pp 5-50.

FAO. (1991). Soil salinity assessment methods and interpretation of electric conductivity measurements. FAO,

Rome, Italy.

FAO. (1999b). *Integrated soil management for sustainable agriculture and food security in southern and east African*. Proceedings of the Expert Consultation . Harare, Zimbabwe 8-12th December, 1997. FAO, AGL Misc/23/99.

FAO. (2000). Agriculture - Land - Water - WAICENT - FAO Search. http://www.fao.org.

FAO. (2001). Food and Agriculture Organization. Annual report. FAO, Rome. Online publication. http://www.fao.org.

FAO. (2003). Food and Agricultural Organization Database. FAO, Rome. Online publication. http://www.fao.org.

FAO. (2006a). *World Reference Base for Soil Resources 2006*. A framework for International classification, correlation and communication. World Soil Resources Reports No.103. FAO, Rome, Italy. http://www.fao.org.

FAO. (2006b). *Management of irrigation-induced-salt affected soils*. SPUSH, CISEAU Project, Food and Agriculture Organization of the United Nations. Rome, Italy. http://www.fao.org.

FAO. (2015). *World reference base for soil resources 2014.* International soil classification system for naming soils and creating legends for soil maps, update 2015. World Soil Resources Reports No.106. FAO, Rome, Italy. http://www.fao.org.

Fenton, M., Alber, C., & Ketterings, Q. (2008). *Soil organic matter. Agronomy Fact Sheet. Fact Sheet 41.* Cornell University Cooperative Extension.

Fernández, F. G., & Hoeft, R. G. (2009). Managing soil pH and crop nutrients. In E.D. Nafziger (ed.) (pp. 91–112). Illinois agronomy handbook. 24th ed. Univ. of Illinois, Urbana.

Ghassemi, F., Jakeman, A. J., & Nix, H. A. (1995). *Salinisation of land and water resources: human causes, extent, management and case studies*. Wallingford: CABI Publishing.

Graaff, R., & Patterson, R. A. (2001). *Explaining the Mysteries of Salinity, Sodicity, SAR and ESP in On-site.* Practice in Proceedings of On-site '01 Conference: Advancing On-site Wastewater Systems, Lanfax Laboratories, Armidale, 361-368.

Hazelton, P., & Murphy, B. (2007). *Interpreting soil test results. What do all the numbers mean?* CSIRO Publishing, 150 Oxford Street, P.O. Box 1139; Collingwood VIC 3066 Australia.

Helfrich, M., Ludwig, B., Buurman, P., & Flessa, H. (2006). Effect of land use on the composition of soil organic matter density and aggregate fractions revealed by solid state 13C NMR spectroscopy. *Geoderma, 136*(1-2), 331-341. http://dx.doi.org/10.1016/j.geoderma.2006.03.048

Hillel, D. (1982). *Introduction to Soil Physics.* Academic Press Inc. New York.

Horneck, D. A., Ellsworth, J. W., Hopkins, B. G., Sullivan, D. M., & Stevens, R. G. (2007). Managing salt affected soils for crop production. A Pacific Northwest Extension publication Oregon State University, University of Idaho, and Washington State University.

Johnston, A. E., & Steén, I. (2000). Understanding Phosphorus and its use in Agriculture. European Fertilizer Manufacturers Association. Avenue E. van Nieuwenhuyse 4 B-1160 Brussels Belgium.

Kanyeka, Z. L., Msomba, S., Penza, M. S. F., & Alluri, K. (1995). *Rice ecosystems in Tanzania. IRRI Notes, 20,* 1,

Kashenge-Killenga, S., Tongoona, P., Derera, J., Kanyeka, Z., & Meliyo, J. L. (2012a). Soil characterisation for salt problems in selected rice irrigation schemes: A key aspect towards improvement of salt Tolerance Rice Genotypes in North Eastern Tanzania. *Journal of Advances in Developmental Research, 3*(1).

Kashenge-Killenga, S., Tongoona, P., Derera, J., & Kanyeka, Z. (2012b). Irrigated rice-based farm characteristics and production constraints in selected salt affected areas of North-eastern Tanzania. *Journal of Advances in Developmental Research, 3*(1).

Keshavarzi, A., & Sarmadian, F. (2012). Mapping of Spatial Distribution of Soil Salinity and Alkalinity in a Semi-arid Region, Annals of Warsaw University of Life Sciences. *Land Reclamation. 44*(1), 3-14.

Ketterings, Q., Czymmek, K. L., Albrecht, G., & Barney, P. (2014). *Starter Phosphorus Fertilizer for Corn. Agronomy Fact Sheet Series. Fact Sheet 8*. Cornell University, Cooperative, Extension

Kiangi, M. F. (2005). *Environmental conservation using rice by-products and reafforestation ndungu village.* (Unpublished master's thesis). Open University of Tanzania and Southern New Hampshire University, UK.

Korsaeth, A., Henriksen, T. M., & Bakken, L. R. (2002). Temperal changes in mineralization and immobilization of N during degradation of plant materials: Implication to plant N supply and nitrogen losses. *Soil Biology and Biochemistry, 34,* 789-799. http://dx.doi.org/10.1016/S0038-0717(02)00008-1

Klute, A. (Ed.).(1986). *Methods of soil analysis, part 1, physical and mineralogical methods (2nd Edition).* American Society of Agronomy, *Agronomy Monographs, 9*(1), Madison, Wisconsin.

Landon, J. R. (Ed). (1991). *Booker Tropical Soil Manual. A handbook for soil survey and agricultural land evaluation in the tropics and subtropics.* Longman Scientific & Technical Publishers, Essex.

Makoi, J., & Ndakidemi, P. (2007). Reclamation of sodic soils in northern Tanzania, using locally available organic and inorganic resources. *African Journal of Biotechnology, 16,* 1926-1931

Marschner, P. (Ed) (2011). *Marschner's Mineral nutrition of higher plant. (3rded).* Academic Press. ISBN 9780123849052.

Metternicht, G. I., & Zinck, J. A. (2003). Remote sensing of soil salinity: Potentials and constraints. *Remote Sensing of Environment, 85,* 1-20. http://dx.doi.org/10.1016/S0034-4257(02)00188-8

Meysner, T., Szajdak, L., & Ku, J. (2006). Impact of the farming systems on the contents of biological active substances and forms of nitrogen in the soils. *Agronomy Research, 4*(2), 531-542.

Mnkeni, P. N. (1996). *Salt affected soils: Their Origin, Identification, Reclamation and Management.* Inter Press. Ltd.

Moberg, J. R. (2000). *Soil and plant analysis Manual.* The Royal Veterinary and Agricultural University, Chemistry Department, Copenhagen, Denmark.

Munsell Color Company. (1992). Munsell Soil Color Charts. Munsell Color Co. Inc. Baltimore.

Nelson, D. W., & Sommers, L. E. (1982). Total carbon, organic carbon and organic matter. In: A.L. Page, R.H. Miller & D.R. Keeney (Eds). *Methods of Soil Analysis, part 2, 2nd edit. (*pp. 539-579). ASA, SSSA Monograph no. 9, Madison, Wisconsin.

Ondrasek, G., Rengel, Z., & Veres, S. (2011). Soil salinisation and salt stress in crop production. In: Shanker, A.K. & Venkateswarlu, B. (Eds). *Abiotic Stress in Plants: Mechanisms and Adaptations,* 171-190. http://dx.doi.org/10.5772/22248

Olsen, S. R., & L. E. Sommers. (1982). Phosphorus. In: A.L. Page, R.H. Miller, and D.R. Keeney (Eds). *Methods of Soil Analysis, part 2. Agronomy monograph No. 9.* (pp. 403-430.). American Society of Agronomy. Madison, Wisconsin.

Qadir, M., Quillérou, E., Nangia, V., Murtaza, G., Singh, M., Thomas, R. J., ... Noble, A. D. (2014). Economics of salt-induced land degradation and restoration. *Natural Resources Forum, 38,* 282-295. http://dx.doi.org/10.1111/1477-8947.12054

Qadir, M., Noble, A. D., Karajeh, F., & George, B. (2015). *Potential business opportunities from saline water and salt-affected land resources.* Colombo, Sri Lanka: International Water Management Institute (IWMI). CGIAR Research Program on Water, Land and Ecosystems (WLE). 29p. (Resource Recovery and Reuse Series 5). http://dx.doi.org/10.5337/2015.206

Qureshi, A. S., Qadir, M., Heydri, N., Turral, H., & Javadi, A. (2007). A review of management for salt prone land and water resources in Iran. Colombo, Sri Lanka. International Water Resources Institute. IWMI Working paper 125.

Rengasamy, P. (2006). World salinization with emphasis on Australia. *Journal of Experimental Botany, 57*(5), 1017-1023. http://dx.doi.org/10.1093/jxb/erj108

Robbins, C. W., & Gavlak, R. G. (1989). Salts- and Sodium affected soils. Bulletin No. 703. Cooperative Extension Services University of Idaho, College of Agric.

Robert, F., & Ulery, A. (2011). An Introduction to Soil Salinity and Sodium Issues in New Mexico. *Las Cruces, Circular, 656,* 1- 6.

Roy, R. N., Fink, A., & Blair, G. J. (2006). *Plant Nutrition for Food Security: A guide for Integrated Nutrient Management.* FAO Fertilizer and Plant Bulletin 16. Rome, Italy.

Seilsepour, M., Rashidi, M., & Khabbaz, B. G. (2009). Prediction of Soil Exchangeable Sodium Percentage Based on Soil Sodium Adsorption Ratio. *American-Eurasian Journal of Agriculture & Environment Science, 5*(1), 1-4.

Shahid, S. A., & Al-Shankiti, A. (2013). Food production in marginal lands – case of GDLA member countries. *International Soil and Water Conservation Research, 1*(1), 24-38. http://dx.doi.org/10.1016/S2095-6339(15)30047-2

Syers, J. K., Johnston, A. E. & Curtin, D. (2008). Efficiency of soil and fertilizer phosphorus use. Reconciling changing concepts of soil phosphorus behaviour with agronomic information. FAO Fertilizer and Plant Nutrition Bulletin 18. Food And Agriculture Organization of the United Nations, Rome.

Thomas, G. W. (1982). Exchangeable cations. In: A.L. Page, R.H. Miller and R.D. Keeney (Eds.).*Methods of Soil Analysis part 2,* 159-165. ASA, SSSA Monograph no. 9, Madison, Wisconsin.

Thomas, R. P. (2010). *Proceedings of the Global forum on Salinization and Climate Change.* (GFSCC2010) Valencia, 25-29.

US Salinity Staff Laboratory. (1954). *Diagnosis and Improvement of Saline and Alkaline Soils.* Agric. Handbook No. 60. USDA, Washington, DC.

Yadav, S., Irfan, M., Ahmad, A., & Hayat, S. (2011). Causes of salinity and plant manifestations to salt stress: A review. *Journal of Environmental Biology, 32*, 667-685.

12

Comparison of Two Harvesting Methods for the Continuous Production of Indigo Plant (*Polygonum Tinctorium*) Leaves in a Closed-Cultivation System

Masaru Sakamoto[1], Dong-An Kim[1], Keiko Imoto[1], Yusuke Kitai[1] & Takahiro Suzuki[1]

[1]Faculty of Biology-Oriented Science and Technology, Kindai University, Wakayama 649-6493, Japan

Correspondence: Masaru Sakamoto, Faculty of Biology-Oriented Science and Technology, Kindai University, Wakayama 649-6493, Japan. E-mail: sakamoto@waka.kindai.ac.jp

Abstract

Although the blue dye indigo has been chemically synthesized for over a century, there is an increasing interest in the indigo plant (*Polygonum tinctorium*) as a source of natural dyes and medicines. To maintain a stable supply of *P. tinctorium* throughout the year, we examined the effect of two harvesting methods on the leaf yield of this plant under a closed-cultivation system. With method 8c-M, all shoot branches >8 cm of the stem bottom were harvested and under method 2b-M, all branches, but two, were harvested at the stem bottom. Both methods enabled sustainable leaf yields from the same plants over 1 year. The total weight of shoot branches harvested by 8c-M was 1.86–3.11 times higher that of shoot branches harvested by 2b-M. Harvesting by 8c-M resulted in shoot branch weights lower than those from plants harvested with the 2b-M. Leaf/shoot ratio was increased in plants harvested by 8c-M. The content of indican, the precursor of indigo, in leaves was not significantly different between the two harvesting methods. Our data may provide a new continuous cultivation method of leaf crops all over years in controlled-cultivation systems.

Keywords: *Polygonum tinctorium*, indican, indigo, year-round cultivation, harvesting methods, closed-cultivation systems

1. Introduction

Indigo is a blue dye that was being extracted from the indigo plant (*Polygonum tinctorium*) until the end of the 19th century, when chemically synthetized indigo almost completely replaced natural indigo (Bechtold et al., 2002). Consequently, the areas covered with *P. tinctorium* decreased on a wide scale and cultivation techniques experienced stagnation. There is currently a renovated needs of indigo plants because of the demand for natural dyes and daily necessities such as soap and plastic products (Velho et al., 2017). Indigo is also used in traditional Chinese medicines to treat systemic psoriasis and leukemia (Koo & Arain, 1998; Xiao et al., 2002).

In Japan, *P. tinctorium* is utilized as a traditional natural dye and is mainly cultivated in the Tokushima Prefecture (Ricketts, 2006). The colorless precursor of indigo, indican, is present in *P. tinctorium* and is chemically converted to the blue dye indigo by hydrolysis and oxidation (Minami et al., 2000). In addition to being the source of the natural dye, *P. tinctorium* is also of pharmacological significance. An increasing number of new pharmaceutical agents with suppressive effects against cancer, inflammation, allergy and microbial growth are obtained from its leaf extracts (Hashimoto et al., 1999; Kunikata et al., 2000; Koya-Miyata et al., 2001; Micallef et al., 2002; Iwaki & Kurimoto, 2002). Thus, a year-round steady supply of *P. tinctorium* leaves is required. However, *P. tinctorium* is an annual short-day plant with a limited leaf harvest period in the field. Moreover, flowering decreases indican concentration in the leaves. Therefore, new cultivation methods for the continuous production of *P. tinctorium* without flowering might be required.

To date, the production of indican and related compounds *in vitro* has been attempted in cell and root-hair cultures of *P. tinctorium* (Chung et al. 1996; Kim et al. 1996; Young-Am et al., 2000), However, these approaches yield indican concentrations that are much lower than those present in fresh leaves (Minami et al., 2000; Campeol et al., 2006). Because indican is mainly present in leaves and is rarely found in other parts, such as stems, roots, and flowers (Minami et al., 2000), to maximize the yield of indican from plants, efficient production strategies of leaves are required. Recently, plant factories have been developed for the sustainable production of many leafy vegetables throughout the year. To our knowledge, there is no experimental attempt to grow *P. tinctorium* in a closed-cultivation system. Therefore, we designed and tested two harvesting methods for the continuous leaf production of this annual plants in a closed-cultivation system.

2. Method

2.1 Plant Material and Growth Condition in a Closed-Cultivation System

Indigo plant seeds (*Polygonum tinctorium* cv. 'Kosenbon' and 'Kojoko') were pregerminated for 3 days at 20°C in the dark. Germinated seeds were sown in sponge cubes of $2 \times 2 \times 2$-cm in size and grown under 200 µmol m^{-2} s^{-1} photosynthetic photon flux (PPF) for 16 h using fluorescent lamps (FL40SBR-A; NEC Co., Japan). The nutrient solution was based on a half-strength culture solution of the Otsuka House A-recipe (Otsuka Chemical Co. Ltd., Japan), as described previously (Sakamoto & Suzuki, 2015). At 13 days after sowing (DAS), seedlings were transferred to a DFT hydroponic container ($34.5 \times 26.5 \times 8.2$-cm high) with continuous aeration under 250 µmol m^{-2} s^{-1} PPF. A total of 12 seedlings were planted per container. The distance from the light source to the plant's bottom was 36 cm. To avoid the diffusion of light out of the cultivation area, light-reflecting plates (MCPET; Furukawa Electric Co., Japan) were placed along all its margins. Nutrient solution was exchanged at every week. Shoot weight was obtained for all harvested shoot branches from a single plant. Leaf weight was obtained by adding the individual weights of all detached leaves from the harvested branches of a single plant. All plant weights were measured as fresh weight. In experiment 1, characteristics of shoot branches were examined. In experiment 2, characteristics of shoot branches, leaves, and leaf indican content were examined. Three plants were subjected to each experimental plot in experiment 1, and 12 plants in experiment 2.

2.2 Harvesting Methods

Harvesting of shoot branches were initiated from 42 DAS in experiment 1, and from 29 DAS in experiment 2. To achieve continuous leaf production in this closed-cultivation system, two cutting methods of shoot branches were examined (Figure 1). The first method, named "8c-M" consisted of the harvesting of shoot branches at >8 cm from the bottom level of the stem. To avoid removing young branches, we only harvested branches containing at least three fully expanded leaves. The second method, "2b-M", consisted of the harvesting of all branches at the bottom of the stem except for two young ones. Harvesting intervals were inconsistent in experiment 1 (7–36 days) and consistent in experiment 2 (7 days).

2.3 Measurement of Indican

The indican content was spectrophotometrically measured following modifications of the method reported elsewhere (Angelini et al., 2004; Gilbert et al., 2004). In brief, leaf disks (100 mg) from fully expanded secondary leaves were placed in 500 µL of distilled water at 100°C for 10 min. The sample was then cooled at room temperature. After the removal of leaf disks, 500 µL of 1% HCl (containing 10 mM FeCl$_3$) were added to the solution and mixed thoroughly. After 30 min, the blue indigo converted from the indican was spectrophotometrically measured at 600 nm. A standard curve was prepared using an indican dilution series prepared by same steps.

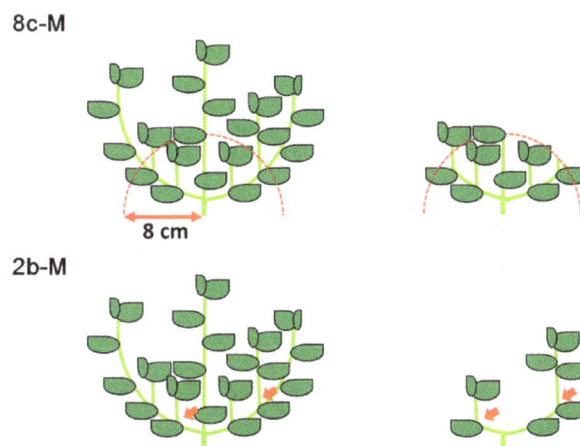

Figure 1. Schematic diagram of two harvesting methods for continuous harvesting of *P. tinctorium*. 8c-M; All shoot branches >8 cm of the stem bottom are harvested except for branches containing less than 3 fully expanded leaves. 2b-M; All shoot branches are harvested at the bottom of stem except for two branches. Arrows show the two shoot branches remained to be harvested

3. Results and Discussion

To achieve the maximum yield of indigo plant leaves by continuously harvesting the same plants, we designed two harvesting methods as described in Materials and Methods (Figure 1). Because *P. tinctorium* is a short-day plant which develops flower buds under prolonged dark periods, we generated a constant long-day environment in our closed-cultivation system to sustain constitutive leaf development without inducing flowering. In our cultivation system, flowering was not induced before 409 DAS (Figure 2). After harvesting branches, new ones emerged from the leaf proximal region, and sometimes from the stem-branching region.

Figure 2. Morphological characteristics of *P. tinctorium* cultivar 'Kosenbon' harvested by two methods after 409 days after sowing in closed-cultivation system. A, plant harvested by 8c-M. B, plant harvested by 2b-M

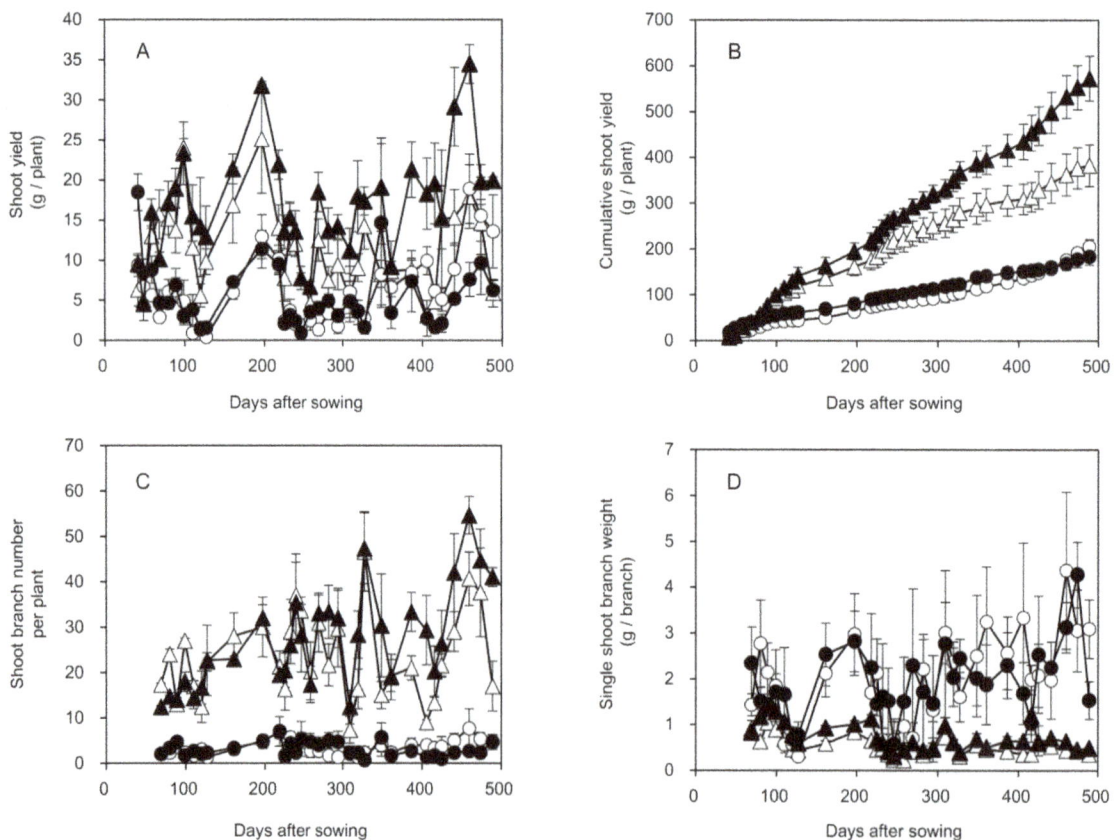

Figure 3. Effect of harvesting methods in *P. tinctorium* cultivars, 'Kosenbon' and 'Kojoko' on shoot yield (A), cumulative shoot yield (B), shoot branch number (C), and single shoot branch weight (D). Closed triangles, 'Kosenbon' harvested by 8c-M; open triangles, 'Kojoko' harvested by 8c-M; closed circles, 'Kosenbon' harvested by 2b-M; open circles, 'Kojoko' harvested by 2b-M. Data are shown by the mean ± SE (*n* = 3)

In experiment 1, both harvesting methods enabled sustainable yield of shoot branches for a total of 489 DAS (Figure 3). Harvested shoot weight was higher with the 8c-M than with the 2b-M (Figures 3A, 3B). The total shoot weight per 'Kosenbon' plants harvested with 8c-M during 489 DAS was 572.90 g, that is highest within in this experiment. The total weight ratio of 8c-M/2b-M was 3.11 and 1.86 in 'Kosenbon' and 'Kojoko' respectively. Similarly, harvested branch number was higher in 8c-M than in 2b-M (Figure 3C). Conversely, the weight of a single shoot branch was lower in plants harvested by 8c-M (Figure 3D). Plant morphologies at 409 DAS were different between plants harvested with the two methods (Figure 2). In plants harvested following 8c-M, the sizes of shoot branches and leaves were uniform compared with those from plants harvested by 2b-M. After harvesting by 8c-M, many branches that could produce new shoot branches remained. In plants harvested by 2b-M, the sizes of the shoot branches and leaves in each branch were variable and were in general larger than those from plants harvested following 8c-M (Figure 3D). Given that the application of 2b-M decreased total stem biomass compared to 8c-M (Figure 2), newly emerging and growing shoot meristems might have been suppressed in plants harvested by 2b-M. To validate these results, we conducted experiment 2 with additional measurements using 'Kosenbon'.

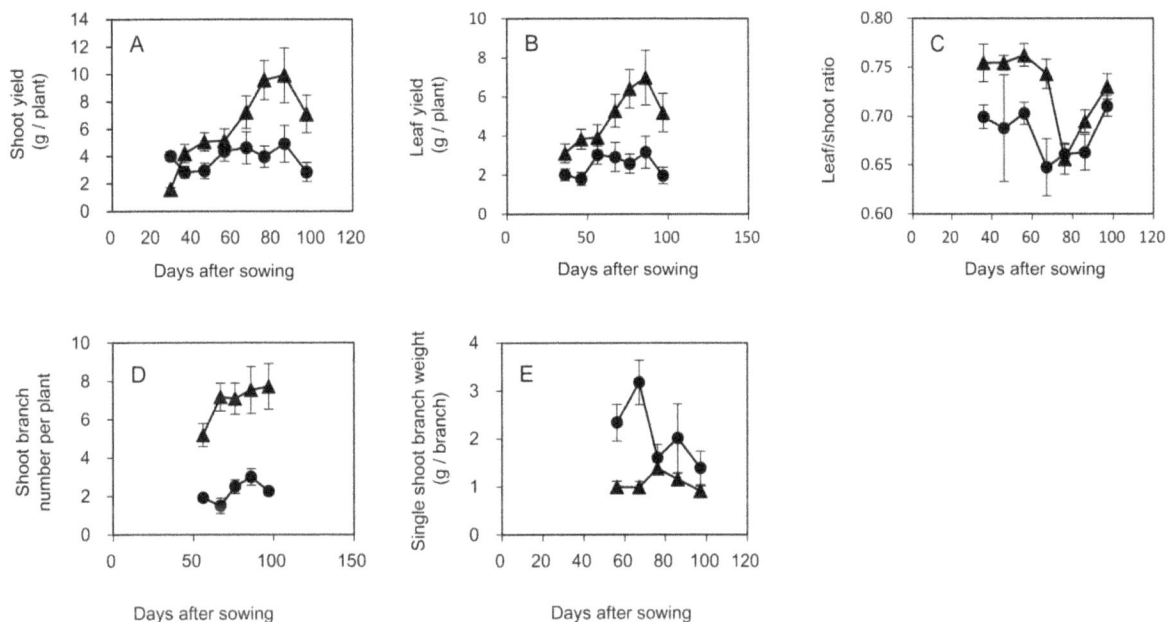

Figure 4. Effect of harvesting methods in *P. tinctorium* cultivar, 'Kosenbon' on shoot yield (A), leaf yield (B), leaf/shoot ratio (C), shoot branch number (D), and single shoot branch weight (E). Closed triangles, 'Kosenbon' harvested by 8c-M; closed circles, 'Kosenbon' harvested by 2b-M. Data are shown by the mean ± SE ($n = 12$)

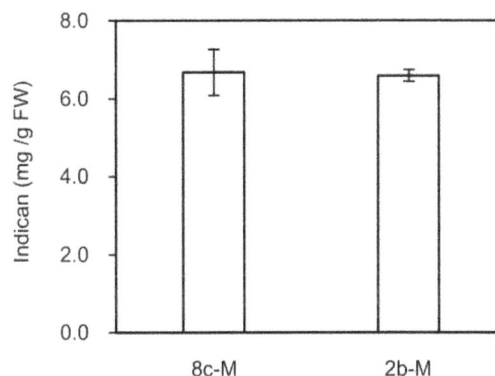

Figure 5. Effect of harvesting methods on the content of indican in leaves of *P. tinctorium* cultivar, 'Kosenbon'. Data are shown by the mean ± SE ($n = 4$)

In experiment 2, the two harvesting methods showed similarities with experiment 1 in shoot weight, shoot branch number, and single shoot branch weight (Figures 4A, 4D, 4E). Leaf weight and shoot yield were increased in plants harvested with 8c-M (Figures 4A, 4B). The leaf/shoot ratio was higher in plants harvested by 8c-M than in plants harvested by 2b-M (Figure 4C).

Because *P. tinctorium* mainly contains indican in leaves and is rarely found in other structures (Minami et al., 2000), a higher ratio of leaf development may result in a higher indican production. To determine whether the applied harvesting methods influenced the production of indican, we measured indican contents of secondary fully expanded leaves from harvested shoot branches. As shown in Figure 5, no significant difference was observed between the two harvesting methods, suggesting that the harvesting method may not have an effect on the metabolic processes leading to indican synthesis. In previous studies, field grown *P. tinctorium* possessed 9.53 mg (32.3 pmol) per g fresh weight of indican in leaves (Minami et al., 2000). In this study, leaf indican contents were 6.68 and 6.60 mg per g fresh weight in 8c-M and 2b-M, respectively (Figure 5). Thus, there was not much difference in leaf indican levels between field and our hydroponically grown leaves, indicating that hydroponics can be alternative cultivation method to gain indican from *P. tinctorium* leaves.

Leaf position is important for the production of specific secondary metabolites in many leaf crops (McConkey et al., 2000; Minami et al., 2000; Koshiishi et al., 2001; Mohanpuria et al., 2009; Fischer et al., 2011). In tea, young leaves have activated caffeine synthetic pathways thereby containing substantially more caffeine than old leaves (Koshiishi et al., 2001; Mohanpuria et al., 2009). Likewise, along the main stem of sweet basil plants, younger leaves have more oil content than older leaves do (Fischer et al., 2011). Accordingly, younger immature leaves contain larger amounts of indican than older ones in *P. tinctorium* (Minami et al., 2000). To obtain indican from leaves in a high concentration, a larger proportion of young immature leaves is needed at harvest. Because the 8c-M yield a higher number of shoot branches than the 2b-M, the development of each branch might be suppressed. The application of the harvest method 8c-M constitutes a reasonable approach to promote larger young immature leaf ratios therefore maximizing the yield of indican throughout the year in our *P. tinctorium* closed-cultivation system. By applying this method in closed-cultivation systems of other leaf crops that contain valuable components in young leaves, a sustainable production of such components may be also attained.

References

Angelini, L. G., Tozzi, S., & Nassi o Di Nasso, N. (2004). Environmental factors affecting productivity, indican content, and indigo yield in *Polygonum tinctorium* Ait., a subtropical crop grown under temperate conditions. *Journal of Agricultural and Food Chemistry, 52*, 7541-7547. https://doi.org/10.1021/jf040312b

Bechtold, T., Turcanu, A., Geissler, S., & Ganglberger, E. (2002). Process balance and product quality in the production of natural indigo from *Polygonum tinctorium* Ait. applying low-technology methods. *Bioresource Technology, 81*, 171-177. https://doi.org/10.1016/S0960-8524(01)00146-8

Campeol, E., Angelini, L. G., Tozzi, S., & Bertolacci, M. (2006). Seasonal variation of indigo precursors in *Isatis tinctoria* L. and *Polygonum tinctorium* Ait. as affected by water deficit. *Environmental Experimental Botany, 58*, 223-233. https://doi.org/10.1016/j.envexpbot.2005.09.006

Chung, I., Kim, T., Bae, G., Lee, H., & Chae, Y. (1996). Stimulation of indirubin production by KNO-3 depletion in indole-supplemented suspension culture of *Polygonum tinctorium*. *Biotechnology Letters, 18*, 947-950. https://doi.org/10.1007/BF00154627

Fischer, R., Nitzan, N., Chaimovitsh, D., Rubin, B., & Dudai, N. (2011). Variation in essential oil composition within individual leaves of sweet basil (*Ocimum basilicum* L.) is more affected by leaf position than by leaf age. *Journal of Agricultural and Food Chemistry, 59*, 4913-4922. https://doi.org/10.1021/jf200017h

Gilbert, K. G., Maule, H. G., Rudolph, B., Lewis, M., Vandenburg, H., Sales, E., Tozzi, S., & Cooke, D. T. (2004). Quantitative analysis of indigo and indigo precursors in leaves of *Isatis* spp. and *Polygonum tinctorium*. *Biotechnology Progress, 20*, 1289-1292. https://doi.org/10.1021/bp0300624

Hashimoto, T., Aga, H., Chaen, H., Fukuda, S., & Kurimoto, M. (1999). Isolation and identification of anti-helicobacter pyrori compounds from *Polygonum tinctorium* Lour. *Nature Medicine, 53*, 27-31.

Iwaki, K., & Kurimoto, M. (2002). Cancer preventive effects of the indigo plant, *Polygonum tinctorium*. *Recent Research Developments in Cancer, 4*, 429-437.

Kim, S. U., Song, K. S., Jung, D. S., Chae, Y. A., & Lee, H. J. (1996). Production of indoxyl derivatives in indole-supplemented tissue cultures of *Polygonum tinctorium*. *Planta Medica, 62*, 54-56.

https://doi.org/10.1055/s-2006-957797

Koo, J., & Arain, S. (1998). Traditional Chinese medicine for the treatment of dermatologic disorders. *Archives of Dermatology, 134*, 1388-1393. https://doi.org/10.1001/archderm.134.11.1388

Koshiishi, C., Kato, A., Yama, S., Crozier, A., & Ashihara, H. (2001). A new caffeine biosynthetic pathway in tea leaves: utilisation of adenosine released from the S‐adenosyl‐L‐methionine cycle. *FEBS Letters, 499*, 50-54. https://doi.org/10.1016/S0014-5793(01)02512-1

Koya-Miyata, S., Kimoto, T., Micallef, M. J., Hino, K., Taniguchi, M., Ushio, S., Iwaki, K., Ikeda, M., & Kurimoto, M. (2001). Prevention of azoxymethane-induced intestinal tumors by a crude ethyl acetate-extract and tryptanthrin extracted from *Polygonum tinctorium* Lour. *Anticancer Research, 21*, 3295-3300.

Kunikata, T., Takefuji, T., Aga, H., Iwaki, K., Ikeda, M., & Kurimoto, M. (2000). Indirbin inhibits inflammatory reactions in delayed-type hypersensitivity. *European Journal of Pharmacology, 410*, 93-100. https://doi.org/10.1016/S0014-2999(00)00879-7

McConkey, M. E., Gershenzon, J., & Croteau, R. B. (2000). Developmental regulation of monoterpene biosynthesis in the glandular trichomes of peppermint. *Plant Physiology, 122*, 215-224. https://doi.org/10.1104/pp.122.1.215

Micallef, M. J., Iwaki, K., Ishihara, T., Ushio, S., Aga, M., Kunikata, T., Koya-Miyata, S., Kimoto, T., Ikeda, M., Hino, K., & Kurimoto, M. (2002). The natural plant product tryptanthrin ameliorates dextran sodium sulfate-induced colitis in mice. *International Immunopharmacology, 2*, 565-578. https://doi.org/10.1016/S1567-5769(01)00206-5

Minami, Y., Nishimura, O., Hara-Nishimura, I., Nishimura, M., & Matsubara, H. (2000). Tissue and intracellular localization of indican and the purification and characterization of indican synthase from indigo plants. *Plant and Cell Physiology, 41*, 218-225. https://doi.org/10.1016/S1567-5769(01)00206-5

Mohanpuria, P., Kumar, V., Joshi, R., Gulati, A., Ahuja, P. S., & Yadav, S. K. (2009). Caffeine biosynthesis and degradation in tea [*Camellia sinensis* (L.) O. Kuntze] is under developmental and seasonal regulation. *Molecular Biotechnology, 43*, 104-111. https://doi.org/10.1007/s12033-009-9188-2

Ricketts, R. (2006). *Polygonum tinctorium*: contemporary indigo farming and processing in Japan. *Indirubin the red shade of indigo. Roscoff, Life in Progress Editions.* 147-156.

Sakamoto, M., & Suzuki, T. (2015). Elevated root-zone temperature modulates growth and quality of hydroponically grown carrots. *Agricultural Sciences, 6*, 749-757. https://doi.org/10.4236/as.2015.68072

Velho, S. R .K., Brum, L. F. W., Petter, C. O., dos Santos, J. H. Z., Simunicd, S., & Kappa, W. H. (2017). Development of structured natural dyes for use into plastics. *Dyes and Pigments, 136*, 248-254. https://doi.org/10.1016/j.dyepig.2016.08.021

Young-Am, C., Yu, H. S., Song, J. S., Chun, H. K., & Park, S. U. (2000). Indigo production in hairy root cultures of *Polygonum tinctorium* Lour. *Biotechnology Letters, 22*, 1527-1530. https://doi.org/10.1023/A:1005668625822

Xiao, Z., Hao, Y., Liu, B., & Qian, L. (2002). Indirubin and meisoindigo in the treatment of chronic myelogenous leukemia in China. *Leukemia & Lymphoma, 43*, 1763-1768. https://doi.org/10.1080/1042819021000006295

13

Land Use Change and Policy in Iowa's Loess Hills

Gaurav Arora[1,2], Peter T. Wolter[3], David A. Hennessy[1,2,4] & Hongli Feng[1,4]

[1]Department of Economics, Iowa State University, USA

[2]Center for Agricultural and Rural Development, Iowa State University, USA

[3]Dept. of Natural Resource Ecology and Management, Iowa State University, USA

[4]Dept. of Agriculture, Food & Resource Economics, Michigan State University, USA

Correspondence: Gaurav Arora, Department of Economics, Iowa State University, USA. E-mail: gaurav88@iastate.edu

Abstract

Land use changes have important implications on ecosystems and society. Detailed identification of the nature of land use changes in any local region is critical for policy design. In this paper, we quantify land use change in Iowa's Loess Hills ecoregion, which contains much of the state's remaining prairie grasslands. We employ two distinct panel datasets, the National Resource Inventory data and multi-year Cropland Data Layers, that allow us to characterize spatially-explicit land use change in the region over the period 1982-2010. We analyze land use trends, land use transitions and crop rotations within the ecoregion, and contrast these with county and state-level changes. To better comprehend the underlying land use changes, we evaluate our land use characterizing metrics conditional on soil quality variables such as slope and erodibility. We also consider the role of contemporary agricultural policy and commodity markets to seek explanations for land use changes during the period of our study. Although crop production has expanded on the Loess Hills landform since 2005, much of the expansion in corn acres has been from reduced soybean acreage. We find that out of the total 258 km^2 increase in corn acreage during 2005-'10, about 100 km^2 transitioned from soybeans. Data also indicate intensifying monoculture with higher percentage of corn plantings for two to four consecutive years during 2000-'10. In addition, crop production is found to have moved away from more heavily sloped land. Cropping does not appear to have increased on lands with higher crop productivity.

Keywords: agro-environmental policy, land use change, land quality, erodibility, monoculture, crop rotations

1. Introduction

This paper seeks to describe land use change in the Iowa Loess Hills landform (ILHL). The Missouri Valley's Loess Hills (figure 1) are comprised of wind-deposited silt hills just east of the Missouri River in Southeast Iowa and Northeast Missouri. Cut through by river tributaries and generally steepest on the west side, hill elevation seldom exceeds 259 feet above the river plain (Note 1). No wider than 15 miles, the landform extends about 320 km from Plymouth County, IA, south through Woodbury, Monona, Harrison, Pottawattamie, Mills and Fremont counties in Iowa as well as Atchison and Holt counties in Missouri. We focus on Iowa's 2800 km^2 portion of the landform, which comprises more than 80% of its total area. The ILHL contains more than 50% of Iowa's remnant prairie (Loess Hills Alliance, 2011).

Mostly under private ownership and largely grass-covered until the 20th Century, row crop production now dominates large patches of this fragmented and erosion-prone landscape (National Resource Inventory (NRI), 2013 p.11). Maize, soybeans and grass are the major land uses while the area also straddles two Metropolitan Statistical Areas (MSAs) from which demand for non-agricultural land uses, such as residences with scenic river overviews, are to be expected. These MSAs are Sioux City to the north and Omaha-Council Bluffs to the south with, respective populations of about 170 000 and 900 000 circa 2014. In addition, fire suppression has led to encroachment by tree species, especially the fire intolerant Eastern Red Cedar, threatening rare native plant species and leaving the loose soil more vulnerable to erosion.

The literature on land use change in western Iowa's Cornbelt is extensive and diverse in direction of inquiry. Secchi, Tyndall, Schulte and Asbjornsen (2008) addressed how high commodity prices can confound conservation efforts, thus placing greater emphasis on the need for targeted practices to obtain highest benefit per unit cost. Secchi, Kurkalova, Gassman and Hart (2010) used the Cropland Data Layer (CDL) to simulate the extent to which biofuels-related expansion may tilt Iowa crop rotations toward more corn intensive rotations. Brown and Schulte (2011) studied aerial photographs to document the decline of small grains and grass agriculture in three Iowa townships between 1937 and 2002. Miller (2006) commented on the roles of urban pressure, topography, erodibility constraints and agro-economic incentives on assembling a remnant prairie, the Broken Kettle Grassland Preserve, at the ILHL's north end. Many technical contributions to our understanding of soil and water conservation on the landform have also been published (e.g. Tomer, Moorman, Kovar & James, 2007).

Specific to the Loess Hills ecoregion, and most relevant to our study, Farnsworth, Schulte and Hickey (2010) connected privately obtained, remotely sensed data on land cover with crop productivity information to develop a conservation priority index that also seeks to account for benefits from tract connectivity. Their inquiry was static with 2006 land uses, just before major changes in United States cropping activities. Arora, Wolter, Feng and Hennessy (2015) used CDL data to quantify land use transitions in the ILHL between 2001 and 2013. They found that grass acres had declined during this period in the ILHL, having moved into wooded categories, and that corn acres had expanded largely at the expense of soybean acres. They expressed surprise, however, at the limited expansion of row-crop production in the region.

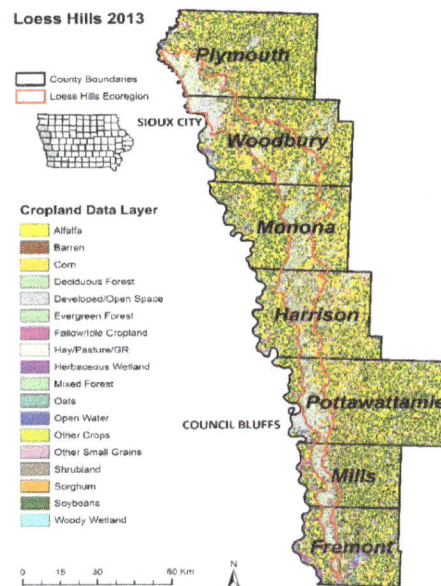

Figure 1. Iowa's Loess Hills landform study area (inside red boundaries, 2797 km^2) and the seven counties in Western Iowa that contain it.

The main objectives of this study are twofold. First, we seek to provide detailed scrutiny of recent land use changes across the ILHL. We characterize regional grassland conversions along with potential factors that impact such conversions such as expansion of cultivated area, urbanization, and invasive wooded species like the Eastern Red Cedar. For this purpose, we utilize longitudinal data from two distinct sources, NRI and CDL, which differ in their scope for spatial as well as temporal dimensions to analyze regional land use. We specify land use trends, land use transition matrices and the structure of corn rotations for the ILHL and contrast these across the counties that skirt the ecoregion. We also conduct a conditional analysis of land use trends on four land quality metrics: erodibility, slope, Corn Suitability Ratings and Land Capability Classification, discussed hereafter under Section 3. Since richer soils are more productive and are expected to offer lower cropping risk, we expect a greater tilt towards crop cultivation on lands with higher productivity. The second objective of this study is to better understand the impacts of commodity markets and agricultural policy on land use decisions in the ILHL. In particular, we determine whether past land use changes in the ILHL are consistent with those in its encircling counties and the state of Iowa. We expect consistency since the private landowners within the landform face a policy and market environment that is very similar to that in neighboring counties. One other contribution of this study is that it provides an approach to identifying spatial errors due to the CDL's land classification scheme and attempts to reduce such errors by employing various remote-sensing techniques.

In Section 2 we provide an overview of relevant policies and the evolving market environment. This is followed by an explanation of our materials and methods; primarily different land use data sources and data processing procedures. After analyzing results, we summarize land use conversion trends with a brief discussion.

2. Policy and Market Environment

The past thirty years has seen a shift in the emphasis of United States agricultural policy away from food and feed production toward energy outputs and also toward environmental outputs that are not generally supported through market incentives. The main agricultural policies of relevance have been those regarding conservation, biofuels and crop insurance.

Although antecedents existed, the Conservation Reserve Program (CRP) was established under the 1985 Farm Bill to incentivize voluntary retirement of environmentally sensitive land from crop production, at least temporarily. Rental contracts with the federal government are typically for ten or more years. The program has proved popular among many land owners and environment advocates, but less popular among agribusinesses and crop producing tenants who identify competition for land and have concerns about lost support for local cropping infrastructure. The program has been renewed in each farm bill through to 2014 although enrollment criteria and maximum enrolled acres have changed over the years. The enrollment cap as well as the enrolled acres have declined under CRP. National enrollment peaked at about 146 000 km^2 in 2007 and declined to about 101 000 km^2 in 2014 (Lubbins and Pease 2014), largely because the offered rental rates were not competitive relative to returns from cropping.

Public funding of conservation easements is another government policy. Easements are legal agreements between a land owner and another party to attenuate owner property rights. Typically the owner obtains monetary compensation and estate tax benefits while the easement has indefinite duration. The *Loess Hills Alliance* seeks to use easements to preserve designated Special Landscape Areas. Other conservation organizations active in the area include *The Nature Conservancy* and the *Iowa Natural Heritage Foundation*. The limited funds available for easement purchases in the ILHL come from private and State of Iowa sources as well as U.S. Department of Transportation's National Scenic Byways Program. In 2014, an initiative to seek National Reserve designation for parts of ILHL, and so open opportunities for additional easement funds from the federal government, failed to gain adequate support among Loess Hills Alliance Board members, where opposition emerged from local land owners. (Note 2)

The 1985 Farm Bill also saw the introduction of conservation compliance provisions whereby those who farm highly erodible lands may be ineligible for some forms of agricultural income support. Growers planting on highly erodible land commit to a conservation plan in order to become compliant. Between 1996 and 2014, eligibility for crop insurance premiums was not conditioned on conservation compliance but linkage was re-established under the 2014 Farm Bill. Recent trends in cropping systems, to be discussed later, have made compliance easier than was the case before the middle 1990s.

For decades preceding the 1996 Farm Bill the commodity-specific income support that growers received depended in large part on cropping choices (Novak, Pease & Sanders, 2015). Some crops, collectively labeled 'program' crops, received support in proportion to acres and yields. Corn was a program crop but soybeans and grass/hay were not. The de-linking of cropping choices and subsidies in the 1996 Farm Bill was motivated by the costs of inflexibility in marketplace response (as growers would lose non-market support upon adapting to market prices) and by International Trade Agreement commitments.

Crop insurance has had at least some federal support since the 1930s, but was not seen as an integral component of income support until the 1990s (Glauber, 2013). In an effort to promote program performance by expanding participation, commencing in 1994 a series of legislative enactments increased premium subsidies and expanded contract choices. Upon passage of the 2014 Farm Bill, crop insurance support had become a firmly established primary pillar of agricultural income support. Although pre-subsidy rates are required to be actuarially fair, as far as is practical, the U.S. Government Accountability Office (U.S. GAO, 2015) has discerned underpricing in production riskier counties where none of these are in Iowa. Nonetheless the growth of crop insurance subsidies, unavailable or less generous for grass-based activities, is likely to promote crop production (Claassen, Cooper & Carriazo, 2011; Feng, Hennessy & Miao, 2013; Miao, Hennessy & Feng, 2014).

Corn-based ethanol has been, indirectly or directly, promoted by the U.S. federal government since the 1970s. Direct support for ethanol production as a renewable fuel came through federal laws passed in 2005 and 2007, which mandated that minimum quantities of certain fuel types be blended with gasoline. As of 2015, more than 200 ethanol plants exist in the United States. Most use corn as feedstock and are located in the Midwest. April 2015 data in Ethanol Producer Magazine identify several plants around the ILHL, including in Council Bluffs (125 mill. gal. capacity), Shenandoah (65 mill. gal.) and Denison (55 mill. gal.) to the south as well as Jackson, Nebraska (50 mill. gal.) and Merrill, Iowa, (50 mill. gal.) to the north. (Note 3). See figure 2 for local cropping infrastructure.

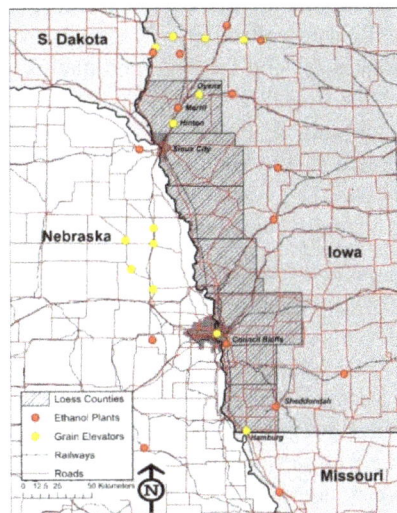

Figure 2. Crop production infrastructure in Iowa's Loess Hills landform study area.

Apart from weather-stressed years, Cornbelt commodity market prices traded in a historically narrow range between 1980 and 2006. A gradual decline in real prices occurred due to technology-driven growth in supplies when compared with slower growth in demands. The aforementioned 2005 and 2007 energy acts changed the output price environment in several ways. The mandates instantly generated higher, assured demand for corn. The corn price range increased from $2-3 in 2004-2007 to $3-7 in 2007-2014. Other commodities also saw price increases so that they remain a competitive use of land resources. A secondary effect was through beef markets. Farm-level beef prices rose in part because feedlot owners needed to cover higher corn input costs or go out of business. The U.S. national beef herd has declined over the 1996-2014 period in the face of higher feed input prices and also adverse weather conditions, where the decline was sharp after 2008. Even so, grassland rental prices also increased as they provide an alternative to corn-based cattle feed.

Marked technological change has occurred in crop production during recent decades. Perhaps most relevant to this study are expanded use of conservation tillage and the advent of genetically modified corn and soybean seeds. Conservation tillage can reduce production costs and can also preserve moisture as a risk management strategy against drought. The Loess Hills area, though far from arid, is among the driest in Iowa. Conservation tillage also protects against soil erosion and is viewed as an acceptable strategy for conservation compliance. However, tillage also provides weed control (Carpenter & Gianessi, 1999) so that growers had been reluctant to adopt conservation tillage. The advent of glyphosate tolerant seeds allowed for cost-effective weed control by use of a single chemical after planting. There is substantial evidence that glyphosate tolerant seed complements less intensive tillage (Perry, Moschini & Hennessy, 2016). In reducing the cost of conservation compliance, these seeds may facilitate corn and soybean production on erodible land. Growers in the seven county area (SCA) have been early and extensive adopters of both conservation tillage and glyphosate tolerant seed. (Note 4)

3. Materials and Methods

To address land use changes in Iowa's Loess Hills and related policy implications, we conducted a comprehensive data analysis at two levels of aggregation: SCA and ILHL. We used two data sources: National Resource Inventory (NRI) data and CDL data. NRI data allow for evaluation of historical land use changes at the county level of aggregation in the SCA (1982-2010), where CDL data are only available for recent years (2000-'14). Given that factors like agricultural policy and agri-infrastructure change incrementally, their impact on regional land use is better understood when longer time-series of land use transitions are used in analysis. For example, inavailability of CDL data prior to year 2000 leaves this data source incapable of informing on change in cropping incentives and regional land use due to the farm bills introduced in 1985 and 1996. On the other hand, and in contrast with NRI data, CDL data are spatially-delineated and so allow us to evaluate changes specific to ILHL.

3.1 National Resource Inventory Data

We utilized National Resource Inventory (NRI 2013) data to evaluate land use/land cover (LULC) trends for the SCA encompassing the ILHL. Focusing on point-level data, the NRI is a survey-based longitudinal database that provides comprehensive information on land characteristics as well as historical uses. In order to conform with

USDA confidentiality protocols, and unlike the CDL, NRI data suppress spatial geo-coordinates (NRI, 2013) although county location is provided. NRI data collection is based on a robust survey methodology that assures reliable and temporally-consistent estimates of land use and land quality parameters. Specifically, the included sample points are intended to represent the overall geographic spread and heterogeneity of natural resources at national and regional levels (Nusser, Breidt & Fuller, 1998). Each sample point is accompanied by its representative weight measured in 100 acre (0.405 km^2) units that differ across sample points. A total of 2,600 NRI points span the SCA's 12 985 km^2 area. Data range from 1982 - 2010, where each NRI point was observed every five years from 1982 to 1997 but annually commencing 2000.

The NRI dataset provides land quality parameters such as land capability classification (LCC) and erodibility index (EI). LCC groups soils into eight classes with regards to limitations for cropping, with higher class codes referring to more severe limitations. LCC classes I and II are most suitable for cropping whereas classes III and IV are more limited in use. Higher LCC classes have severe limitations and are considered to be unsuitable for cultivation (p. 65 in Helms 1992). EI measures soil's erosion potential whereby soils with higher index values are costlier for cropping as they entail pertinent management costs to limit erosion and preserve crop productivity (NRI, 2013). About 1% of NRI points had non-constant LCC and/or EI values during the 1982-2010 period. These points were excluded when assessing land use by LCC and/or EI status. Within this domain of constant LCC lands; 6.6% of points have missing LCC for all years, 40.2% were in classes I-II, 48.3% were in classes III-IV, and 4.9% were assigned higher classes.

Locations with EI values that were either missing or temporally varying accounted for 31.1% of NRI points, where a disproportionate fraction were in the urban category. Among the non-missing, constant EI points, the index ranges from 0.8 to 160.2 where 32.2% had EI $\geq$ 8. The USDA's Natural Resource Conservation Service (NRCS) defines soils with EI $\geq$ 8 as highly erodible. Since the 1996 Farm Bill, soils with EI $\geq$ 8 have been eligible for CRP regardless of other attributes.

3.2 Cropland Data Layers

Pre-processing: Spatially explicit raster CDL data were downloaded from the CropScape portal of the National Agricultural Statistics Service (United States Department of Agriculture National Agricultural Statistical Service (USDA NASS), 2012) and clipped by the SCA encompassing the ILHL (figure 1). These CDL data are produced annually for the 48 contiguous states (since 2000 for Iowa) using a combination of (1) multiple satellite imagery dates each year to capture crop phenology differences, (2) concurrent USDA Farm Service Agency's (FSA) training/validation data, and (3) augmented versions of the raster-based National Land Cover Data (NLCD) from 2001 and 2006 (Fry, Coan, Homer, Meyer & Wickham, 2009; Boryan, Yang, Meller & Craig, 2011). Landsat-based CDL data (30 m pixels) from 2001, 2005, 2010, and 2013 were selected to quantify LULC change trends for the ILHL. Advanced Wide Field Sensor-based (AWiFS) CDL data (56 m pixels) from years 2006-'09 were excluded since differences in spatial resolution are known to negatively affect area estimation precision in remote sensing-based studies (Lunetta, Knight, Ediriwickrema, Lyon & Worthy, 2006; Wright & Wimberly, 2013).

Downloaded CDL data could not be analyzed directly due to inherent difference in (1) classes from 2001 to 2013 (table 1); (2) missing classes (e.g., roads in 2001 and 2005), (3) class confusion between dominant crop types and non-crop types (figure 3), and (4) because later improvements in the CDL processing stream reduced spatial errors due to single-pixel salt-and-pepper effects (D.M. Johnson [NASS], personal communication) that remain in earlier CDL years (figure 3). Similar preprocessing was necessary for early versions of NLCD (1992 and 2001) so that class structure and pixel-wise effects were equivalent (Wolter, Johnston & Niemi, 2006). For instance, the 2001 and 2005 CDLs did not specifically include a 'developed' class that contained roads as did later years (figure 3). Also, spatial inconsistencies within the 'Fallow/Idle Cropland' class through time have been identified in prior studies (Kline, Singh & Dale, 2013). Laingen (2015) has recommended that all data be treated with circumspection, placing emphasis on the data generation processes.

NASS warns against the use of non-agricultural classes such as Grass/Pasture and Fallow/Idle as these classes have low classification accuracy (USDA NASS, 2013). (Note 5) Rather, NASS suggests substituting NLCD's non-agricultural classes. Hence, non-agricultural classes from the 2001 CDL were used to clip the recently improved 2001 NLCD layer. Resulting NLCD classes were recoded to fit the CDL class structure and then overlain back on the 2001 CDL. We also used the 2001 CDL agricultural classes to clip the 2001 NLCD under the suspicion that 2001 CDL agricultural classes were being confused with the non-agricultural classes, especially 'Forest' (figure 3). Again, resulting NLCD classes were recoded and overlain in the 2001 CDL. A similar approach, using 2006 NLCD, was used to reconcile such errors in the 2005 CDL (figure 4).

Table 1. CDL cover type classes pertinent to Iowa and the disparity between years. An 'X' denotes that the class existed and was adequately represented, while 'U' denotes that the class existed but was grossly under-represented. No value '-' indicates that the class did not exist or had zero total area.

CDL Cover Type	2001	2005	2010	2013
Corn	X	X	X	X
Soybeans	X	X	X	X
Barley	-	-	X	X
Oats	X	X	X	X
Rye	-	-	X	X
Flaxseed	-	-	X	-
Spring Wheat	-	-	X	-
Winter Wheat	-	X	X	X
Other Small Grains	X	X	-	-
Other Crops	X	X	U	-
Alfalfa	X	X	X	X
Other Hay/Non-alfalfa	-	-	X	X
Fallow/Idle Cropland	X	X	U	U
Forest	X	X	X	-
Deciduous Forest	-	-	X	X
Evergreen Forest	-	-	X	X
Mixed Forest	-	-	X	X
Developed	U	X	-	-
Dev/Open Space	-	-	X	X
Dev/Low intensity	-	-	X	X
Dev/Med Intensity	-	-	X	X
Dev/High Intensity	-	-	X	X
Grass Pasture	X	X	X	X
Non-Ag/Undefined	-	X	-	-
Shrubland	-	-	U	U
Barren	-	-	X	X
Wetlands	-	X	X	-
Herbaceous Wetland	-	-	X	X
Woody Wetlands	-	-	X	X
Water	X	X	X	X

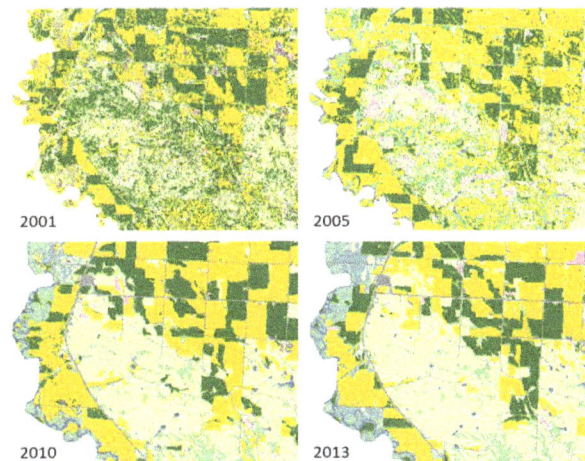

Figure 3. Speckling and misclassifications between old (top) and newer (bottom) CDL data processing for a 127 km^2 area in western Iowa. Soybeans (dark green) and Corn (orange) are confused with Grass/Pasture (light green) in 2001, while Corn, Forest (intermediate green), and Alfalfa (magenta) are confused with Grass/Pasture in 2005. Also note that roads (gray) are missing from 2001 and 2005.

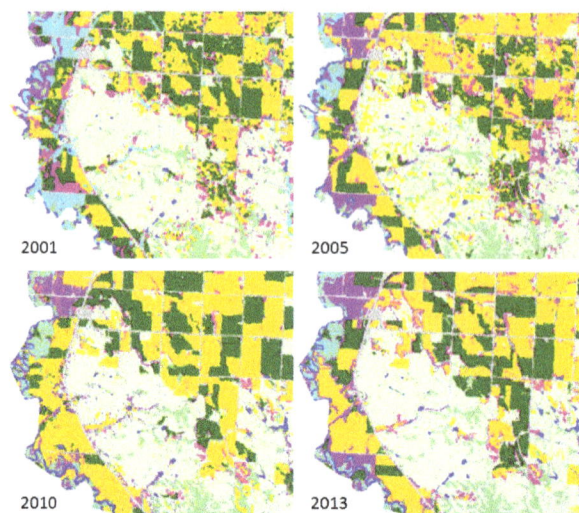

Figure 4. Results of class parity correction using 2001 and 2006 NLCD and subsequent, simultaneous, despeckling of the 2001, 2005, 2010, and 2013 CDLs. Despeckling was achieved via a 3 x 3 majority filter applied to the union of the four respective CDLs into one thematic image with a four-year, change vector, attribute assigned to each pixel. This change vector attribute was used afterward to recreate the individual CDLs in figure 3.

Once CDL classes were equivalent among the {2001, 2005, 2010, 2013} time steps, we then used a matrix union function to combine the information into one physical image. This matrix union function, when applied to two consecutive CDL years (e.g., Y_1-Y_2), provides a "from-to" attribute in the resulting thematic output attribute table with which one may track land cover changes between those years. Here, we simply applied this function three consecutive times (Y_1 -Y_2, Y_{1-2} -Y_3, and finally Y_{1-2-3} -Y_4) to achieve the four-year ($Y_{1-2-3-4}$) "from-to" vector of change for each pixel. A 3 x 3 pixel majority despeckling function was then applied to this resulting image to simultaneously weed out spurious and illogical single pixels of changes through time (e.g. water-corn-forest-soy). After the multi-temporal despeckling operation was complete, the four-year "from-to" change vector attribute was then used to guide recoding of the single, thematic image back into the four respective CDL years as separate, thematic, image layers.

To allow for further consideration on the sorts of land that have seen changing use, we have linked CDL data to land quality. Corn Suitability Ratings (CSR, Miller 2005) and slope data were obtained from the SSURGO database maintained by USDA NRCS. (Note 6)

Quantifying Land Use/Land Cover Change: Quantifying change among CDL crop types and other land uses for the four selected years was performed within the SCA and the ILHL. We considered four slope percent classes [0-5, 6-10, 11-15, ≥16] and four CSR classes [1-69, 70-79, 80-89, >90] across the SCA and within the ILHL. Given that CDL and NRI are two distinct data sources with different data generating processes, we did not expect the comparable land use numbers to match exactly. We did expect to see reasonably similar trends for the SCA among the common land use categories and this is largely true. Surprisingly though, while in table 2 the NRI data show an increase in corn acreage and decrease in hay and pasture over the 2001-'05 interval, the CDL data in table 3 provide a reverse trend.

The pre-processing exercise and land use analysis that uses CDL were performed using the ERDAS Imagine software, and the analysis that used NRI data was conducted with SQL in SAS software and the Pivot Tables application in Microsoft Excel.

4. Results and Discussion

We present land use trends and land use transition matrices to characterize conditional (on land quality) and unconditional land use changes for SCA and ILHL. We report comparisons among the two independent datasets, as well as among the two aggregation levels within the unconditional land use trends. In addition, linkages have been established, when applicable, between the observed land use change within the landform and agricultural policy. A comparison with state and national level land use change is also considered. Last, but not least, to facilitate further scrutiny we empirically specify the structure of rotations for corn over the 2000-'10 period to assess whether cropping in this region is moving toward monoculture. We present our findings below.

4.1 Unconditional Land Use Trends

To utilize the full extent of NRI data while emphasizing more recent available data, table 2 provides summary data on cultivated area under crops, hayland and pastureland for 1982, 1992, 2001, 2005 and 2010 (NRI, 2013). It is evident that area under crop cultivation, at about 70% of total area in the SCA, dominates 'other' land uses. Corn and soybeans account for almost all land under tilled crops. Area under corn fell by 734 km^2 from 1982 to 2001, before returning to 1982 levels in 2010, trends that are broadly consistent with the nation as a whole. Soybean area has largely counterbalanced corn area where total acres under either crop increased by 5% over 1982-2010 and by 3% over 2001-2010. Soybean acres did decline in area between 2005 and 2010 whereas they increased by 7.5% nationally over the same period. This contrast may have been due to limited opportunities to substitute other crops out in favor of corn in the SCA. 'Other crops' declined over the interval but mainly before the 1996 Farm Bill. Area allocated to pasture saw a declining trend over the three decades, losing 551 km^2 between 1982 and 2010, whereas the decline in area under hay was more modest.

Table 2. Land use trends for Seven County Area (in km^2).

Land Use Category	1982	1992	2001	2005	2010
Corn	5,372	5,077	4,638	4,858	5,655
Soybeans	3,527	3,475	4,472	4,449	3,727
Other Crops	402	109	102	26	62
Hay	392	280	413	358	358
Pasture	1,544	1,369	1,140	1,022	993
Forest	563	591	640	657	645
CRP	0	746	331	316	130
Urban	297	309	353	379	405

Source: National Resource Inventory data.

NRI records acreage under the CRP general sign-up scheme, introduced in the 1985 Food Security Act. (Note 7) Acreage under such CRP sign-ups had fallen substantially (83%) by 2010, when compared to 1992. Table 2 also shows a 60% decline in the CRP general-sign up acreage in the SCA between 2001 and 2010. Note that acres reported by the USDA Farm Service Agency (USDA FSA) under continuous sign-up in Iowa almost doubled in this period and their contribution towards total CRP acres also grew substantially from 16.7% in 2001 to 42.5% in 2013. However, as much of the ILHL landform is characterized as highly erodible, we expect continuous sign-ups to be a small fraction of all sign-ups for this region. In addition, at 1.1% growth per year, land under urban uses grew at a definite but modest rate during 1982-2010. (Note 8)

Turning to CDL data, corn acres have increased by about 75 km^2 in the ILHL between 2001 and 2013, see bottom right panel in table 3. Soybean acres have declined over the period by about 100 km^2 so that, perhaps surprisingly, there has been no net change in total acres devoted to row crops over the period. There has been expansion of the forest (+67 km^2) and fallow/idle cropland (+70 km^2) categories. The ILHL is topographically variable, especially in regard to slope and land quality. Within the region there have been some subtle changes. Crop production has expanded toward the north, especially in Woodbury and Monona counties where cropping is extensive. Crop production has declined in the more southerly counties where the landform is thin and cropping is limited. Crop production has generally moved from soybeans and toward corn overall along the hills but corn has not expanded in Plymouth County to the north or Pottawattamie, Mills and Fremont counties along the landform's narrowing tail. Forests have expanded uniformly over the area.

Land use in marginal growing areas should, by definition, be more sensitive to price movements than should other areas. Over 2005-'10 Iowa's planted corn acres increased by 4.7% while planted soybean acres decreased by 2.5%. Over 2010-'13 Iowa's planted corn acres increased by 1.5% while planted soybean acres decreased by 4.1%. Table 3 shows that over 2005-'10 the ILHL's planted corn acres increased by 62% while planted soybean acres declined by 2%. Over 2010-'13 the ILHL's planted corn acres decreased by 8% while planted soybean acres decreased by 20%. The area does appear to have had more variable responses to external conditions than the state as a whole, likely due to the challenging production environment that it poses for growers.

Because CDL classification protocols for developed acres have evolved substantially through these years, CDL data do not directly allow for an assessment of change to this category. In a separate query that appropriately adjusted for the redefinitions, we found a 2.6% increase in ILHL development acres (from 103 km^2 acres to 106 km^2) over the 2001-'13 period. We had expected a larger increase.

Table 3. County level and aggregate changes on Loess Hills landform (in km^2).

	Plymouth				Woodbury			
	2001	2005	2010	2013	2001	2005	2010	2013
Corn	42.5	28.3	53.2	41.8	235.0	191.7	287.0	278.2
Soybeans	27.7	33.4	35.4	27.5	196.4	211.3	219.7	182.0
Fallow/Idle Cropland	26.6	17.2	8.0	14.8	77.7	77.8	32.1	53.4
Hay/Pasture/Grass	115.4	127.3	125.7	137.0	158.6	184.9	132.6	156.9
Forest	41.7	48.2	40.1	42.3	47.1	55.1	53.9	52.3
Developed (all years)	19.4	19.4	19.4	19.4	135.4	135.4	135.4	135.4
	Monona				**Harrison**			
Corn	102.4	88.4	150.4	138.9	121.0	88.1	144.3	130.1
Soybeans	95.6	92.2	88.7	71.5	104.4	104.9	82.4	64.2
Fallow/Idle Cropland	24.9	28.8	18.7	31.5	25.2	40.3	26.5	44.7
Hay/Pasture/Grass	129.9	134.7	86.2	99.1	69.6	80.1	62.4	73.5
Forest	120.5	129.2	132.0	133.0	97.3	104.4	102.6	106.9
Developed (all years)	38.3	38.3	38.3	38.3	40.7	40.7	40.7	40.7
	Pottawattamie				**Mills**			
Corn	74.5	50.7	79.8	74.0	25.0	11.7	20.5	18.8
Soybeans	52.5	50.6	52.8	37.4	17.6	17.1	20.5	14.9
Fallow/Idle Cropland	14.7	31.6	17.6	27.3	9.3	17.9	9.3	12.5
Hay/Pasture/Grass	72.0	77.0	57.9	67.0	46.7	49.4	41.7	46.9
Forest	62.2	66.6	67.3	69.4	45.9	49.0	51.0	51.2
Developed (all years)	68.3	68.3	68.3	68.3	19.7	19.7	19.7	19.7
	Fremont				**Loess Hills Polygon**			
Corn	11.4	3.7	11.6	5.5	611.7	462.4	746.8	687.4
Soybeans	9.3	8.8	7.4	8.5	503.4	518.2	506.9	406.1
Fallow/Idle Cropland	2.4	11.2	5.7	7.7	180.7	224.8	118.0	192.0
Hay/Pasture/Grass	28.8	30.8	26.6	29.0	621.1	684.1	533.1	609.4
Forest	51.9	52.8	55.4	55.0	476.6	515.7	507.3	517.1
Developed (all years)	9.6	9.6	9.6	9.6	331.2	331.2	331.2	331.2

Source: Cropland Data Layer data.

Note: Acres in the 'Developed' category were fixed at 2001 levels.

4.2 Conditional Land Use Trends

<u>Seven-County-Area</u>: Table 4 presents land use trends conditional on being in LCC classes I-II and on being in classes III-IV. Classes I-IV contain almost all of the region's corn and soybean acres. Furthermore, 75% or more of hay and pasture acres also lie on LCC I-IV lands. Although, we find relatively higher acreage-shares of hay, pasture, forests and CRP lands on LCC III-IV lands. Corn acres have seen a slight migration to better land, perhaps because of incentives that the CRP provides to more limited land, while hay and pasture acres have shifted away from LCC classes I-II.

Table 4. Land use trends for Seven County Area (in km^2), conditional on Land Capability Class. Percent of all land in that use is in parentheses.

LCC I, II Land Use Categories					
	1982	1992	2001	2005	2010
Corn	2,500 (46.5)	2,472 (48.7)	2,253 (48.6)	2,447 (50.4)	2,750 (48.6)
Soybeans	1,712 (48.5)	1,841 (53.0)	2,207 (49.4)	2,172 (48.8)	1,812 (48.6)
Other Crops	120 (29.9)	30 (27.5)	26 (25.5)	9 (34.6)	19 (30.6)
Hay	117 (29.8)	56 (20.0)	88 (21.3)	33 (9.2)	53 (14.8)
Pasture	443 (28.7)	391 (28.6)	324 (28.4)	251 (24.6)	245 (24.7)
Forest	105 (18.7)	93 (15.7)	103 (16.1)	102 (15.5)	105 (16.3)
CRP	0 (-)	81 (10.9)	13 (3.9)	13 (4.1)	1 (0.8)
LCC III, IV Land Use Categories					
	1982	1992	2001	2005	2010
Corn	2,790 (52)	2,558 (50.4)	2,330 (50.2)	2,355 (48.5)	2,866 (50.7)
Soybeans	1,747 (49.5)	1,563 (45.0)	2,245 (50.2)	2,258 (50.8)	1,895 (50.8)
Other Crops	263 (65.4)	79 (72.5)	77 (75.5)	17 (65.4)	43 (69.4)
Hay	268 (68.4)	213 (76.1)	304 (73.6)	311 (86.9)	294 (82.1)
Pasture	777 (50.3)	693 (50.6)	548 (48.1)	516 (50.5)	499 (50.3)
Forest	196 (34.8)	200 (33.8)	227 (35.5)	247 (37.6)	234 (36.3)
CRP	0 (-)	629 (84.3)	302 (91.2)	294 (93.0)	119 (91.5)

Source: National Resource Inventory data.

Table 5. Land use trends for Seven County Area (in km^2), conditional on Erodibility Index.

EI < 8					
Land Use Category	1982	1992	2001	2005	2010
Corn	2,799	2,732	2,431	2,568	3,211
Soybeans	2,178	2,251	2,660	2,613	1,857
Other Crops	134	35	26	9	19
Hay	114	78	86	40	56
Pasture	10	6	17	17	0
CRP	0	51	4	4	0
EI ≥ 8					
Land Use Category	1982	1992	2001	2005	2010
Corn	2,151	1,761	1,615	1,679	1,789
Soybeans	944	962	1,319	1,351	1,397
Other Crops	227	73	77	17	43
Hay	224	120	243	212	215
Pasture	0	11	5	5	0
CRP	0	593	294	288	108

Source: National Resource Inventory data.

Note: Urban land and forest land are generally not assigned EI values.

Table 5 provides trends for the EI <8 and EI ≥ 8 categories. Although almost 40% of land in the SCA had to be excluded from evaluating the trend statistics, due to missing and transitioning EI values, 88-91% of cropped acres lie within the land parcels that had constant, non-missing EI values for all years. About 40% of the region's cropped acres with reported, constant EI values were on land with EI ≥ 8, along with only 13-23% of total hay and pasture lands over the period of our analysis. Acres under hay, pasture and CRP categories generally had higher erodibility index values. It is noteworthy that corn acres on EI ≥ 8 land was lower in 2010 when compared with its 1982 counterpart. This may be, in part, due to the advent of CRP, and partly due to conservation compliance constraints.

ILHL: Table 6 shows that corn presence and recent corn expansion have been concentrated on less steeply sloped river valley tracts that cut through the landform, see figure 5. Total acres to corn and soybeans on slopes >10% has changed minimally over the period. A moderate decline in the hay/pasture/grass category over 2001-'13 occurred mainly on shallow slopes while expansion in forest occurred throughout. Expansion in the fallow/idle cropland category occurred mainly on lower (≤ 5%) and higher slopes (> 15%) where the highest proportional expansion of this category has been on higher slopes, a notable observation given that overall CRP acres on the landform likely declined over the 2001-'13 period.

Table 7 reports land use changes by four corn suitability rating (CSR) categories: ≤ 69, or least suitable lands, 70-79, 80-89 and ≥90, or most suitable lands, see Miller (2005). Corn expanded and soybean contracted in each category over the 2001-'13 period. Total land in either corn or soybeans increased slightly on lower quality land but decreased slightly on better quality land, a perplexing finding that also applies for the SCA (table not shown). Land area under forest has expanded or trended sideways for all land quality categories while the data indicate that area in the hay/pasture/grass category has declined slightly for better land categories. Fallow/idle cropland is found to have expanded on higher quality land and to have contracted on lower quality land, perhaps due to high error in the category (as previously mentioned) or forest encroachment.

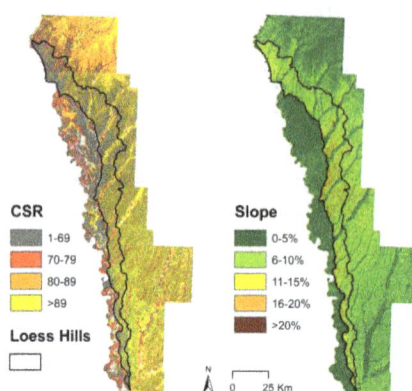

Figure 5. Landscape stratification classes for the seven-county study area. Corn suitability rating (CSR) on the left side and percent slope of terrain classes on the right side.

Table 6. Change in crop and other land use/cover areas on the ILHL, by four slope categories (km^2).

	0-5 % Slope				6-10 % Slope			
	2001	2005	2010	2013	2001	2005	2010	2013
Corn	257	216	317	298	206	152	252	236
Soybeans	233	235	219	181	167	176	177	142
Other Crops	1	2	0	0	1	3	0	0
Alfalfa	6	4	6	5	9	7	8	6
Fallow/Idle Cropland	49	66	34	62	68	85	42	65
Hay/Pasture/Grass	136	151	107	129	215	238	184	210
Forest	71	81	78	80	116	128	126	129
Developed (all years)	124	124	124	124	110	110	110	110
	11-15 % Slope				**>15 % Slope**			
	2001	2005	2010	2013	2001	2005	2010	2013
Corn	92	60	113	101	43	22	49	36
Soybeans	67	70	74	57	24	24	25	17
Other Crops	0	2	0	0	0	1	0	0
Alfalfa	6	5	4	3	3	2	2	1
Fallow/Idle Cropland	42	48	25	37	20	24	16	25
Hay/Pasture/Grass	160	176	144	161	106	116	96	106
Forest	117	126	127	128	159	168	169	171
Developed (all years)	55	55	55	55	36	36	36	36

Source: Cropland Data Layer data.

Note: Acres in the 'Developed' category were fixed at 2001 levels.

Table 7. Change in crop and other land use/cover areas within the ILHL, by four CSR categories (km^2).

	1-69 CSR2				70-79 CSR2			
	2001	2005	2010	2013	2001	2005	2010	2013
Corn	309	211	371	342	38	36	52	47
Soybeans	227	243	256	201	39	37	30	26
Other Crops	14	6	1	1	0	0	0	0
Alfalfa	17	14	12	9	1	0	1	1
Fallow/Idle Cropland	127	146	73	108	5	5	4	8
Hay/Pasture/Grass	455	504	410	456	13	14	11	14
Forest	390	415	417	422	6	8	6	7
Developed (all years)	194	194	194	194	14	14	14	14
	80-89 CSR2				**90-100 CSR2**			
	2001	2005	2010	2013	2001	2005	2010	2013
Corn	158	128	197	178	99	79	117	111
Soybeans	139	141	131	106	91	91	83	68
Other Crops	2	2	0	0	10	0	0	0
Alfalfa	4	3	5	4	2	1	2	2
Fallow/Idle Cropland	34	44	26	49	14	28	15	25
Hay/Pasture/Grass	104	114	79	97	44	47	32	39
Forest	45	54	50	51	22	26	25	26
Developed (all years)	64	64	64	64	53	53	53	53

Source: Cropland Data Layer data.

Note: Acres in the 'Developed' category were fixed at 2001 levels.

4.3 Temporal Transition Matrices

We present temporal transition matrices for the SCA using NRI data as well as CDL land cover data. Pivot tables such as table 8 provide a matrix of land use transitions over time. They allow us to identify interesting transitions among the land use categories under study, which may have important policy implications for this region. We analyze pivot tables using NRI data for 2001-'05, 2005-'10 and using CDL data for 2001-'05, 2005-'10, and 2010-'13 periods. Whereas CDL data allows us to capture the most recent transitions, NRI data helps in quantifying conversions into urbanization.

Table 8 provides further perspective on cropping pressure in the ILHL. The table is a transition matrix that categorizes CDL data to be broadly consistent with NRI categories. (Note 9) Net movement from corn and soybeans into hay/pasture/grass was large during 2001-'05, a period of stable corn planted acres in the United States, at about 324 000 km^2, and of comparatively low commodity prices. Strong net movement in the other direction occurred between 2005 and 2010, consistent with the national movement toward more cropland to meet growing demand for commodities. The shift back to hay/pasture/grass and fallow/idle cropland over 2010-'13 as well as contraction in both corn and soybean acres is not consistent with national trends. In 2013, corn and soybean area planted in the United States were, respectively, 384 000 km^2 and 308 000 km^2. In 2010 the corresponding numbers were 348 000 km^2 and 312 000 km^2. Many of the additional corn acres have come from outside the traditional Cornbelt, especially from Great Plains states such as North Dakota and Kansas.

Table 8 shows that much of the corn acreage that moved out of cropping between 2001 and 2005 in the ILHL likely went into grass and fallow cropland, a pattern that was reversed in the subsequent five years so that corn acres were 138 km^2 larger in 2010 than in 2001. High corn acreage was sustained in the 2010-'13 period through declining soybean acres. While shifts occurred into grass and fallow cropland categories, net grass acres have declined due to outward transitions into the forests category (not included here) where invasive eastern red cedar is a problem in the area.

Table 9 reveals that corn and soybean areas are sources of urbanized land in the SCA with 10 km^2 converted during 2001-'05 and 15 km^2 converted during 2005-'10. CRP in 2001-'05 (6 km^2) and pasture in 2005-'10 periods (7 km^2) are other sources of conversion. Also, CRP lands transitioned into corn and soybean production (116 km^2) at a very high rate during 2005-'10, relative to the 2001-'05 period (6 km^2 CRP lands into corn and soybean). Hay and pasture have also lost considerable areas to corn and soybean production, i.e., 176 km^2 in 2001-'05 and 133 km^2 in 2005-'10.

Table 8. Pivot table for ILHL using CDL data but NRI specific land use categories (in km2), 2001-'05, 2005-'10 & 2010-'13.

		2005				
		Corn	Soy	Hay/Pasture/Grass	Fallow/Idle Crop	*Grand Total*
2001	Corn	335	129	53	70	588
	Soy	106	349	1	42	498
	Hay/Pasture/Grass	1	16	559	18	594
	Fallow/Idle Crop	16	17	50	83	167
	Grand Total	457	512	663	213	1,846
		2010				
		Corn	Soy	Hay/Pasture/Grass	Fallow/Idle Crop	*Grand Total*
2005	Corn	168	262	4	18	452
	Soy	362	113	12	20	506
	Hay/Pasture/Grass	113	69	415	17	613
	Fallow/Idle Crop	68	48	49	47	213
	Grand Total	710	492	480	102	1,784
		2013				
		Corn	So	Hay/Pasture/Grass	Fallow/Idle Crop	*Grand Total*
2010	Corn	255	341	68	61	726
	Soy	397	50	22	31	501
	Hay/Pasture/Grass	15	5	494	0	514
	Fallow/Idle Crop	12	7	1	89	110
	Grand Total	680	404	585	181	1,850

Source: Cropland Data Layer data.

Table 9. Pivot table for Seven County Area using NRI specific land use categories, 2001-'05 & 2005-'10.

| | | Specific Land Use 2005 (km^2) | | | | | | |
		Corn	Soy	Hay	Pasture	CRP	Urban	Grand Total
Specific Land Use 2001 (km^2)	Corn	3,904	680	20	0	0	7	4,611
	Soybeans	741	3,677	15	0	0	3	4,435
	Hay	74	15	324	0	0	0	413
	Pasture	57	30	0	1,020	0	5	1,113
	CRP	6	0	0	2	316	6	331
	Urban	0	0	0	0	0	353	353
	Grand Total	4,782	4,402	358	1,022	316	361	11,272
		Specific Land Use 2010 (km^2)						
		Corn	Soy	Hay	Pasture	CRP	Urban	Grand Total
Specific Land Use 2005 (km^2)	Corn	1,448	3,263	42	0	0	14	4,766
	Soybeans	3,974	398	7	0	0	1	4,379
	Hay	53	58	243	0	0	0	354
	Pasture	22	0	0	993	0	7	1,022
	CRP	107	9	66	0	130	0	312
	Urban	0	0	0	0	0	379	379
	Grand Total	5,605	3,727	358	993	130	383	11,230

Source: National Resource Inventory data.

4.4 Structure of Rotations

As all indicators hold that corn acreage has expanded in the region dominated by corn-soybean rotations while cropped acreage has seen very limited change, it is certain that more corn intensive rotations are being used. Table 10 provides confirmation using CDL data for the ILHL. The table shows that the percent of all corn land in a given year that returns to corn the next year has trended upward over the years. The pattern is more obvious when three and four year corn sequences are viewed. Figure 6 depicts CDL data that provide evidence of more intensive corn rotations toward the landforms thick north end. Our finding corroborates Plourde, Pijanowski & Pekin (2013), who used CDL data across much of the Greater Mississippi watershed to discern an intensification of corn in rotations during 2003-2010.

Table 10. Cropping sequence as evidence on narrowing cropping patterns on the ILHL, 2000-2010. The sequences are characterized as acreages, measured in km^2.

Year	2000	2001	2002	2003	2004	2005	2006	2007	2008	2009
C_t	668	773	545	614	522	529	628	582	571	642
C_tC_{t+1}	172	115	111	109	113	191	259	184	180	197
% CC sequence	26	15	20	18	22	36	41	32	31	31
$C_tC_{t+1}C_{t+2}$	46	50	37	45	59	135	107	108	108	-
% CCC sequence	7	6	7	7	11	26	17	19	19	-
$C_tC_{t+1}C_{t+2}C_{t+3}$	19	14	16	26	39	49	66	72	-	-
% CCCC sequence	3	2	3	4	7	9	11	12	-	-

Source: Cropland Data Layers.

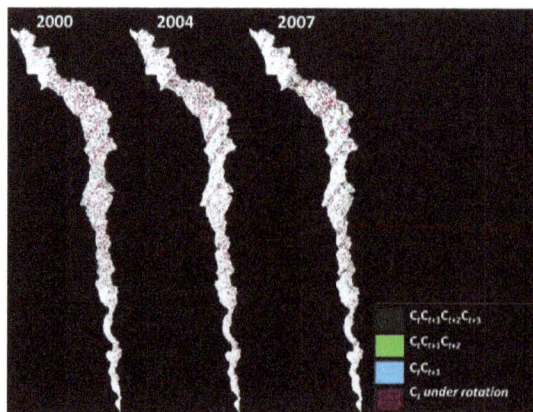

Figure 6. Evidence of more intensive corn rotations in the ILHL.

Notes: For 2000, 'C_{2000}' gives the are in corn that year, '$C_{2000}C_{2001}$' gives the area in corn both that and the following year, while '% CC sequence' gives ($C_{2000}C_{2001}/C_{2000}$)×100. To compute the above statistics we apply multi-temporal despeckling and re-coding operations, as discussed under 'Materials and Methods', on CDL years 2000-'10.

5. Concluding Remarks

Several factors have led to growth in tilled acres across the United States since 2006. In the Loess Hills we conclude that corn production has increased where much of the expansion has been through displacement of soybeans. The evidence does point to grassland loss in the area where we remind the reader that pixel-level misclassification rates is very high for CDL data on grass and fallow/idle categories. Tables 2 and 3 show that the hay/pasture/grass category declined by about 12% and 17%, respectively, over 2005-2013 while the respective figures for the fallow/idle category are declines of about 15% and 21%. Even if evidence on grassland loss is discarded, there are adverse implications for environmental services as corn production in rotation is believed to improve soil quality (Karlen et al., 2006) as well as reduce demand for chemicals that improve fertility (Stanger & Lauer, 2008) and manage pests (Gassmann et al. 2014).

Some parts of the ecoregion have seen cropland expansion, most notably southeast of Sioux City, while crop production has declined toward the less heavily cropped south. Both corn and forest acres have expanded everywhere in the hills over 2005-'13 but it should be noted that separating forested land and grass cover is problematic, especially in the presence of invasive shrubs. There is little evidence that cropping has moved to better quality land although there is some evidence that it has moved away from steeper slopes. The limited evidence available does not point to urban development as a major factor in the area but our view is that the matter warrants further inquiry. Land identified as fallow/idle has declined, likely due to a net decline in CRP acres.

Given the various forces that have aligned in recent years to incentivize row crop production and given national trends, it is not clear why row cropping has not expanded by more along the hills. Perhaps, for some reason, trends toward mechanization have favored less hilly land. Perhaps too, notwithstanding the growth in reduced tillage methods throughout the United States, conservation compliance regulations and targeted CRP sign-ups have proven to be more effective in protecting grass and wooded land in the area than elsewhere? Whether the Loess Hills are distinctive or our finding reflects a more general pattern of comparative constraint in row crop activity on hill terrain in recent years is a matter that warrants further inquiry and will be an important issue for policy design.

References

Arora, G., Wolter, P. T., Feng, H., & Hennessy, D. A. (2015). Characterizing and comprehending land use change in the Loess Hills Region. *Agricultural Policy Review Winter Issue*, available at http://www.card.iastate.edu/ag_policy_review/display.aspx?id=33.

Boryan, C., Yang, Z., Mueller, R., & Craig, M. (2011). Monitoring US Agriculture. The US Department of Agriculture, National Agricultural Statistics Service. Cropland Data Layer Program. *Geocarto International, 26*(5), 341-358. http://dx.doi.org/10.1080/10106049.2011.562309

Brown, P. W., & Schulte, L. A. (2011). Agricultural landscape change (1937-2002) in three townships in Iowa, USA. *Landscape and Urban Planning, 100*(3), 202-212. http://dx.doi.org/10.1016/j.landurbplan.2010.12.007.

Carpenter, J., & Gianessi, L. (1999). Herbicide tolerant soybeans: Why growers are adopting Roundup Ready varieties. *AgBioForum, 2*(2), 65-72. Retrieved from http://www.agbioforum.org/v2n2/v2n2a02-carpenter.htm.

Claassen, R., Cooper, J. C., & Carriazo, F. (2011). Crop insurance, disaster payments, and land use change: The effect of sodsaver on incentives for grassland conversion. *Journal of Agricultural and Applied Economics, 43*(2), 195-211. http://dx.doi.org/10.1017/s1074070800004168

Farnsworth, D. A., Schulte, L. A., & Hickey, S. (2010). Evaluation of current and alternative spatial patterns in the Loess Hills. *Proceedings of the 22ⁿᵈ North American Prairie Conference*. D. Williams, B. Butler, D. Smith, Eds. (pp. 9-17). University of Northern Iowa, Cedar Falls, Iowa.

Feng, H., Hennessy, D. A., & Miao, R. (2013). The effects of government payments on cropland acreage, conservation reserve program enrollment, and grassland conversion in the Dakotas. *American Journal of Agricultural Economics, 95*(2), 412-418. http://dx.doi.org/10.1093/ajae/aas112

Fry, J. A., Coan, M. J., Homer, C. G., Meyer, D. K., & Wickham, J. D. (2009). Completion of the National Land Cover Database (NLCD) 1992-2001 Land cover change retrofit product. U.S. Geological Survey Open-File Report 2008-1379, 18 p. Retrieved from http://www.mrlc.gov/nlcdrlc.php.

Gassmann, A. J., Petzold-Maxwell, J. L., Clifton, E. H., Dunbar, M. W., Hoffmann, A. M., Ingber, D. A., &

Keweshan, R. S. (2014). Field-evolved resistance by western corn rootworm to multiple *Bacillus thuringiensis* toxins in transgenic maize. *Proceedings of the National Academy of Sciences, 111*(14), 5141-5146. http://dx.doi.org/10.1073/pnas.1317179111

Glauber, J. W. (2013). The growth of the Federal Crop Insurance Program, 1990-2011. *American Journal of Agricultural Economics, 95*(2), 482-488. http://dx.doi.org/10.1093/ajae/aas091.

Helms, D. (1992). The development of the Land Capability Classification. In (D. Helms, Ed.) *Readings in the History of the Soil Conservation Service*, Washington, DC: Soil Conservation Service. Retrieved from http://www.nrcs.usda.gov/Internet/FSE_DOCUMENTS/stelprdb1043484.pdf.

Karlen, D. L. Hurley, E. G., Andrews, S. S., Cambardella, C. A., Meek, D. W., Duffy, M. D., & Mallarino, A. P. (2006). Crop rotation effects on soil quality at three Northern corn/soybean belt locations. *Agronomy Journal, 98*(3), 484-495. http://dx.doi.org/10.2134/agronj2005.0098

Kline, K. L., Singh, N., & Dale, V. H. (2013). Cultivated hay and fallow/idle cropland confound analysis of grassland conversion in the Western Corn Belt. *Proceedings of the National Academy of Sciences USA, 110*(31), E2863-E2863. http://dx.doi.org/10.1073/pnas.1306646110.

Laingen, C. (2015). Measuring cropland change: A cautionary tale. *Papers in Applied Geography, 1*(1), 65-72. http://dx.doi.org/10.1080/23754931.2015.1009305

Loess Hills Alliance. (2011). The Loess Hills of Western Iowa: Common Vision and Comprehensive Plan 2011. http://www.loesshillsalliance.com/pdfs/loesshillsalliance-comprehensiveplan-lq.pdf, last accessed 8/7/2015.

Lubbins, B., & Pease, J. (2014). Conservation and the Agricultural Act of 2014. *Choices, 29*(2), 142-154, available online at http://www.choicesmagazine.org/magazine/pdf/cmsarticle_374.pdf, visited 8/7/2015.

Lunetta, R. S., Knight, J. F., Ediriwickrema, J., Lyon, J. G., & Worthy, L. D. (2006). Land-cover change detection using multi-temporal MODIS NDVI data. *Remote Sensing of Environment, 105*(2), 142-154. http://dx.doi.org/10.1016/j.rse.2006.06.018.

Miao, R., Hennessy, D. A., & Feng, H. (2014). Sodbusting, crop insurance, and sunk conversion costs. *Land Economics, 90*(4), 601-622. http://dx.doi.org/10.3368/le.90.4.601.

Miller, G. (2005). Corn suitability ratings. PM 1168. Ames, IA: Iowa State University Extension (revised October 2012). Retrieved from file:///C:/Users/Gaurav/Downloads/PM1168%20(2).pdf.

Miller, J. R. (2006). Restoration, reconciliation, and reconnecting with nature nearby. *Biological Conservation, 127*(3), 356-361. http://dx.doi.org/10.1016/j.biocon.2005.07.021.

Novak, J. L., Pease, J., & Sanders, L. (2015). Agricultural Policy in the United States: Evolution and Economics. Routledge Textbooks in Environmental and Agricultural Economics, New York. http://dx.doi.org/10.1007/s41130-016-0013-6

NRI. (2013). Summary Report: 2010 National Resources Inventory, Natural Resources Conservation Service, Washington, DC, and Center for Survey Statistics and Methodology, Iowa State University, Ames, Iowa. Retrieved from http://www.nrcs.usda.gov/Internet/FSE_DOCUMENTS/stelprdb1167354.pdf, last accessed 8/7/2015.

Nusser, S. M., Breidt, F. J., & Fuller, W. A. (1998). Design and estimation for investigating the dynamics of natural resources. *Ecological Applications, 8*(2), 234-245. http://dx.doi.org/10.2307/2641063.

Perry, E., Moschini, G., & Hennessy, D. A. (2016). Testing for complementarity: Glyphosate tolerant soybeans and conservation tillage. *American Journal of Agricultural Economics, 98*(3), 765-784. http://dx.doi.org/10.1093/ajae/aaw001

Plourde, J. D., Pijanowski, B. C., & Pekin, B. K. (2013). Evidence for increased monoculture cropping in the Central United States. *Agriculture, Ecosystems and Environment, 165*(January), 50-59. http://dx.doi.org/10.1016/j.agee.2012.11.011

Secchi, S., Tyndall, J., Schulte, L. A., & Asbjornsen, H. (2008). High crop prices and conservation – Raising the stakes. *Journal of Soil and Water Conservation, 63*(3), 68A-73A. http://dx.doi.org/10.2489/jswc.63.3.68a.

Secchi, S., Kurkalova, L., Gassman, P. W., & Hart, C. (2010). Land use change in a biofuels hotspot: The case of Iowa, USA. *Biomass & Bioenergy, 35*(6), 2391-2400. http://dx.doi.org/10.1016/j.biombioe.2010.08.047.

Stanger, T. F., & Lauer, J. G. (2008). Corn grain yield response to crop rotation and nitrogen over 35 years. *Agronomy Journal, 100*(3), 643-650. http://dx.doi.org/10.2134/agronj2007.0280

Tomer, M. D., Moorman, T. B., Kovar, J. L., James, D. E., & Burkart, M. R. (2007). Spatial patterns of sediment and phosphorus in a riparian buffer in Western Iowa. *Journal of Soil and Water Conservation, 62*(5), 329-338.

USDA National Agriculture Statistics Service. (2012). CropScape-cropland data layer. U.S. Department of Agriculture, National Agricultural Statistics Service, Washington. Retrieved from http://nassgeodata.gmu.edu/CropScape/

USDA, National Agriculture Statistics Service. (2013). Cropland data layer metadata. Retrieved from www.nass.usda.gov/research/Cropland/metadata/meta.htm. Last accessed 5/31/2015.

U.S. GAO. (2015). Crop Insurance: In Areas with Higher Crop Production Risks, Costs are Greater, and Premiums may not Cover Expected Losses. GAO-15-215 U.S. Government Accountability Office, Washington, D.C., February. Retrieved from http://www.gao.gov/products/GAO-15-215

Wolter, P. T., Johnston, C. A., & Niemi, G. J. (2006). Land use land cover change in the U.S. Great Lakes Basin 1992 to 2001. *International Journal for Great Lakes Research, 32,* 607-628. http://dx.doi.org/10.3394/0380-1330(2006)32[607:lulcci]2.0.co;2

Wright, C. K., & Wimberly, M. C. (2013). Recent land use change in the Western Corn Belt threatens grasslands and wetlands. *Proceedings of the National Academy of Sciences USA, 110*(10), 4134-4139. http://dx.doi.org/10.1073/pnas.1215404110

Notes

Note 1. Source: http://iowa.sierraclub.org/LoessHills/LoessHillsHome.htm.

Note 2. See 'National Park Reserve designation causes controversy in western Iowa's Loess Hills' by Brianna Clark, posted at KTIV News Channel, February 18, 2014, http://www.ktiv.com/story/24755375/2014/02/18/controversy-surrounding-the-national-park-reserve-designation -of-the-loess-hills, last visited 8/7/2015.

Note 3. See www.ethanolproducer.com/plants/listplants/US/Existing/Sugar-Starch/.

Note 4. Over the 1998-2001 period 47% of soybean growers in the SCA adopted both conservation tillage and glyphosate tolerant soybean seed, when compared with 37.4% nationwide. For the 2007-2011 period the comparable figures were 72.5% and 65.3%. See table 1 and supporting text in Perry et al. (2015) for explanations of data. Ed Perry kindly made SCA summary data available to us.

Note 5. Land assigned to the 'Fallow/Idle' is considered to be cropped but is not currently under a discernible crop. Much of this land may be in CRP.

Note 6. The spatially-delinated soil data can be retrieved from http://www.nrcs.usda.gov/wps/portal/nrcs/detail/soils/survey/geo/?cid=nrcs142p2_053627. Farnsworth et al. (2010) also used CSR data to measure productivity. They use the original rating whereas we use CSR2, as made available in 2013 and 2014 by the USDA NRCS.

Note 7. High erodibility is the only eligibility criterion for the CRP general sign-up. The continuous sign-up, introduced in the 1996 Farm Bill, also targets land that adopt certain conservation practices, such as wetland restoration and conserving riparian buffers. Continuous sign-up contracts with high priority conservation practices can be enrolled any time during the year. The NRI records CRP lands under continuous sign-up in their respective categories like cropland, forest, grassland, etc. (p. 12, NRI 2013).

Note 8. Perceptions among concerned observers in the area are that developed areas are expanding more rapidly than these data would suggest (personal correspondence with Susan Hickey, *The Nature Conservancy*).

Note 9. Numbers are somewhat smaller than in table 3 as minor categories have been removed.

14

Effect of Inoculation with Arbuscular Mycorrhizal Fungi on Selected Spring Wheat Lines

Yonaisy Mujica Pérez[1], Christiane Charest[3], Yolande Dalpé[2], Sylvie Séguin[2], Xuelian Wang[2] & Shahrokh Khanizadeh[2]

[1] Instituto Nacional de Ciencias Agrícolas, San José de las Lajas, Cuba

[2] Ottawa Research and Development Centre, Agriculture and Agri-Food Canada, Ottawa, ON, Canada

[3] Department of Biology, Faculty of Science, University of Ottawa, Ottawa, ON, Canada

Correspondence: Shahrokh Khanizadeh, Ottawa Research and Development Centre, Agriculture and Agri-Food Canada, Ottawa, ON, Canada. E-mail: Shahrokh.Khanizadeh@agr.gc.ca, http://khanizadeh.info

Abstract

An experiment was performed in a completely randomized split-plot design using five lines of spring wheat (*Triticum aestivum* L.) (AW-774, AC Carberry, HY-162, Major and AAC Scotia) and two arbuscular mycorrhizal fungi (AMF) strains (*Rhizoglomus irregulare* and *Glomus cubense*). Two different inoculant forms (solid and liquid) for the *G. cubense* strain were evaluated. The main plot was AMF, and the subplot was spring wheat lines. Data on heading date, plant height, fresh, and dry biomass, yield, grain quality (chemical composition of the seeds, gluten, and sugar), root structure, and colonization by AMF were collected. The results show a positive effect of inoculation in comparison with the control treatment. The liquid and solid *G. cubense* inoculants provided better results than inoculation with *R. irregulare*. Fungus indicators were in agreement with root morphological parameters because of the effect induced by AMF activity. Yield increased significantly in the mycorrhizal treatments.

Keywords: fungus, colonization, yields, cereal

1. Introduction

Wheat (*Triticum aestivum* L.) is the world's third produced cereal after maize and rice. It is estimated that by 2015, the production of this cereal will increase by 70% (Tilman et al., 2011) because of the growing needs of the human population for food, feed, fibre, and fuel (Fedoroff et al., 2010). To meet this demand, and because of the limited availability of uncultivated land (Garnett et al., 2013), agriculture faces a problem related to the transformation of many ecosystems to intensive grain production (Mueller et al., 2012).

Strategies such as the use of more environmentally friendly alternatives, natural processes, and environmental conservation play a vital role in current agricultural production. A large number of microorganisms exert positive effects on the growth and development of plants in the rhizosphere region, and are involved in various activities, including dynamic resources availability to plants and preservation of soil fertility (Priyadharsini & Muthukumar, 2015).

Microorganisms play an important role in agricultural systems. Specifically, arbuscular mycorrhizal fungi (AMF) are potential components of sustainable management systems (Adesemoye & Kloepper, 2009). These fungi are biotrophic because they depend on their host to complete their life cycle. They establish symbiotic associations with the majority of terrestrial plants in a diversity of ecosystems (Graham, 2008; Smith & Read, 2008; Neumann & George, 2010), improve plant tolerance to both biotic stresses (e.g., pathogens) and abiotic stresses (e.g., drought, soil salinity, and pollution) (Banuelos et al., 2014; Ruiz-Lozano et al., 2012; Cicatelli et al., 2014; Songachan et al., 2011), and play a role in remediation (Mrnka et al., 2012).

Over the past few decades, companies throughout the world have manufactured and commercialized AMF inoculants using either single AMF species or mixtures of AMF species that may include plant-growth-promoting rhizobacteria or other symbiotic and/or biocontrol fungi (Gianinazzi & Vosátka, 2004). The industrial manufacturing of AMF as crop inoculants is relatively new, and, despite practical demonstrations of the efficiency of AMF, and crop producers have been slow to adopt them. Inoculation with effective microorganisms could lead to enhanced crop productivity and higher incomes for farmers.

The effectiveness of the AM symbiosis is highly dependent on the host plant genotype (Yücel et al., 2009). Differences in the level of association of AMF with particular genotypes were found in four Canadian spring wheat cultivars (Xavier & Germida, 1998). These hexaploid Canadian wheat genotypes differed in AM root mycorrhizal development levels and in their response to inoculation, which ranged from positive to negative. The selection of crop cultivars that have strong associations with AMF may be important in providing adequate soil fertility to the crop (Singh et al., 2012).

The aim of this study was therefore to evaluate the influence of AMF on biomass, root morphology and yield in spring wheat lines.

2. Materials and Methods

2.1 Arbuscular Mycorrhizal Fungi Strain

Two granular inoculants were applied at seeding as follows: 20 g of *Rhizoglomus irregulare* (MYKE PRO commercial inoculants; 1-propagule/g), or 1 g of *Glomus cubense* (living culture of the type-specimen DAOM 241198; 1000-spores/g) per pot. Liquid *G. cubense* inoculant (20-spores/mL) obtained from the National Institute of Agricultural Science in Cuba (Fernández et al., 2004) was applied, through water irrigation, 7 d after seed germination (25-mL per pot). Controls plants were not inoculated.

2.2 Experimental Design and Plant Material

This experiment was conducted in a greenhouse at Agriculture and Agri-Food Canada (Ottawa Research and Development Centre), in Ottawa, Ontario, Canada, under controlled conditions during the period of February to June 2015. The experiment was performed in a completely randomized design with a split- plot arrangement. Five lines of spring wheat (*Triticum aestivum* L.) (AW-774, AC Carberry, HY-162, Major and AAC Scotia), two granular AMF strains (*R. irregulare* and *G. cubense*), and liquid *G. cubense* mycorrhizal treatments were used. The main plot was AMF, and the subplot was spring wheat lines. Each treatment had six replicates for a total 120 pots. The soil used was sieved, homogenized, and sterilized (twice on successive days), and then mixed with pure washed sand (1:1, v/v). Seeds were sown by hand at a depth of 2.5 cm after a germination test had been performed. Six seeds were sown per pot, and one month after planting, two plants per pot were removed. After seeding, irrigation with tap water was applied to maintain soil moisture near the maximum water-holding capacity. Urea was applied two times at a rate of 5-g per pot after 30 days of plant growth.

2.3 Measurements and Analysis

At 120 days, two random test plants were uprooted carefully from each pot treatment. The roots were washed with tap water, and a fresh root portion (200 g) was used to estimate root colonization levels by the grid line intersect method (Giovannetti & Mosse, 1980) after the roots had been bleached with 10% KOH using the microwave oven (Dalpé & Séguin, 2013) and stained with acid fuchsin (Phillips & Hayman, 1970). The frequency and intensity of colonization indicators were determined according to the methodology described in Trouvelot et al. (1986).

Plant height, fresh and dry roots and plants biomasses were measured. Total length (cm) and width (cm^2) of roots were measured using WinRHIZO Pro image analysis software. The number of grains per spike, number of grains per plant, grain weight, and yield were determined. Grain protein was evaluated using Kjeldahl digestion and total Kjeldahl nitrogen analysis.

2.4 Statistical Analysis

Data were analysed using the GLM procedure of the SAS software package (SAS Institute, 1989), and the means were separated using the least significant difference (LSD) method at the 5% level.

3. Results and Discussion

The effects of mycorrhizal inoculation on different parameters of the spring wheat lines AW-774, AC Carberry, HY-162, Major, and AAC Scotia are shown in Tables 1 to 5, respectively. Mycorrhizal root colonization was different between spring wheat lines and between AMF treatments. The Major and AW-774 lines showed a higher level of colonization (58% - 59%), whereas AC Carberry and HY-162 showed a lower level (46% - 47%). The AAC Scotia line showed the lowest mycorrhizal colonization level (35%) in comparison with two other wheat lines.

Arbuscular mycorrhizal fungi interact at the root–soil interface in a coordinated manner. Their hyphae absorb and translocate water and nutrients from the soil to the plant to increase its growth and development (Dwivedi, 2015; Priyadharsini & Muthukumar, 2015). Wheat is colonized by mycorrhizae, but colonization depends on AMF strains and soil conditions. Studies conducted in India showed that the interaction between mycorrhizae and wheat depends on the variety, and thus it is important to evaluate mycorrhizal dependency in the crops being grown (Solaiman et al., 2014).

Table1. Effects of arbuscular mycorrhizal fungal inoculation of spring wheat line AW-774 on plant parameters.

Treatment	Yield (g/pot)	Plant height (cm)	Spike weight (g)	Tiller number	Fresh biomass (g)	Dry biomass (g)	Fresh root biomass (g)	Dry root biomass (g)	Root colonization frequency (%)	Root colonization intensity (%)	Root total length (cm)	Root total width (cm²)
Control	4.77 c	50.87	6.52 c	6.33 c	1.21	1.19	14.55 c	0.40	1.00 c	0.01 c	413.50 b	16.82 c
Glomus cubense (Liquid)	7.70 a	57.47	9.73 a	9.67 a	1.30	1.29	23.72 a	0.47	58.00 a	2.45 a	923.30 a	42.52 a
Rhizoglomus irregulare (Solid)	6.75 b	52.88	8.55 b	7.50 b	1.23	1.23	17.58 b	0.44	35.50 b	0.58 b	836.90 a	37.23 a
Glomus cubense (Solid)	8.48 a	57.25	10.67 a	9.67 a	1.33	1.30	23.19 a	0.42	59.17 a	2.48 a	592.50 b	30.06 ab
F-value	27.94	1.52	30.45	28.67	1.54	1.45	61.50	1.40	379.39	66.70	9.15	10.22
LSD	0.89	7.82NS	0.96	0.91	0.13NS	0.13NS	1.67	0.07NS	4.13	0.46	226.59	10.27
SE	0.30	2.65	0.32	0.31	0.04	0.04	0.57	0.02	1.40	0.16	76.81	3.48

Note: Different letters within each block indicate significant differences ($p \leq 0.05$) according to the least significant difference (LSD) test. NS: not significant.

SE: standard error

Table 2. Effects of arbuscular mycorrhizal fungal inoculation of spring wheat line AC Carberry on plant parameters.

Treatment	Yield (g/pot)	Plant height (cm)	Spike weight (g)	Tiller number	Fresh biomass (g)	Dry biomass (g)	Fresh root biomass (g)	Dry root biomass (g)	Root colonization frequency (%)	Root colonization intensity (%)	Root total length (cm)	Root total width (cm²)
Control	3.66 b	42.83 b	4.96 b	6.33 c	0.90 b	0.87 b	11.05 d	0.25 c	1.00 c	0.01 d	391.55b	16.51 b
Glomus cubense (Liquid)	5.22 a	49.40 a	7.06 a	10.17 a	1.17 a	1.15 a	19.52 b	0.34 a	47.50 a	2.23 a	383.84b	15.89 b
Rhizoglomus irregulare (Solid)	4.46 ab	44.12 b	6.35 a	7.00 c	1.05 ab	1.02 ab	14.16 c	0.31 b	36.67 b	0.60 c	494.60 a	23.26 a
Glomus cubense (Solid)	5.22 a	49.43 a	7.12 a	8.67 a	1.19 a	1.16 a	20.89 a	0.37 a	46.83 a	1.83 b	559.05 a	26.33 a
F-value	6.26	6.74	7.51	20.37	5.24	5.67	125.50	12.25	369.87	99.28	7.10	5.14
LSD	0.88	3.94	1.08	1.13	0.17	0.17	1.21	0.04	3.36	0.31	93.67	6.67
SE	0.30	1.34	0.37	0.38	0.06	0.06	0.41	0.01	1.14	0.10	31.75	2.26

Note: Different letters within each block indicate significant differences (p≤ 0.05) according to the least significant difference (LSD) test. NS: not significant.

SE: standard error

Table 3. Effects of arbuscular mycorrhizal fungal inoculation of spring wheat line HY-162 on plant parameters.

Treatment	Yield (g/pot)	Plant height (cm)	Spike weight (g)	Tiller number	Fresh biomass (g)	Dry biomass (g)	Fresh root biomass (g)	Dry root biomass (g)	Root colonization frequency (%)	Root colonization intensity (%)	Root total length (cm)	Root total width (cm²)
Control	3.48 b	31.35c	4.79 b	6.33b	0.75 b	0.74 b	9.53b	0.19b	1.00 c	0.01 d	348.75	16.12
Glomus cubense (Liquid)	4.70 a	42.10a	5.90 a	8.33a	0.98 a	0.96 a	13.71a	0.26a	45.67 a	6.52a	384.23	18.82
Rhizoglomus irregulare (Solid)	3.68 b	36.50b	5.01 b	6.33 b	0.87 ab	0.85 ab	10.25 b	0.24 a	36.50b	0.60 c	390.50	15.82
Glomus cubense (Solid)	4.64 a	42.03a	5.94 a	8.50 a	1.00 a	0.98 a	13.05 a	0.24 ab	47.50 a	2.23 b	354.38	16.43
F-value	7.10	15.87	5.49	18.33	4.26	3.78	20.11	2.67	519.32	1.56	0.46	0.92
LSD	0.71	3.81	0.75	0.83	0.16	0.17	1.35	0.05	2.80	6.93	90.78NS	4.23NS
SE	0.24	1.29	0.25	0.28	0.06	0.06	0.46	0.02	0.95	2.35	30.67	1.44

Note: Different letters within each block indicate significant differences (*p*≤ 0.05) according to the least significant difference (LSD) test. NS: not significant.

SE: standard error

Table 4. Effects of arbuscular mycorrhizal fungal inoculation of spring wheat line Major on plant parameters.

Treatment	Yield (g/pot)	Plant height (cm)	Spike weight (g)	Tiller number	Fresh biomass (g)	Dry biomass (g)	Fresh root biomass (g)	Dry root biomass (g)	Root colonization frequency (%)	Root colonization intensity (%)	Root total length (cm)	Root total width (cm²)
Control	4.73 c	50.21c	7.09 c	6.17b	1.31 b	1.28 b	15.60 c	0.41 c	1.00 c	0.01 c	433.06 c	17.18c
Glomus cubense (Liquid)	7.45 a	79.37a	9.78 a	9.17 a	1.59 a	1.58 a	36.25 a	0.73 a	58.17 a	2.47 a	988.21 a	43.00a
Rhizoglomus irregulare (Solid)	6.38 b	56.90b	8.42 b	6.83 b	1.39 b	1.37 b	24.46 b	0.47 b	38.33 b	0.79 b	736.96b	33.76b
Glomus cubense (Solid)	7.87 a	79.62a	9.60 a	10. 17 a	1.60 a	1.59 a	35.95 a	0.74 a	58.00 a	2.45 a	1010.02a	48.04a
F-value	16.10	119.38	10.59	28.04	11.99	11.95	163.71	154.00	429.18	48.11	31.24	47.05
LSD	1.03	4.11	1.13	1.05	0.12	0.13	2.29	0.04	3.87	0.52	142.25	5.83
SE	0.35	1.39	0.38	0.36	0.04	0.04	0.78	0.01	1.31	0.18	48.22	1.98

Note: Different letters within each block indicate significant differences (p≤ 0.05) according to the least significant difference (LSD) test. SE: standard error

Table 5. Effects of arbuscular mycorrhizal fungal inoculation of spring wheat line AC Scotia on plant parameters.

Treatment[a]	Yield (g/pot)	Plant height (cm)	Spike weight (g)	Tiller number	Fresh biomass (g)	Dry biomass (g)	Fresh root biomass (g)	Dry root biomass (g)	Root colonization frequency (%)	Root colonization intensity (%)	Root total length (cm)	Root total width (cm)
Control	5.28b	40.67	7.81b	5.50	1.21	1.19	8.73 c	0.21 b	1.00 b	0.01 b	403.75ab	16.73ab
Glomus cubense (Liquid)	6.08a	41.87	9.09a	5.67	1.22	1.20	12,05 a	0.25 a	35.66 a	0.58 a	488.80 a	23.44a
Rhizoglomus irregulare (Solid)	5.62ab	41.00	7.69b	5.33	1.19	1.17	10.63 b	0.22 ab	35.50 a	0.56 a	485.05 a	21.66ab
Glomus ubense (Solid)	5.98ab	40.80	8.22ab	5.50	1.21	1.19	12.43 a	0.24 ab	36.50 a	0.59 a	285.10b	15.55b
F-value	1.91	0.19	3.82	0.39	0.05	0.05	15.25	1.86	298.69	197.06	3.79	2.65
LSD	0.77	3.67NS	0.95	0.64NS	0.13NS	0.14NS	1.26	0.04	2.98	0.06	144.69	6.90
SE	0.26	1.24	0.32	0.22	0.05	0.05	0.43	0.01	1.01	0.02	49.05	2.34

Note: Different letters within each block indicate significant differences (p≤ 0.05) according to the least significant difference (LSD) test. NS: not significant.

SE: standard error

The mycorrhizal colonization results showed differences between *G.cubense* and *R.irregulare,* which could be related to infectivity. Some studies demonstrated that fungal infectivity can be associated with differences in inoculum level and in the ability of the fungi to colonize roots (Solaiman et al., 2014). In our case, the inoculum concentration was different for both strains; such a difference can induce variations in fungal colonization and interfere with the relationship between inoculum level and infectivity. However, other research demonstrated a positive effect of the *G.cubense* strain in a different crop grown in red soil with low to high fertility (Rivera et al., 2007). The mycorrhizal colonization levels obtained in the present study are similar to those reported for rice cultivation in saline conditions (Fernández et al., 2011).

The root morphology results showed significant differences between wheat lines, but for the HY-162 line, no positive effect to AMF was found. A comprehensive analysis of root morphology variables for the Major wheat line showed that *G.cubense* was more effective than *R.irregulare*. The results for these variables were in agreement with those for fungal colonization. Although root morphological variables for the AAC Scotia line showed significant differences, the level of mycorrhizal colonization was low, with no significant differences from the levels for the other lines, and exceeded only the level for the control treatment.

Significant differences in plant growth indicators (plant height, fresh and dry weights of roots and biomass) between all five wheat lines under study were determinated. In two lines (AW-774 and AAC Scotia), inoculation with AMF strains showed no positive effect on plant height, but in the rest of the lines, significant differences were obtained in relation to the control treatment.

The tallest plants were achieved with the Major wheat line, at 79 cm, in the treatments inoculated with *G.cubense,* whereas inoculation of Major with *R.irregulare* produced a height of 56 cm. Inoculation with liquid and solid *G.cubense* stimulated plant height for the AC Carberry and HY-162 wheat lines, with values of 49 and 42 cm, respectively. Root dry weight differed between wheat lines and between inoculation treatments. The AW-774 wheat line did not show significant differences for this indicator, but levels were variable and exceeded the value for the control treatment.

Plant growth indicators can be used to evaluate benefits when microorganisms are applied, given that plant growth and development are stimulated significantly. Our results show differences depending on the AMF strains and spring wheat lines, but inoculant treatment responded to fungal colonization. The results were higher when *G.cubense* (liquid and solid) inoculum was applied. This response could be related to mycorrhizal effectiveness (Bonfante & Genre, 2010; Fernández et al., 2011). The positive effect of AMF on the height and development of the aerial part and root system of plants was previouly reported in various crops, including maize (*Zea mays*) (Sheng et al., 2011), tomato (*Solanum lycopersicum*) (Hajiboland et al., 2010), wheat (*Triticum aestivum*) (Stonor et al., 2014), rice (*Oryza sativa*) (Fernández et al., 2011) and pepper (*Capsicum annuum*) (Çekiç et al., 2012).

The yield of a crop is the end result of the interaction of several factors. In this study, differences between AMF strains and between wheat lines were found for yield components. Inoculation with liquid and solid *G.cubense* increased tiller number and spike weight in four of the wheat lines (AW-774, AC Carberry, HY-162, and Major). Inoculation with *R.irregulare* increased the levels of those indicators as well, but to a lower extent than inoculation with *G.cubense* did. Wheat yield was stimulated by AMF inoculation. Inoculation with liquid and solid *G.cubense* was more effective than inoculation with *R.irregulare* was, but both species showed better results than control treatment.

The effects of mycorrhizal inoculation on total nitrogen and grain protein in the spring wheat lines are shown in Table -6. In this case, AMF inoculation did not have a positive effect for both indicators, and only the HY-162 line showed significant differences. This response could be related to the fertilization and mineral requirements of each wheat line.

Table 6. Effects of arbuscular mycorrhizal fungal inoculation of spring wheat lines on total nitrogen (N) and grain protein.

Treatment[a]	AW-774		AC Carberry		HY-162		Major		AAC Scotia	
	Total N (%)	Grain protein (%)	Total N (%)	Grain protein (%)	Total N (%)	Grain protein (%)	Total N (%)	Grain protein (%)	Total N (%)	Grain protein (%)
Control	3.27	18.63	3.50b	20.03	3.87ab	22.10ab	3.80	21.66	3.51	19.96
Glomus cubense (Liquid)	3.27	18.63	3.70a	21.10	3.98a	22.70a	3.67	21.00	3.70	21.10
Rhizoglomus irregulare (Solid)	3.29	18.73	3.60ab	20.53	3.60c	20.56c	3.37	19.20	3.60	20.53
Glomus cubense (Solid)	3.43	19.56	3.64a	20.80	3.68bc	21.00bc	3.49	19.90	3.45	19.70
F-value	0.61	0.62	2.45	2.24	6.88	6.94	1.78	1.77	0.39	0.37
LSD	0.32NS	1.87NS	0.17	0.98NS	0.21	1.21	0.46NS	2.70NS	0.58NS	3.35NS
SE	0.10	0.58	0.05	0.30	0.07	0.37	0.14	0.83	0.18	1.03

Note: Different letters within each block indicate significant differences ($p \leq 0.05$) according to the least significant difference (LSD) test. NS: not significant. SE: standard error

The results showed a positive effect of inoculation in comparison with the control treatment. Liquid and solid *G.cubense* inoculant produced better results than inoculation with R.*irregulare*. Fungus indicators were in agreement with root morphological parameters because of the effect induced by AMF activity. Yield increased significantly in the mycorrhizal treatments.

Acknowledgements

This work was supported by a Government of Canada international scholarship (Emerging Leaders in the Americas Program). The authors would like to acknowledge Agriculture and Agri-Food Canada greenhouse staff members Art Smith and Kathie Upton for their help maintaining the set-up.

References

Adesemoye, A. O., & Kloepper, J. W. (2009). Plant-microbes interactions in enhanced fertilizer-use efficiency. *Applied Microbiology and Biotechnology, 85*, 1-12. http://dx.doi.org/10.1007/s00253-009-2196-0

Banuelos, J., Alarcón, A., Larsen, J., Cruz-Sánchez, S., & Trejo, D. (2014). Interactions between arbuscular mycorrhizal fungi and Meloidogyne incognita in the ornamental plant Impatiens balsamina. *Journal of Soil Science and Plant Nutrition, 14*, 63-74.

Bonfante, P., & Genre, A. (2010). Mechanisms underlying beneficial plant–fungus interactions in mycorrhizal symbiosis. *Nature Communications, 1*, 48. http://dx.doi.org/10.1038/ncomms1046

Çekiç, F. Ö., Ünyayar, S., & Ortaş, I. (2012). Effects of arbuscular mycorrhizal inoculation on biochemical parameters in *Capsicum annuum* grown under long term salt stress. *Turkish Journal of Botany, 36*, 63-72.

Cicatelli, A., Todeschini, V., Lingua, G., Biondi, S., Torrigiani, P., & Castiglione, S. (2014). Epigenetic control of heavy metal stress response in mycorrhizal versus non-mycorrhizal poplar plants. *Environmental Science and Pollution Research, 21*, 1723-1737. http://dx.doi.org/10.1007/s11356-013-2072-4

Dalpé, Y., & Séguin, S. M. (2013). Microwave assisted technology for the clearing and staining of arbuscular mycorrhizal fungi in roots. *Mycorrhiza, 23*, 333-340. http://dx.doi.org/10.1007/s00572-012-0472-9

Dwivedi, O. P. (2015). Distribution and association of arbuscular mycorrhizal fungi in different cultivars of wheat from Lalganj Pratapgarh District of Utter Pradesh, India. *Advances in Bioscience and Biotechnology, 6*, 353-357. http://dx.doi.org/10.4236/abb.2015.65034

Fedoroff, N. V., Battisti, D. S., Beachy, R. N., Cooper, P. J. M., Fischhoff, D. A., Hodges, C. N., & Zhu, J. K. (2010). Radically rethinking agriculture for the 21st century. *Science, 327*, 833-834. http://dx.doi.org/10.1126/science.1186834

Fernández, F., Dell'Amico, J. M., Angoa, M. V., & de la Providencia, I. E. (2011). Use of a liquid inoculum of the arbuscular mycorrhizal fungi *Glomus hoi* in rice plants cultivated in a saline Gleysol: A new alternative to inoculate. *Journal of Plant Breeding and Crop Science, 3*, 24-33.

Fernández, F., Dell'Amico, J. M., & Pérez, Y. (2004). Producto inoculante micorrizógeno líquido. Cuban Patent 23479.

Garnett, T., Appleby, M. C., Balmford, A., Bateman, I. J., Benton, T. G., Bloomer, P., & Godfray, H. C. J. (2013). Sustainable intensification in agriculture: Premises and policies. *Science, 341*, 33-34. http://dx.doi.org/10.1126/science.1234485

Gianinazzi, S., & Vosátka, M. (2004). Inoculum of arbuscular mycorrhizal fungi for production systems: science meets business. *Canadian Journal of Botany, 82*, 1264-1271. http://dx.doi.org/10.1139/b04-072

Giovannetti, M., & Mosse, B. (1980). An evaluation of techniques for measuring vesicular–arbuscular mycorrhizal infection in roots. *New Phytologist, 84*, 489-500. http://dx.doi.org/10.1111/j.1469-8137.1980.tb04556.x

Graham, J. H. (2008). Scaling-up evaluation of field functioning of arbuscular mycorrhizal fungi. *New Phytologist, 180*, 1-2. http://dx.doi.org/10.1111/j.1469-8137.2008.02608.x

Hajiboland, R., Aliasgharzadeh, N., Laiegh, S. F., & Poschenrieder, C. (2010). Colonization with arbuscular mycorrhizal fungi improves salinity tolerance of tomato (*Solanum lycopersicum* L.) plants. *Plant and Soil, 331*, 313-327. http://dx.doi.org/10.1007/s11104-009-0255-z

Mrnka, L., Kuchár, M., Cieslarová, Z., Matějka, P., Száková, J., Tlustoš, P., & Vosátka, M. (2012). Effects of endo- and ectomycorrhizal fungi on physiological parameters and heavy metals accumulation of two species from the family Salicaceae. *Water, Air and Soil Pollution, 223*, 399-410. http://dx.doi.org/10.1007/s11270-011-0868-8

Mueller, N. D., Gerber, J. S., Johnston, M., Ray, D. K., Ramankutty, N., & Foley, J. A. (2012). Closing yield gaps through nutrient and water management. *Nature, 490*, 254-257. http://dx.doi.org/10.1038/nature11420

Neumann, E., & George, E. (2010). Nutrient uptake: The arbuscular mycorrhiza fungal symbiosis as a plant nutrient acquisition strategy. In: H. Koltai & Y. Kapulnik (Eds) *Arbuscular mycorrhizas: Physiology and function* (pp. 137-167). Dordrecht, Netherlands: Springer. http://dx.doi.org/10.1007/978-90-481-9489-6_7

Priyadharsini, P., & Muthukumar, T. (2015). Insight into the role of arbuscular mycorrhizal fungi in sustainable agriculture. In: P. Thangavel & G. Sridevi (Eds), Environmental sustainability: Role of green technologies (pp. 3-37). New Delhi:Springer. http://dx.doi.org/10.1007/978-81-322-2056-5_1

Rivera, R., Fernández, F., Fernández, K., Ruiz, L., Sánchez, C., & Riera, M. (2007). Advances in the management of effective arbuscular mycorrhizal symbiosis in tropical ecosystems. In: C. Hamel & C. Plenchette (Eds), Mycorrhizae in crop production (pp. 151-188). Binghampton, NY: Haworth Press.

Ruiz-Lozano, J. M., Porcel, R., Azcón, C., & Aroca, R. (2012). Regulation by arbuscular mycorrhizae of the integrated physiological response to salinity in plants: New challenges in physiological and molecular studies. *Journal of Experimental Botany, 63*, 4033-4044. http://dx.doi.org/10.1093/jxb/ers126

SAS, Institute. (1989). SAS user's guide: Statistics. Version 6. SAS Institute Inc., Cary, NC, USA.

Sheng, M., Tang, M., Zhang, F., & Huang, Y. (2011). Influence of arbuscular mycorrhiza on organic solutes in maize leaves under salt stress. *Mycorrhiza, 21*, 423-430. http://dx.doi.org/10.1007/s00572-010-0353-z

Singh, A.K., Hamel, C., DePauw, R.M., & Knox, R.E. (2012). Genetic variability in arbuscular mycorrhizal fungi compatibility supports the selection of durum wheat genotypes for enhancing soil ecological services and cropping systems in Canada. *Canadian Journal of Microbiology, 58*, 293-302. http://dx.doi.org/10.1139/w11-140

Smith, S. E., & Read, D. J. (2008). Mycorrhizal symbiosis. (3rd ed.). Academic Press: New York.

Solaiman, Z. M., Abbott, L. K., & Varma, A. (Eds.). (2014). Mycorrhizal fungi: Use in sustainable agriculture and land restoration. Berlin: Springer.

Songachan, L. S., Lyngdoh, I., & Highland, K. (2011). Colonization of arbuscular mycorrhizal fungi in moderately degraded sub-tropical forest stands of Meghalaya, Northeast India. *Journal of Agricultural Technology, 7*, 1673-1684.

Stonor, R. N., Smith, S. E., Manjarrez, M., Facelli, E., & Smith, A. (2014). Mycorrhizal responses in wheat: Shading decreases growth but does not lower the contribution of the fungal phosphate uptake pathway. *Mycorrhiza, 24*, 465-472. http://dx.doi.org/10.1007/s00572-014-0556-9

Tilman, D., Balzer, C., Hill, J., & Befort, B. L. (2011). Global food demand and the sustainable intensification of agriculture. *Proceedings of the National Academy of Science, 108*, 20260-20264. http://dx.doi.org/10.1073/pnas.1116437108

Trouvelot, A., Kough, J., & GianinazziPearson, V. (1986). Mesure du taux de mycorhization VA d'un systeme radiculaire. Recherche de méthodes d'estimation ayantune signification fonctionnelle. In: V. Gianinazzi-Pearson & S. Gianinazzi (Eds.), Physiological and genetical aspects of mycorrhizae: Proceedings of the 1st European Symposium on Mycorrhizae (pp. 217-222). Paris: INRA

Xavier, L. J. C., & Germida, J. J. (1998). Response of spring wheat cultivars to *Glomus clarum* NT4 in a P-deficient soil containing arbuscular mycorrhizal fungi. *Canadian Journal of Soil Science, 78*, 481-484. http://dx.doi.org/10.4141/S97-106

Yücel, C., Özkan, H., Ortaş, I., & Yağbasanlar, T. (2009). Screening of wild emmer wheat accessions (*Triticum turgidum* subsp. *dicoccoides*) for mycorrhizal dependency. *Turkish Journal of Agriculture and Forestry, 33*, 513-523.

15

Dates Palm Farming Systems Sustainability and Risk Efficiency in Oman

Kheiry Hassan M. Ishag[1]

[1]Dhofar Cattle Feed Company, P.O. Box 1220, PC 211, Sultanate of Oman

Correspondence: Kheiry Hassan M. Ishag, Dhofar Cattle Feed Company, P.O. Box 1220, PC 211, Sultanate of Oman. E-mail: kheiryishag@hotmail.com

Abstract

Date Palm is the most important cultivar in Oman and occupies 35% of total cultivated area and 78% of the total fruit trees area. The Ministry of Agriculture and Fisheries attempted to improve Date production but due to climate change and environmental constrains agricultural production has shown instability in production and generated a desire to build a sustainable farming systems. The study used stochastic model to analyze Date Palm Farming Systems in Oman and identify the most sustainable and risk efficient one. Stochastic Efficiency with Respect to Function (SERF) technique use certainty equivalent concept to rank a set of risk efficient alternatives. The study found Batinah Region is the most risk efficient region and got a positive NPV with a probability of 88% followed by Dakhiliyah Region 77% and Dahirah Region 67%. The study also indicates Batinah and Dakhiliyah Regions which cultivate 52% of total Date Palm area in Oman is located in risk efficient farming system areas. Risk premium analysis performed and shows Dakhiliyah Region farmers can pay up to RO 59 for replanting Date Palm tree and move to Batinah Region farming practices with less risky farming system. The farmer in Dahirah Region is willing to pay RO 144 per Date Palm tree to shift to more efficient farming system practices such as Batinah Region. Date Palm Farming System analysis should be used as a basis and foundation for replanting program and resources and environmental constrains need be considered. A right economic incentives to be given to encourage production of selected Date Palm varieties for each region to increase ecosystem resilience and economic benefit.

Keywords: Date Palm, simulation model, stochastic efficiency with respect to a function (SERF), risk efficient, resilience ecosystem, sustainability, farming systems

1. Introduction

The Date Palm cultivation is considered as the most important agricultural activities in Oman with a large number of varieties and spread over a large integrated ecological and farming system. It occupies 35% of total cultivated area and 78% of the total fruit trees in Oman as per Ministry of Agriculture Statistics (2013).The Date Palm trees occupies about 57,430 acres and includes 6,79 Million productive trees. The total date production at year 2013 reached 308 thousand ton 53% for direct human consumption, 20% used as animal feed, 16% for industry use and 11% for export.

Oman has about 200 date palm varieties and 30 types of these varieties are recognized as good varieties and have commercial and high market absorption demand. Top good varieties such as Khalas produce about 8% of total production and second top varieties date such as Zabad , Khanizi , Barni and Madlozi are commercial dates varieties and can be stored and sale off the season with a reasonable price. Oman benefited from different environmental and climate zones and adopted farming systems which extend the harvesting season to six months from May to November, (Kheiry Hassan M. Ishag et al., 1994). Accordingly date palm varieties can also be grouped as early, intermediate and late mature varieties. The early mature date palm varieties such as Nagal which is 11% from total dates production in Oman is cultivated at Batinah i.e. 38% and Dakhiliyah regions i.e. 30% of total Nagal production. Farad, Khanizi and Khalas are intermediate mature varieties and cultivated at Batinah and Sherqiah regions. A late mature Date variety such as Khasab is cultivated at Batinah and Dakhiliyah regions. However, farmers are diversifying date palm cultivated varieties to manage risk associated with price and yield uncertainty in term of extending harvesting period and producing different type of dates for different use i.e. human, industrial, animal feed. Farmers should manage to develop resilience agricultural systems by introducing affordable technologies and strategies such that ecosystem functions and services can be maintained and livelihoods can be protected, (Brenda B. Lin, 2011).

The date Palm farming systems in Oman can be grouped to six farming systems. Batinah Region considers as the leading region and produces 136 thousand tons in year 2013, which represent about 44% of the total date's production in Oman. The second region is Dakhiliyah Region and produces 63 thousand tons i.e. 20% followed by Sherqiah Region 14% and Dahirah Region which produce 11% of the total Oman production of Dates. Dakhiliyah Region got the highest Date Palm average productivity per tree with 62 kg/tree followed by Dahirah Region 47Kg/tree and Batinah 46 Kg/tree and Sherqiah Region 39 Kg/tree, and Muscat Region 30 Kg/tree. Region production and Date Palm tree productivity depend on many factors such as date palm varieties, cropping pattern, soils and soil fertilities, irrigation system, water quality and availability and farming system resilience and survivals.

Social sustainability in term of consumers' preference of Dates products, short supply chain systems and availability of local Dates marketplace are required. In addition to Farm waste management, multifunctional farm's activities such as selling manure and feed dates to animal have important role in farming system sustainability, (Anna Gaviglio et al., 2016).

However, comparing Date Palm Regions in term of region production, number of date palm trees at each region and tree productivity ignoring farming systems profitability and economic sustainability will not give a full picture to understand problems and develop strategy for Date Palm sector in Oman, (Kheiry Hassan M. Ishag et al., 1997). As a result, farming system economic sustainability and risk efficiency is the most important issue need to be studied to provide data for policy advisers.

Problem statement

Date Palm is the most important cultivar in Oman and occupies 35% of total cultivated area and 78% of the total fruit trees area. The Government Authority attempted to improve Date production through supporting Research, Extension and Farmers. The production increased during 1999-2001 and reduced due to environmental constrain, water resource shortages and drought climate in 2002 and 2004. The Ministry of Water Resources indicted that the annual ground water recharge quantity in Oman reaches (1295) millions M^3 which is less than the total country water consumption of about 25%. However, this unbalance water supply and demand is the main resources constrain facing Date Palm development program.

The Ministry of Water Resources stated that aquifers are recharged by rainfall and surface water of a bout (70%) of water demand. The renewable water recharge shows that Batinah Region comes first in aquifer annual recharge, followed by Al Dakhliya and Al Sharqiya Regions. Due to water shortage and environmental constrains the Date Palm cultivated areas reduce recently by 36% from 93,534 acres in year 2000 to 59,602 acres in 2013. The unsustainable date production situation encourages Government to announce for long term Date Palm development program and to cultivate One Million good varieties Date Palm trees. The study investigates and aims to identify the right land and sustainable farming system region that can accommodate this project and improve resilience systems.

Literature review

Farming system sustainability concern about system capability to maintain production in spite of major constrains and disturbance. The system resilience and ability to continue in the future in term of financial viability and farm resources degradation can be taken as a measurement to choice between alternative Date Palm farming systems, (Lien G. et.al., 2007). Stochastic and dynamic nature of the farming systems can be model and probability of getting positive return for each farming system can be calculated.

Monte Carlo Simulation Model used for project appraisal by (Savvakis C. Savvides, 1994). He outline that dynamic integrated analysis provided a range of outcomes that can decrease risk of uncertainty variables and give more reliable results to investor. Additional information regarding water management policy analysis and aquatic ecosystems can be found in (Folkes et al., 2002) ; (MA , 2005; Sanders and Lewis, 2003) ; (Blumenfeld et al., 2009; Carpenter, Brock and Hanson, 1999; Chen et al., 2009).

In this study we used Monte Carlo Simulation Models to quantify risk and uncertainty associated with Date Palm farming system. The quantitative risk analysis will estimate the probability of getting positive NPV for reach Region and provide decision makers a tool to improve Date Palm cultivation and achieving One Million Date Trees Project aims simultaneously: sustaining irrigated agriculture farming to achieve food security and maintain resources associated natural environment. (S. Quiroga, 2010) used Monte Carlo Simulations technique to estimate crop yield risk at different water resources variability.

The stochastic efficiency of alternative Date Palm farming system can rank risky alternative over a range of risk aversion. The stochastic efficiency with respect to a function (SERF) technique developed recently by (Hardaker et al., 2004) and based on ranking risky alternatives in terms of expected utility function and certainty equivalents (CE). Certainty Equivalents is defined as the sure sum of the money with the same utility can be accepted to compensate risk alternative (Hardaker et al., 2004). The Stochastic Efficiency with Respect to a Function (SERF) used by (Lien et al., 2006) to assess organic and conventional cropping systems sustainability. (Eihab M. F. et al., 2011), used (SERF) methodology to analyze conventional and conservation tillage system from experimental plots data at Iowa State University. He argued (SERF) method is useful tool to assist policy makers and advisers on solving problems involving agricultural risk. In this study, the (SERF) technique is applied to assess a set of alternative risky farming systems. The SERF method ranked regional farming systems in terms of the CE over a range of risk aversion levels. SERF can compare any level of decision makers" preferences including different level of risk. The aggregated Date Palm area and production data at Regional level were calculated. These values were multiplied by average date price for each date varieties cultivated at each region. The stochastic simulation models were employed to examine NPV distribution for the six different Regions and their Date Palm farming system alternatives. The research aimed to investigate Date production regions and their Date Palm farming system sustainability and risk efficiency over a range of risk aversion level.

2. Methodology

The farming system evaluation and viability is depending on estimating the future values of the crops return and operation costs by using available information from a specific situation in the past. The normal approach used in investment appraisal is to calculate a "best estimate" of the variables based on the available data and use it as an input in farming appraisal model. The regional (NPV) is calculated as the most likely outcome of the farming system.

2.1 Net Present Value

The NPV of regional farming system was used to evaluate economic sustainability. The net cash flow for reach region is calculated by subtracting the cost from the revenue and discounted by the interest rate to obtain the NPV of the Region. The variables identified in term of deterministic and stochastic variables and NPV obtained as a range of values instead of a single value in a conventional financial evaluation.

2.2 Monte Carlo Simulation

Monte Carlo Simulation Model is a computerized algorithm model designed to evaluate the variability of the input variables of a model. The model can be used to estimate the effects of main variables on project outputs. The process started with the identification and assessment of the key variables. Then the probability density function is estimated to fit the best range of uncertainty around the expected value. The historical Date Palm data at each region obtained from MAF statistics (2013) and Date Palm Production Survey (2013). The Monte Carlo Simulation model including these variables is then run using random generated input values taken from the probabilistic distribution function. The model combines inputs and generate the estimated outcome value for (NPV). Monte Carlo Simulation Model is currently recorded as the most powerful technique can be used to incorporate risk and uncertainty. The model is suitable for complex project and the more risky and uncertainty variables associated within the project.

Monte Carlo dy-namic simulation model is used in this research for the evaluation of farming system economic sustainability and risk efficiency within different level of risk aversion. The stochastic budgeting methods are used to incorporate risk and uncertainty variables in Date Palm farming system in Oman. The study used @Risk 7.5 program from (Palisade Corporation, Ithaca, New York) and add-in for Excel was utilized to account for stochastic nature of key farming system variables in the model. Parameters in Table 1 and input distribution were obtain from historical data and used in the model to estimate cash flow for each region.

2.3 Stochastic Efficiency with Respect to a Function (SERF)

SIMETAR program (2011) was used to carry on SERF analysis to assess Date Palm farming systems risk efficiency and sustainability under different level of risk aversion. The whole region stochastic simulation model was used and model is run for 10 years in the future to calculate NPV for each region and assess the economic sustainability of different farming system alternatives. The negative NPV indicates project failure and less risk efficient alternative (Hansen and Jones, 1996).

The Certainty Equivalent graphs were also constructed to display and rank Date Palm farming systems according to risk efficiency across the specified range of ARAC values and farming system with higher CE is preferred to those with lower CE. The risk premiums were also calculated for each region by subtracting less preferred alternative regional CE values from preferred CE values at given ARAC values. Given a utility function u(·), a random wealth variable X, and an initial level of wealth w0, the certainty equivalent is :

$$CE = u-1\{E[u(X + w0)]\} - w0,$$

The risk premium measure the minimum amount that would have to be paid to a farmers and decision maker to justify a switch from alternative present farming system to other less risky Date Palm farming system. An analysis of Date Palm farming system of six regions was conducted using a ten year farm level data and simulation model. Total number of Date Palm trees and varieties for each region, yield, price, investment and operation cost are collected from Date Palm Production Survey and Ministry of Agriculture Statistics (2013). The model simulates the costs and returns of the farm for six regions and the NPV probability distributions generated by the simulation model and used to rank the best alternative region across a full range of RACs.

2.4 Data Collection

Historical data were collected to perform partial budgeting analysis for six alternatives Date Palm farming systems in Oman and consider parameters such as date palm trees varieties, number of date palm panted trees, regional date yield, sale price, cost of inputs and other operation cost for each region. The study used simulation analysis to identified main stochastic variables need to be incorporated in the model such as Date Palm trees varieties cultivated for each region, Date yields, inputs cost, and sale prices for each. The probability distributions of the risky input variables (triangle – normal - bionomial) identified and Cumulative Distribution Function (CDF) of the output (NPV) for each region were calculated.

The Date Palm farming system sustainability was investigated by performing Stochastic Efficiency with Respect to a Function (SERF) Analysis for six regions. SIMETAR (2011) program was used to calculate and generate Certainty Equivalent (CEs) for each region and rank alternatives regions as per their risk efficiency. The data used in this study obtained from different sources such as Ministry of Agriculture statistics (2013), Date Palm Production survey (2013) and previous studies and can be grouped in to two categories as under :

Regional level data for Date Palm farming systems in six regions :

- Regional data and current farming systems data such as intercropping and cropping pattern.
- Regional level data such as Date Palm varieties (yields, price, operation costs) for each region.
- Irrigation system, water quantity available for each region, water and soil salinity.
- Regional Date Palm varieties use (human use, industrial use, animal feed) and consumptions.
- Regional Dates marketing, harvesting time and duration and market price.
- Regional level historical data for Date Palm Farms area development at different region.

Farm level data and capital cost of cultivation one Acre :

- Cost of land preparation and Date offshoots planting trees.
- Operation cost for each Date Palm varieties at each region.

2.5 Model Structure

The study started modeling process by defining inputs and parameters effecting Date Palm cultivation income and return. The qualitative risk analysis in this study is aim to provide a high level of understanding of risks facing farming system sustain Date Palm growing in Oman. Such analysis may increase attention of old Date Palm trees replacement policy adviser to the top risks they need to manage effectively, (Qiu Ling Guo, 2001) and (James, 2007).

Percentage of annual yield increase, sale price and cost increase and total Date production for each region are collected and calculated from regional historical data. The market absorption Date varieties rate for each region is calculated from Date Palm production and marketing survey historical data.

The main risk and uncertainty variables identified in the models were:

- Low market absorption Date varieties.
- Cost of establishing Date farm and capital cost increase for each region.
- Total Date Palm trees and varieties grown at each region.
- Dates products selling price volatility for each varieties grown at each region.
- Cost of production per ton for each region.

- Annual increase in sales price and unit cost.
- Total Date saleable volume for each variety at each region.
- Date Palm trees yield variation at six regions.

The region financial performance and cash flows is performed after selecting key parameters and the probability of all individuals risk combined on parameters. The model run and result of the analysis generate the probability that region will meet its quantitative objectives and cash flow projection. The probability distributions of the parameters are incorporated in to simulation model to quantified risks range for each region, Table 1. Six models formed and constructed to represent six regions and their Date Palm farming system. The models run with @Risk add-in software to test economic viability and sustainability for each region and farming system.

Table 1. The main input parameters distribution used in Batinah Region Faming System Model

Risk	Affects	Distribution	Absolut/ percentage	Impacts		
				Min	Most likely	Max
1st year yield	Revenue	Normal	Absolut	60 000		78 203
Low market absorption var.	Revenue	Normal	Percentage	23%		
Increase in yield ton	Revenue	Triangular	Percentage	2%	2.5%	3%
Sale Price/ton	Revenue	Triangular	Absolut	450	650	850
1st year unit cost/ton	Cost	Triangular	Percentage	29%	35%	60%
Increase in sales price	Revenue	Triangular	Percentage	1%	2%	3%
Increase in cost	Cost	Triangular	Percentage	1%	2%	3%

Latin hypercube sampling technique and procedure used calculate NPVs for each region and statistic results generated after 5000 number of iterations runs. In the simulation, values of parameters entering into the model were chosen from their respective probability distributions by Latin hypercube sampling technics and were combined according to functional relationships in the model to determine whole region outcome and return i.e. NPV.

The Cumulative Distribution Functions (CDFs) of NPVs for each region is generated by repeating the process for a large number of times to give estimates of the output distributions of farming system performance and economic viability. SIMETAR program was used to calculate Stochastic Efficiency with Respect to a Function (SERF) and evaluate farming system risk efficiency and rank farming system regions within different level of risk preferences. The study finally performed CE analysis to estimate risk premium price and subsidy should be given to farmers to replant Date Palm trees in a less risky farming system practices and utilizing their poor farm land, saline soil and poor water quality in a sustainable manner.

2.6 Farming System Risk Allocation

The risks of Date Palm farming systems are normally shared by farmers within the region in term of irrigation system, water availability, water and soil salinity and Date Palm pests and diseases. However, some farmers are better able to cope with certain specific risks than others and grow water salinity and drought tolerance varieties. Resilience of farming system could be achieved by Date Palm diversification and risk-sharing. A reasonable risk-taking offset and optimizing risk allocation within the farmer in the same region can help farmers to cope with un-control risk facing some region and mitigate water pumping restriction in some areas.

The main risks cannot be control by Date Palm farmers:

- Risk of yield reduction: The risk of low yield that could not cover operation and investment cost.
- Risk of Dates sale price reduction: The sale price during the season is low and farmers need to sell their products to recover operation cost and loans repayment.
- High price vulnerability and total volumes of un-marketable Dates for each region (markets absorb rate).
- Risk of increasing operation cost: Increase of raw material, operation and maintenance cost.
- Risk of increasing capital cost: Risk of Dates trees replanting cost increase due to offshoots price increase.

Date Palm farming system risk efficiency and economic sustainability could be increased and enhanced by farmers' attitude and perception of risk sharing strategies.

3. Result and Discussion

3.1 Date Palm Farming Systems in Oman

Oman is located within Date Palm Belt area (15-27°) north latitude with long summer season extended for six months and warm winter and low rain fall. Date Palm farming system in Oman can be grouped into two main farming systems spread over six regions. The country date production in year 2013 reached 308 427 tons, 44% from Batinah Region, 21% Dakhiliyah Region, 15% Sherqiah Region, 11% Dahirah Region, 3% Muscat Region, 2% Musandam Region and 4% from others area. Farming system for each region can be described as under.

3.1.1 Coastal Farming System

The Coastal Farming System is coastal oasis farming system type located near the sea in low lying land and regular floods from mountains allowed considerable aquifer recharge and leaching salt from soils. The land located far distance from sea has sweeter and deeper irrigation water and crop is irrigated with wells, Tariq M. Alzidgali et al. (1993). The Coastal Farming System includes 57% of total Date Palm trees and produce 54% of Oman production. However, this farming system can be represented by Batinah Region, Muscat Region and Musandam Region.

3.1.1.1 Batinah Region

Batinah Region is a top region in Date Palm area, production and number of cultivated Date Palm trees i.e. 2.5 Million trees. Batinah Region can be grouped to three different farming systems zone:

Coastal area: The area is located in a tropical climate zone and characterized by high temperatures and humidity with high saline water and soil and low yield date palm trees. Date Palm is cultivated in intercropping with Alfalfa and lime trees and small area of vegetables and cereal crops is grown. The rapid expansion of cultivated area created water deficit in aquifer recharge and compensated by sea saline water and vegetable and Alfalfa crops moved inland. Date Palm cultivated varieties dominated by low yield, water and soil tolerance and stable varieties such as Um-Silla and Mabsali. Fishing activities is practices and dried dates are exported.

Sahel area: The area is located next to coastal area with a medium farm size (5-15 acres) and well-organized farms and use modern irrigation systems. Irrigation water is less saline and water quality determined the cropping pattern in this area. The climate is dry and Rhodes Grass crop is dominant cultivated crop. Farmers use fertilizers and pesticide and chemicals and modern bumps for irrigation. Date Palm cultivated varieties dominated by good quality and high yield varieties such as Khalas and Khasab and low yield variety such as Um-Silla which is used for human and animal feed.

Large Modern Farms area: The area is located next to Sahel area near mounting area and established recently as a large commercial farm. Date Palm cultivated with Rhodes Grass crop and Alfalfa and fruit trees. Date Palm cultivated in this area is of good quality varieties and cropping density is low compared to other regions.

Batinah Region cropping pattern dominated by fruit trees i.e. 60% such as dates, mango and lime trees, 20% forage crops and 20% vegetable crops. All agricultural operations are cared out by unskilled expatriates labour. Batihah Region has three Date cultivation zones and grows 31 Date Palm sustainable varieties. Date Palm area decreased from 28 560 acres in year 2000 to 20 278 acres in 2013 to maintains resilience farming system through resources management and crop diversification and recorded as resilience and sustainable Region in Oman.

3.1.1.2 Muscat Region

Muscat Region is the fifth region in term of production area and number of trees i.e. 289 303 Date Palm trees. Farms in Muscat region are small and Date Palm area start declining due to urbanization and farm land reclassification. The average Date Palm tree production in year 2013 recoded as 31 Kg/tree, Table 2.

3.1.1.3 Musandam Region

Musandam Region is dominated with mounting area and low production varieties are grown with an average of 30 Kg/trees. The total Date Palm cultivated area in year 2013 is 1497 acres and produce 6503 ton of dates.

3.1.2 Interior Farming System

The Interior Farming System located inland and experience with small farm areas and date grown in intercropping systems and high density plantation is recorded. The farming system has a dry climate area with low humidity. Farms irrigate crops with Al Falaj irrigation system and with good water quality. The Interior Farming System includes 43% of total Date Palm trees and produce 46% of Oman production. However, this farming system can be represented by Dakhiliyah Region, Dahirah Region and Sherqiah Region.

3.1.2.1 Dakhiliyah Region

The Dakhiliyah Region is the third top region in term of cultivated number of Date Palm trees i.e. 905 thousand trees and fourth in area and second in production. Water resources in this region are limited and farmers have a good experience of date palm cultivation and high yield Date Palm varieties such as Khalas, Fardh and Nighal are cultivated with the highest yield per tree i.e. 62.5 Kg/tree.

3.1.2.2 Dahirah Region

Dahirah Region farming system is the third Region in term of Date Palm cultivated area (6134 acres) and fourth in term of production and number of Date Palm trees i.e. 670 thousand trees with average production of 47 Kg/tree. Farahd and Khalas Date Palm varieties are cultivated at this region, table 2.

3.1.2.3 Sherqiah Region

Sherqiah Region is the second top region in Date Palm area and number of Date Palm trees i.e. 1.04 Million trees and recorded as the third region in production in year 2013. Date Palm trees are grown in Oasis agro-ecosystem area with loamy and sandy soil. The Date Palm is grown in intercropping and irrigated with Al Falaj Irrigation system and good quality of dates are grown such as Khalas, Hilali and Khasab varieties and Mabsaly for export.

Table 2 shows that in Dakhiliyah and Batinah regions farmers are grown more date palm varieties to extend harvesting time and mitigate risk and cope with yield uncertainty and price volatility. The total number of Date Palm varieties grown in Dakhiliyah and Batinah regions is 22 and 31 respectively. The Research Center in Oman recommends plant density of 65 date palm trees per Acre as the correct and appropriate planting density. Table 2 shows Date Palm planting density in Dakhiliyah and Batinah regions are low compare to other regions, and this may explain the reasons for high yield and production of date per acre in these regions and adaptation with climate changes. Crop diversification and inter cropping practices is also observed at Dakhiliyah Region and improve ecosystems.

Table 2. Date Palm Number of varieties, tree density and production for each Region in Oman (2013)

Location	Muscat	Musandam	Dakhiliyah	Batinah	Sherqiah	Dahirah
Date Varieties	17	13	22	31	19	13
Density/acre	91	143	34	53	90	121
Production kg/tree	31.18	30.39	62.53	46.41	39.09	46.98
Production ton/acre	2.99	4.34	6.67	6.71	3.87	5.69
Number of trees	289 303	191 124	905 126	2 546 071	1 036 361	670 265
Production ton	10 145	6 503	63 237	136 132	44 360	34 947
Establishment cost	1 456	2 145	850	318	2 070	2 299
Revenue	1 944	2 821	3 335	4 362	2 089	4 268
Operation cost/acre	682	1 382	1 501	1 527	1 149	1 920
Net return	1 262	1 439	1 834	2 835	940	2 348

3.2 Monte Carlo Simulation Models Run Results

The study investigated the regional economic performance and sustainability and calculates NPV by using regional data level. Batinah Region recoded the highest NPV i.e. 158 million Rials followed by Dakhiliyah 104 million and Dahirah 61 million as shown in Table 3. However, this indicates that net present value calculation is useful for investigating economic performance measurements, but it is also important to obtain the variation between NPVs and calculate the CVs and SDs as a risk measurement tools and decision support factors.

The Coefficient of Variation of the probability distribution of NPV was low for Batinah and Dakhiliyah regions and indicates regional farming system sustainability and risk efficiency. The smaller CVs and variance between NPVs i.e. (Batinah & Dakhiliyah) makes CDF curve steeper and less risky, whereas larger variance and CVs makes CDF curve flatter Figure 2. The result also shows that Batinah Region will get a positive NPV with 88% probability and Dakhiliyah Region get positives NPV with 75% probability, Table 3. Musandam and Dahirah Region model show a low Skewness figure which indicates downside risk control farming systems.

Table 3. Date Palm Farming System in Oman-Statistics for NPVs (000) for each Region

Models Location	Model (1) Muscat	Model (2) Musandam	Model (3) Dakhiliyah	Model (4) Batinah	Model (5) Sherqiah	Model (6) Dahirah
Mean (000)	5 127	5 396	104 474	157 960	53 048	61 189
SD (000)	38 174	144 834	157 040	146 286	135 335	158 766
CV	744%	2 684%	150%	93%	255%	259%
Skewness	-0.048	0.007	0.226	0.509	0.433	-0.071
Kurtosis	-0.107	0.081	0.501	0.816	1.029	0.357
Min(000)	(103 033)	(467 998)	(366 014)	(250 903)	(381 082)	(583 940)
Max	134 357	417 779	735 539	751 035	609 637	539 521
Range	237 390	885 777	1 101 553	1 001 938	990 719	1 123 461

The cultivated Date Palm varieties have an important and major effect on NPV for each region. The early and late maturity dates has a high market absorption demand and can be sold with high price off season. Moreover, dates for human consumption such as (Khalas, Khasab, Hilali) has high market absorption rate and can be sold with high price within the season, whereas dates used for industrial purposes such as (Farad, Mabsili) and for animal feed such as (Um-Silla) has a low market price.

The Batinah Region produces 93% of total Um-Silla country production. This variety is sustainable and water salinity tolerance variety, as a result, region production from this variety increased from 25 885 tons in 1997 to 28 791 tons in 2013 due to resilience and propensity of region farming system and its ability to retain productivity and cope with water deficits and water salinity risk increase. However, this variety contributes about 24% of region date production and is a late maturity date variety and can be stored and consumed as dry date off the season. The second varieties in Batinah region is Shahl i.e. 13% followed by early maturity variety Naghal i.e. 11% and consumer favorites and premium price late maturity variety Khasab i.e.10% of date production.

The cropping pattern analysis is performed for all regions and shows that fruit crops (including Dates) occupied 54% of the total cultivated area followed by forage crops 27%, vegetables 12% and field crops 6%, Figure 1. Date Palm area for the country has been reduced from 93,534 acres in 2000 to 59,602 acres in 2013 to cope with water shortage problems. Batinah Region reduced fruits trees, field crops, forage areas and Date Palm area by 29% from year 2000 to year 2013. Dakhiliyah and Sherqiah Regions use intercropping system and increased forage crops and fruit trees areas by cultivating Alfalfa under fruit and Date Palm trees. Farmers get multiple benefits and obtain additional income from forage, Dates, fruits and vegetables, and improve soil fertility from leguminous forage and enhancing water productivity (yield per unit of water consumed) when irrigation is used for intercropped fields.

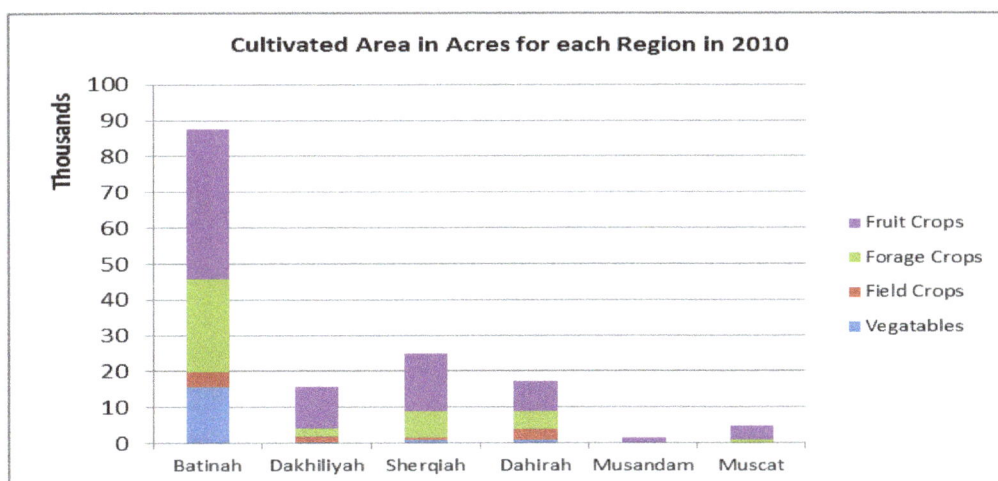

Figure 1. Cultivated Area in Acres and Cropping Pattern for each Region in Oman 2010.

3.3 Date Palm Regions and Cumulated Distribution Function Analysis

To test Date Palm Farming System sustainability the Cumulated Distribution Function CDF graphs performed to illustrate the range and probabilities of NPV for combinations of farming system for each region. Due to CDF lines cross in the graph we could not ranked regions and their Date Palm Farming System sustainability by using first degree stochastic dominance, and Stochastic Efficiency with respect to a Function (SERF) is performed to have a good ranking analysis and most risk efficient region and farming system alternatives. The analysis indicates Batinah Region farming system distribution line on the right is preferred to those on the left.

Muscat Region Date Palm farming system could not manage downside risk and were not viable and less efficient due to low profit and probability of getting positive return which is 54% only. Figure 2 shows Batinah Region Farming System with a curve on the right side of the graph is a less risky and most sustainable farming system. The Dakhiliyah Region is the second most risk efficiency and got a probability of 75% to achieve a positive NPV. The study indicates Batinah and Dakhiliyah Regions which cultivate 52% of total Date Palm area in Oman is located in risk efficient farming system areas.

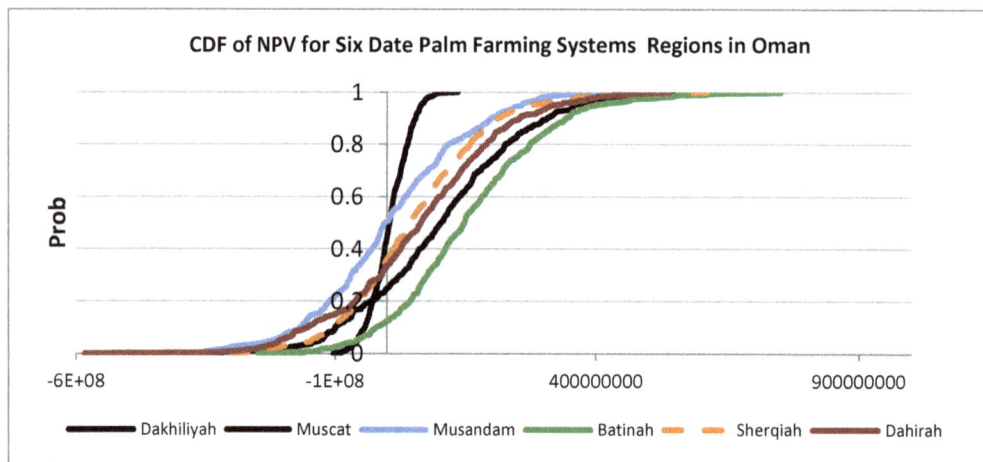

Figure 2. Comparison of 6 CDF of NPVs of Regions' Dates Palm Farming Systems in Oman.

3.4 SERF Analysis and Farming System Certainly Equivalent

The SERF analysis performed and calculated Certainly Equivalent values over a range of absolute risk aversion coefficients and represent different decision makers' degree of risk aversion. Decision makers are risk averse if ARAC below (0), risk neutral if ARAC equal (0) and risk preferring if ARAC more than (0) figures. The ARAC values used in this study represent risk neutral, rather risk and extremely risk averse as shown in Figure 3.

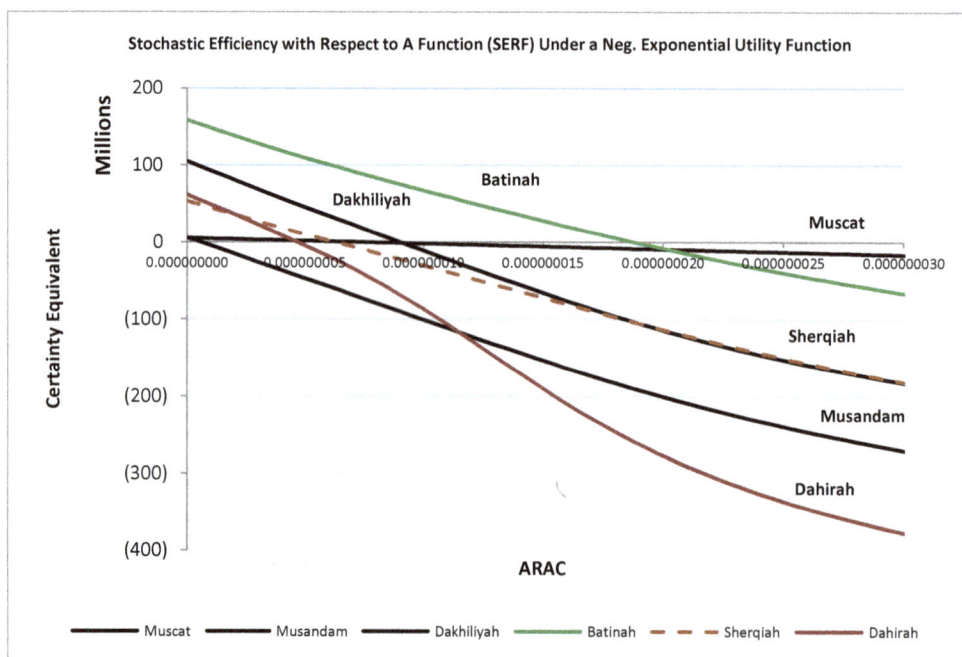

Figure 3. SERF for NPVs of six Date Palm farming system in Oman.

In Figure 3 above the SERF analysis is performed by using SIMETAR software program to compare six farming system alternatives simultaneously for all ARAC values in the range from (0.000000) to (0.000000018), and identified Batinah Region alternative as most risk efficient under all risk aversion. Under risk rather with (0.000000008) ARAC Batinah and Dakhiliyah Farming systems are risk efficient and Batinah farming system is the only efficient farming system under extremely risk aversion level. Muscat Region ranked as second most efficient farming system extremely risk aversion level as shown in Table 4.

Table 4. Ranking of Risky Alternatives by Risk aversion using CE for NPV (000,000) of Regions

Risk degree	Normal Risk		Rather Risk		Rather Risk		Extreme Risk	
ARAC	0.0000000		0.000000005		0.000000008		0.000000018	
Rank	Alternative	CE	Alternative	CE	Alternative	CE	Alternative	CE
1	Batinah	158	Batinah	108	Batinah	85.77	Batinah	5.15
2	Dakhiliyah	104	Dakhiliyah	43	Dakhiliyah	14.05	Muscat	-8.07
3	Dahirah	61	Sherqiah	8.9	Muscat	-0.529	Dakhiliyah	-96.79
4	Sherqiah	53	Muscat	1.4	Sherqiah	-12.14	Sherqiah	-97.95
5	Musandam	5.4	Dahirah	-6.88	Dahirah	-46.56	Musandam	-183.49
6	Muscat	5.1	Musandam	-48.9	Musandam	-79.57	Dahirah	-249.90

The study ranks six alternative farming systems using SERF, over the range of risk normal, rather risk and extremely risk averse. The results are presented graphically in Figure 3 and numerically in Table 4. The study reveals that under normal risk aversion all Date Palm Farming Systems NPV are positive and Batinah, Dakhiliyah and Dahirah got the first, second and third ranking alternatively. Under rather risk (8.08E) only Batinah and Dakhiliyah got a positive CE values and under extreme risk only Batinah Region got a positive CE values. Table 4 also shows that Batinah Region is the most sustainable farming system in Oman and got a positive CE value even under extremely risk aversion.

3.5 Risk premium and Date Palm replacement policy

Risk premiums measure the value of the preferred alternative farming system over a less preferred alternative. The risk premium can be calculated by subtracting the CE value of less-preferred farming system alternative from CE value of the preferred one at each RAC level. The SERF analysis can help Decision Maker's with deferent risk preferences to select risk efficient farming system alternatives as per their risk aversion level.

Figure 3 show the difference between farming systems' CEs value which represents Decision Makers willingness to exchange the preferred (Date Palm Farming System) risky alternative for another less-preferred risky farming system alternative. The value of WTP is calculated as the difference between the CE for a risky farming system alternative and represents the date palm replacement payment and subsidy necessary to make the farmers and investors indifferent between the less-preferred farming system alternative and the preferred farming system alternative (Batinah Region):

$$WTP = CE_{preferred} - Ce_{alternative}$$

The SERF rankings methodology and WTP calculation are used to examine farming system sustainability and Date Palm replacement policy and strategies needed in Oman. Figure 3 shows that Batinah Region is most preferred farming systems under normal risk aversion, followed by Dakhiliyah, Sherqiah and Dahirah under normal risk aversion. From table 4, it is evident that Decision Makers for Date Palm replacement program should give more attention and support to risky regions i.e. Dahirah, Muscat and Musandam Regions. The study also indicates that Batinah Region benefit from date variety and crop diversification and increase ecosystem resilience and economic benefits. Dakhiliyah and Sherqiah Regions benefit from intercropping system and enhance water productivity.

4. Conclusion

The main objective of this paper is to identify Date Palm Farming System sustainability and ranked them over the range of risk aversion levels. The study also evaluate Date Palm cultivation viabilities in Oman and estimate economic sustainability through calculation of future values of different Date Palm Farming System and projected net cash flows by estimating yield and price and other main key variables which effect NPV and farming system sustainability. The whole region stochastic model using historical data of Date Palm cultivated area, number of trees, yield and price for each region is constructed. The risk and uncertainty associated with Date Palm farming systems and farming practices at different regions in Oman incorporated in the analysis. The stochastic character and shape were also incorporated by identifying probability distribution for each uncertain variable.

The study reverses that all Date Palm Farming System in Oman are viable and got a positive NPV at normal risk aversion level. The Batinah Date Palm Farming System recorded as the most risk efficient and economic sustainable system with 88% probability of achieving positive NPV, whereas Musandam farming system got a positive NPV with low probability of 49% only. Dakhiliyah farmer grows high yield and Date quality varieties such as Khalas, Khasab, Hilali for human consumption, whereas Sherqiah farmers grows Mabsili Date variety for export as dried boiled Dates. Farmers in these regions are manage to develop resilience agricultural systems by introducing tolerance Date Palm varieties and using affordable technologies and strategies such that ecosystem functions and services can be maintained and livelihoods can be protected. Although Batinah Region has a highest water deficit consist of 69% of total country deficit but the region mange to cope with the risk and ranked as the most risk efficient region in Oman by adjusting cropping pattern water requirement with annual underground water recharge and Region water supply.

Batinah and Dakhiliyah Regions farmers are diversifying date palm cultivated varieties to manage risk associated with price and yield loss uncertainty and extended harvesting period to six months to mitigate market access risk. Moreover, Batinah and Dakhiliyah managed to cope with water shortage and environment constrains by cultivating and producing different type of dates for different economic use i.e. direct human consumption, industrial, export and animal feed use. Batinah Region produces 93% of Oman production from Um-Silla variety to cope with water salinity risk and compensate animal feed shortage.

The study indicates that local Date Palm verities cultivated in Oman are well adapted to environment and underground water limitation and other social economic attributes and constrains. The Date Palm Farming System shows its resilience, persistence and survivals under risk and uncertainty shocks in the future and Batinah Region ranked as the most risk efficient farming system under all risk aversion levels.

Risk premium analysis indicates that Dakhiliyah Region farmers can pay up to RO 59 for replanting Date Palm tree with a sustainable variety to avoid price and yield reduction risk he is facing and to move to Batinah Region farming practices with less risky farming system. The farmer in Dahirah Region is willing to pay RO 144 per Date Palm tree to shift to more efficient farming system such as Batinah Region farming system.

Date Palm Farming System analysis and finding of this study should be used as a basis and foundation for replanting program and selecting of Date Palm varieties, planted area, cropping pattern at each region. Crop and livestock activates interaction and resources and environmental constrains should be considered. Date Palm replacement program and policy makers should give economic incentives needed to encourage production of selected sustainable Date Palm varieties and crop diversification that suit and mitigate risk for each region.

Acknowledgements

The author thanks Mr. Al Hag Bakhit from Directorate General of Agriculture Dhofar Region-Ministry of Agriculture and Fisheries for providing Data and comments that greatly improved the manuscript.

Competing interests

The author declared that he has no competing interest and declared that the research was conducted in the absence of any commercial or financial relationships that could be construed as a potential conflict of interest.

References

Anna, G., Mattia, B., Maria, E. M., Eugenio, D., & Alberto, Pirani. (2016). The social pillar of sustainability: a quantitative approach at the farm level. *Agricultural and Food Economics, 4,* 15. http://dx.doi.org/10.1186/s40100-016-0059-4.

Brenda, B. L. (2011). Resilience in Agriculture through Crop Diversification: Adaptive Management for Environmental Change. *BioScience, 61*(3), 183-193. http://dx.doi.org/10.1525/bio.2011.61.3.4

Blumenfeld, S. L., Christophersen, C. T., &Coates, D. (2009), Water, Wetlands and Forests. *A Review of Ecological, Economic and Policy Linkages.* CBD Technical Series No. 47. Montreal/Gland, Switzerland, Secretariat of the Convention on Biological Diversity and Secretariat of the Ramsar Convention on Wetlands.

Carpenter, S., Brock, W., & Hanson, P. (1999). Ecological and social dynamics in simple models of ecosystem Management. *Conservation Ecology, 3*(2), 4. http://www.consecol.org/vol3/iss2/art4/

Chen, X. D., Lupi, F., He, G. M., & Liu, J. G. (2009). Linking social norms to efficient conservation investment in payments for ecosystem services. *Proceedings of the National Academy of Sciences of the United States of America (PNAS), 106*, 11812-17. http://dx.doi.org/10.1073/pnas.0809980106

Eihab, M. Fathelrahman, J. C., Ascough, II, Dana, L. H., Robert, W., Malone, P. H., Lori, J. W., & Ramesh, S. K. (2011). Economic and Stochastic Efficiency Comparison of Experimental Tillage Systems in Corn and Soybean Under Risk. *Expl Agric, 47*(1), 111-136. http://dx.doi.org/10.1017/S0014479710000979

Folke, C., Carpenter, S., Elmqvist, T., Gunderson L., Holling, C. S., & Walker, B. (2002). Resilience and Sustainable Development: Building Adaptive Capacity in a World of Transformations. Scientific Background Paper on Resilience for the process of the World Summit on Sustainable Development on behalf of the Environmental Advisory Council to the Swedish Government. *Interdisciplinary Center of Natural Resources and Environmental Research*, Stockholm University, Sweden. http://dx.doi.org/10.1579/0044-7447-31.5.437

Hardaker, J. B., Richardson, J. W., Lien, G., & Schumann, K. D. (2004). Stochastic efficiency analysis with risk aversion bounds: a simplified approach. *Australian Journal of Agricultural and Resource Economics, 48*, 253-270. http://dx.doi.org/10.1111/j.1467-8489.2004.00239.x

Hansen, J. W., & Jones, J. W. (1996). A system framework for characterizing farm sustainability. *Agriculture Systems, 51*, 185-201. http://dx.doi.org/10.1016/0308-521X(95)00036-5

James, W. R., Brian, K. H., Joe, L. O., & Chope, G. II. R. (2007). Including Risk in Economic Feasibility Analyses: The Case of Ethanol Production in Texas. *Journal of Agribusiness, 25*, 2.

Kheiry, H. M. I., & Al-Marzogi, M. H. (1997). *Date Palm Development Strategy in Oman*. Directorate General of Agricultural Research, Ministry of Agriculture & Fisheries, Sultanate of Oman, Muscat.

Kheiry, H. M. I., Al-Marzogi, M. H., & Riad, S. A. D. (1994). *Technical and economic constrains of date palm sector and farming systems in Sultanate of Oman*, Report submitted to AFESD, IFAD and IDB. Syria.

Lien, G., Hardaker, J. B., & Flaten, O. (2006). Economic sustainability and risk efficiency of organic versus conventional cropping systems. *Aspects of Applied Biology*, 79, 2006.

Lien, G., Hardaker, J. B., & Flaten, O. (2007). Risk and economic sustainability of crop farming systems. *Agricultural Systems, 94*(2), 541-552. http://dx.doi.org/10.1016/j.agsy.2007.01.006

Qiu, L. G. (2001). *Development of risk analysis models for decision-making in project management*. PhD Thesis submitted to School of the Built Environment, Napier University, Edinburgh, UK.

Savvakis, C. S. (1994). Risk analysis in investment appraisal" *Project Appraisal Journal, 9*(1), 3-18.

Saunders, J. F., & Lewis, W. M. (2003). Implications of climatic variability for regulatory low flows in the South Platte River basin. Colorado. J. Am. *Water Resource Association, 39*, 33-45. http://dx.doi.org/10.1111/j.1752-1688.2003.tb01559.x

Tariq, M., Alzidgali, M. A. H., Rajinder, K. S., & Kheiry, H. M. I. (1993). South Batinah Integrated Study. *Soil survey and land classification project, 3*.

Quiroga, S., Fernandez-Haddad, Z., & Iglesias, A. (2010). Risk of water scarcity and water policy implications for crop production in the Ebro Basin in Spain. *Hydrology and Earth System Sciences Discussions Journal, 7*, 5895-5927. http://dx.doi.org/10.5194/hessd-7-5895-2010

16

Profitability of Organic Vegetable Production via Sod Based Rotation and Conventional Versus Strip Tillage in the Southern Coastal Plain

Mona Ahmadiani[1], Chun (Cathy) Li[1], Yaqin Liu[1], Esendugue Greg Fonsah[2], Christine M. Bliss[3], Brent V. Brodbeck[3] & Peter C. Andersen[3]

[1]Department of Agriculture and Applied Economics, University of Georgia, Corner Hall, Athens, GA 30206, USA

[2]Department of Agriculture and Applied Economics, University of Georgia, Tifton, GA, 31793, USA

[3]University of Florida, Institute of Food & Agricultural North Research and Education Center 155 Research Road, Quincy, FL 32315, USA

Correspondence: Mona Ahmadiani, Department of Agriculture and Applied Economics, University of Georgia, Corner Hall, Athens, GA 30206, USA. E-mail: monaah@uga.edu

Abstract

There are little economic data concerning the profitability of organic vegetable crops in the Southern Coastal Plain, especially in reference to sod-based rotation and tillage alternatives. A three-year experiment was conducted at the North Florida Research and Education Center-Quincy involving a crop rotation sequence of oats and rye (winter), bush beans (spring), soybean (summer) and broccoli (fall). Bush beans and broccoli were the cash crops. This paper presents analyses of the riskiness of organic production utilizing years in bahiagrass prior to initiating the crop rotation sequence and conventional tillage (CT) versus strip tillage (ST). Methods of "Risk-rated enterprise budget" and "Analyses of Variance-Covariance Matrix (ANOVA)" were utilized for determining relative profitability, and coefficient of variation was applied for measuring riskiness of each treatment. Three years of bahiagrass prior to initiating the crop rotation sequence, in combination with conventional tillage, had the highest profitability and ranked as the least risky scenario. The second most profitable treatment was conventional tillage with four years of bahiagrass. Focusing on strip tillage, four years of bahiagrass with strip-tillage ranked third in term of profitability.

Keywords: economic returns, broccoli, bush bean, sod-based rotation, tillage treatment

1. Introduction

The demand for organically-grown products in the United States has increased substantially such that the previously supply-driven organic sector is now market-driven. As a result, the marketing boom has increased retail sales of organic foods from $3.6 billion in 1997 to $21.1 billion in 2008 (Dimitri & Oberholtzer, 2009). Consequently, the organic sector is now the fastest growing sector of the U.S. agricultural economy. As of 2007, the 48-state total certified organic farms was over 4 million acres (USDA Census; 2009). However, transition to organic production has great regional variation. California has the most organically-certified acreage in production, with Washington, Wisconsin, Minnesota, Iowa, Pennsylvania, Ohio, New York, Vermont and Maine rounding out the top 10 (Greene & Kreman, 2003).

In contrast, the Southern Coastal Plain (defined in this paper as Florida, Georgia and Alabama) have less than 15,000 acres of organic production (USDA Census, 2009), or less than 0.3% of the national total acreage. Moreover, there has been a great reduction in the number of commercial farms in the southeastern U.S. based on agronomic crops due to higher costs of inputs, climatic conditions and low commodity prices. Until recently, there has been little incentive for the industry to develop and utilize sustainable farming systems such as sod-based crop rotation or by utilizing organic agro-ecosystems (Drost et al., 1996; Katsvairo et al., 2007). Individual organic producers in the Southern Coastal Plain have taken the initiative, and are already organically producing a variety of crops. Winter vegetables are the most prominent due to lower pest pressures and marketing windows. However, the acreage of organic vegetables remains small, off-farm inputs are high and marketing is primarily local. Current local markets for organic vegetables and other products are not meeting consumer demand. Thus, development of more efficient production systems that increase the profitability and assist in transition to organic systems is critical in the Southern Coastal Plain.

Economic sustainability increasingly depends on selecting profitable enterprises, sound financial planning, proactive marketing, risk management, and good overall management (Sullivan, 2003). Previous studies have focused on crop rotation and tillage methods individually. For instance, Gebremedhin and Schwab (1998) reviewed the economic literature on crop rotation systems, and reported that rotation-based cropping systems usually perform as well or better than continuous monocultures under most environmental and economic conditions. Zwingli et al. (1989) analyzed the economic potential of vegetable production using simple crop rotations and report that rotations are generally more stable with respect to markets. Martin and Hanks (2009) evaluated the combination of crop rotations and tillage, and calculated net return of each treatment from the enterprise budget. Minimum tillage and a corn rotation was illustrated to have the largest net return and lowest risk in the every other year rotation.

Crop rotation and minimum tillage practices can be economically profitable and environmentally sustainable. Other studies applied expected value-variance analysis to three years data of center-pivot irrigated corn production and tested three tillage methods (conventional, conservation and minimum tillage practices). They found that the conservation tillage methods consistently produced the highest profitability (Tew et al., 1986; Archer et al., 2008).

Klonsky (2012) found expected yields of several crops (alfalfa, tomato, field corn, broccoli/lettuce, raisins, walnuts) in California were comparable for organic and conventional production, with an exception of lower expected output for organic strawberries and almonds; profitability of these crops depended on price premiums. Crop yield and economic performance of organic, low-input and conventional farming system were compared over an eight-year period on research from the "Sustainable Agriculture Farming System Project" in California (Clark et al. 1999). They applied four farming-system treatments including 2 and 4 years rotation (Note 1) and compared conventional, low input and organic systems. The organic treatments had the highest return and were the most profitable among four year rotations. Also, bean yields and profit over the eight years indicate good short-term and long-term potential for low-input and organic systems and make it a good candidate as a transition crop even without a premium price. Marois and Wright (2003) used a working business model to predict the potential income of continuous cropping of cotton and peanuts compared to a 4-year rotation consisting of 2 years of grazing cattle on bahiagrass followed by peanuts and cotton. The model predicted that a 200-acre farm would yield $5,000 per year growing continuous cotton compared to $22,000 per year under the 4-year rotation that included two years of bahiagrass rotation. Marois et al. (2002) showed the average peanut yield in the Southeast U.S. was about 2,500 lbs. per acre, but yields after bahiagrass were often 3,500–4,500 lbs. per acre. Economic modeling showed that profits for cotton and peanuts in a sod-based rotation were about two times greater than those for a peanut-cotton-cotton rotation.

The benefits of bahiagrass include increased soil porosity, soil organic matter, soil moisture retention, and reduced plant disease and soil nematode populations (Gallaher et al., 2003; Johnson et al., 1999; Marois et al., 2003; Marois & Wright, 2003; Sumner et al., 1999). It is inexpensive to establish and maintain compared to similar perennial grasses such as hybrid bermudagrass (Hancock et al., 2013). For example, bahiagrass establishment costs and annual expenses were approximately $30 to $200 per acre, and $50 to $150 less than bermudagrass respectively. Balkcom et al. (2007) compared crop yields, production costs, revenue and net economic returns in the conventional peanut-cotton-peanut vs. bahiagrass-bahiagrass-peanut-cotton rotations, and conventional tillage (CT) versus strip tillage (ST). Net returns for the two crop rotations and tillage systems were determined using enterprise production budget for each year. Rotation system did not affect yield, revenue or return. Tillage was the most important factor in the study, and CT was superior to ST both in terms of yield, revenue and net return.

This study evaluates an organic vegetable production system utilizing variable years of sod-based rotation and bahiagrass, and CT versus ST in the Southern Coastal Plain. The main production limitations in the Southern Coastal Plain are infertile, compacted, droughty soils, limited water retention, loss of valuable top soils and pest suppression. Sod-based crop rotation and organic crop production greatly improve the physical properties of soils (Drinkwater et al., 1995; Waldon et al., 1998). Furthermore, the benefits of crop rotation systems in organic production may be enhanced if used in combination with ST. Strip tillage may decrease production expenses for labor, fuel, and machinery, reduce water loss from soil due to evaporation, and positively affect yields by minimizing soil loss and soil compaction (Gebremedhin & Schwab, 1998). Although utilization of ST in organic systems may require intensive management to alleviate weed control problems, severe erosion and depleted soils in the Southern Coastal Plain dictates that reduced or no systems tillage should be utilized whenever possible.

The objective of this paper is to analyze profitability and economic viability of different combination in sod-based rotation coupled with tillage treatments in organic vegetable production (Note 2). Yields and input costs data during the long term field experimental period was collected and analyzed using enterprise budget approach. A comparison of variable years of sod-based rotation coupled with different tillage techniques (CT and ST) were carried out to determine which system provides optimal solution in terms of financial lucrativeness to growers. There are many attendant economic, environmental and ecological components of these agricultural systems; however, it is clear that the adoption of agricultural technologies by farmers are closely related to economic profitability (Hoorman et al., 2009; Garcia-Préchac et al., 2004; Siri-Prieto & Ernst, 2009; Ikemura and Shukla, 2009; Johnson et al., 1999; Sumner et al., 1999)

2. Material and Methods

The primary goal of this project was to integrate advances made on sod-based rotation and strip tillage into organic systems for vegetable production in the Southern Coastal Plain. In this study we combined; 1) two tillage treatments of conventional tillage (CT) and strip tillage (ST), and 2) five rotation treatments including 0, 1, 2, 3 and 4 years of bahiagrass preceding annual vegetable cropping rotations. All treatment combinations were utilized in similar annual crop rotation with oats/rye (winter), bush beans (spring), soybean (summer) and broccoli (fall). Experiments were conducted for 3 years of field trials (from 2010-2013) at the University of Florida, North Florida Research and Education Center, Quincy, Florida. The experiment had four replications for all ten treatments (Note 3). For the purpose of simplicity, the above-mentioned treatments will be denoted by conventional tillage with zero year of bahiagrass (CT0), conventional tillage with one year of bahiagrass (CT1), conventional tillage with two years of bahiagrass (CT2), conventional tillage with three years of bahiagrass (CT3), conventional tillage with four years of bahia grass (CT4), strip tillage with zero year of bahiagrass (ST0), strip tillage with one year of bahiagrass

(ST1), strip tillage with two years of bahiagrass (ST2), strip tillage with three years of bahiagrass (ST3) and strip tillage with four years of bahiagrass (ST4) in the rest of the paper (Table 1; Note 3).

Table 1. Summary Codes for the different treatments in the sod-based rotation studies in Florida, 2011-2013.

Treatment	Explanation
CT0	Conventional tillage with zero year of bahiagrass
CT1	Conventional tillage with one years of bahiagrass
CT2	Conventional tillage with two years of bahiagrass
CT3	Conventional tillage with three years of bahiagrass
CT4	Conventional tillage with four years of bahiagrass
ST0	Strip tillage with zero year of bahiagrass
ST1	Strip tillage with one years of bahiagrass
ST2	Strip tillage with two years of bahiagrass
ST3	Strip tillage with three years of bahiagrass
ST4	Strip tillage with four years of bahiagrass

The climatological and environmental conditions, detailed experimental design, soil characteristics and treatments are described in Bliss et al., (2016).   Economics and statistical analysis were adopted using a risk-rated enterprise budget approach and the classic two-way factorial ANOVA. Fundamental economic measures for conducting the economic analysis of alternative production scenarios are net revenue of each cropping system and approach of risk-rated budget enterprise were utilized for comparing net revenue and riskiness of above treatment.

3. Risk-Rated Enterprise Budget Approach

Preparation of an enterprise budget and analyzing the preferable measure of return requires identification of cultural operations and their associated costs, identification and valuation of production inputs, and the proper valuation of output (Gebremedhin and Schwab, 1998; Fonsah & Maltag Group, 2009a; 2009b; Fonsah & Chidelebu, 2012). A risk-rated enterprise budget is the broadest assessment of profitability compared to a deterministic enterprise budget; the expected returns of each treatment will be computed by taking different scenarios of price and yield into account. Considering broccoli and bush bean as two cash crops in this experiment, 5 scenarios of "pessimistic", "worst", "median", "best" and "optimistic", are assigned for constructing risk-rated budget enterprise. A 15% and 35% percentage decrease and increase in broccoli yield and 21% and 28% decrease and increase in its price were adopted for these scenarios, respectively, based on historic yield information and USDA data (Fonsah et al., 2008; Ferrer et al., 2011).

A 25% and 50% increase and decrease in bush bean yield and 25% and 46% increase and decrease in its price were also adopted for afore-mentioned scenarios.   Variable costs of each production systems were broken down into pre-harvest, harvesting and marketing costs and the revenue was calculated by multiplication of median value of each cash crop (broccoli and bush bean) by its median price (Fonsah et al., 2008; Fonsah et al., 2007; Fonsah & Hudgins, 2007; Byrd et al., 2006 & 2007).

4. Results

Table 2 presents the average yield of two cash crops in this experiment for the ten treatments. More years of treatment with bahiagrass resulted in higher average yield of bush bean for both tillage treatments (Appendix A). Although CT produced higher average yield than ST for the same years of bahiagrass treatment for both crops, average yield of broccoli did not show significant increase in subsequent years of using bahiagrass. Comprehensive yield data for each year for bush bean and broccoli are illustrated in Bliss et al. (2016).

Table 2. Average yield of bush bean and broccoli in the different sod-based rotation study in Florida, 2011-2013 (lbs./acre)

Crops	CT0	CT1	CT2	CT3	CT4	ST0	ST1	ST2	ST3	ST4
Bush beans Avg Yields	7,012	6,774	6,141	12,222	12,104	5,971	5,566	6,323	9,154	9,640
Broccoli Avg Yields	5,004	4,833	6,230	7,505	6,638	4,568	3,100	5,314	4,563	4,780

Enterprise budgets were developed by The UGA Fruits and Vegetable Extension Ag-Economics Specialist, and value of net revenue for broccoli and bush bean for different treatments were calculated. Net return measure that was considered as the main variable for decision-making process that revealed profitability, were calculated under five possible risk-rated pricing and yield outcome (best, optimistic, median, pessimistic, and worst). The expected net return of each treatment in each year with its probability of occurrence is presented in appendix B (Fonsah & Hudgins, 2007; Fonsah et al., 2008; Ferrer et al., 2011).

Table 3 illustrates that, for bush bean, three years of bahiagrass was the most profitable among the CT treatments ($4,843/Ac), and net revenue increased substantially in second year of cultivation ($7,199/Ac) for this treatment. The second best scenario after CT3 in term of profitability was CT4 with value of $5,610/Ac. Second year of vegetable production for three year of bahiagrass (ST3) had the highest profit ($5,105/Ac) among strip-tillage treatments. The third year of vegetable cultivation for ST0 treatment yield net revenue of $4,150/Ac that was induced by its high yield equivalent to 8,524 pounds per acre.   Total variable cost as a component of total cost is a function of yield, therefore, substantial variation in yield that might be induced by exogenous agricultural shocks like climate condition, lead to variation in variable cost. Total variable cost shows larger variation than total fixed cost (standard deviation is 864 vs 35).

Table 3. A breakdown of all the key economic parameters and net revenue for the different sod-based rotation of bush bean, 2011-2013 (Dollar/acre)

Treatment	Year	Pre-harvest variable costs	Total Harvest and marketing	Total Variable costs	Total Fixed costs	Total costs	Total Revenue	Net revenue
CT0	2011	1,851	1,976	3,827	804	4,631	7,480	2,849
	2012	2,066	1,587	3,652	836	4,488	5,656	1,168
	2013	2,364	1,769	4,134	881	5,014	7,187	2,172
CT1	2012	2,066	1,820	3,885	836	4,721	6,487	1,766
	2013	2,364	1,593	3,957	881	4,838	6,470	1,633
CT2	2011	1,851	867	2,719	804	3,522	3,284	(239)
	2012	2,066	1,863	3,928	836	4,764	6,639	1,875
	2013	2,364	1,903	4,267	881	5,148	7,730	2,582
CT3	2012	2,066	3,020	5,085	836	5,921	10,764	4,843
	2013	2,364	3,411	5,775	881	6,656	13,854	7,199
CT4	2013	2,364	2,892	5,256	881	6,137	11,747	5,610
ST0	2011	1,851	884	2,735	769	3,503	3,345	(158)
	2012	2,066	1,392	3,458	801	4,259	4,963	704
	2013	2,364	2,404	4,768	845	5,613	9,763	4,150
ST1	2012	2,066	989	3,055	801	3,855	3,525	(330)
	2013	2,364	1,753	4,117	845	4,963	7,121	2,158
ST2	2011	1,851	752	2,603	769	3,372	2,846	(525)
	2012	2,066	2,055	4,121	801	4,921	7,325	2,404
	2013	2,364	1,737	4,101	845	4,947	7,055	2,108
ST3	2012	2,066	2,282	4,348	801	5,148	8,134	2,986
	2013	2,364	2,716	5,080	845	5,925	11,031	5,105
ST4	2013	2,364	2,294	4,658	845	5,504	9,319	3,815

Applying the classic two-way factorial ANOVA (Afifi & Azen, 1973) for assessing the treatment effect on bush bean net revenue illustrates that more years of bahiagrass with ST lead to statistically significant increase in net-revenue of bush bean. The corresponding p-value of F statistic for testing hypothesis of effectiveness of both treatments was $P<0.0001$ (Table 4). ANOVA table shows that separate main effect of more years of bahiagrass and using ST treatment have statistically significant power in explaining change in net revenue ($P < 0.0001$ and $P < 0.038$, respectively). Interaction term in two-way ANOVA matrix captures the change in effect of treatment 1 over different values of treatment 2. The interaction term for two treatments was not significant ($P< 0.23$) and this means that effects of treatment 1 and treatment 2 are independent.

Table 4. Statistical analysis of treatment effect on net revenue of bush bean.

Source	Partial sum of square	Degree of freedom	Mean sum of square	F - stat	Prob < F
Model	229 834 126	9	25 537 125.1	7.11	0.0001
Years of bahiagrass	201 877 686	4	50 469 421.4	14.06	0.0000
Tillage*	15 975 386.4	1	15 975 386.4	4.45	0.0378
Bahiagrass*tillage	20 380 679.9	4	5 095 169.98	1.42	0.2345
Residual	308 799 202	86	3 590 688.39		
Total	538 633 328	95	5 669 824.5		

* For designing the ANOVA analysis, code 1 and 2 are assigned for conventional tillage and strip tillage respectively and same codes as years of bahiagrass are assigned for bahiagrass treatments.

Table 5 shows average economic parameters for bush bean over three years of experiment. Three and four years of bahiagrass lead to maximum value of net revenue for bush bean respectively. Also, for a given year of bahiagrass, CT treatment leads to higher net revenue compared to ST treatments. Comparing "Total fixed cost" and "Total variable cost" across treatments shows that CT is more costly than ST for the same number of years of bahiagrass. Considering CT3 and ST3 as the most profitable treatments for the bush bean for two alternative tillage treatments, opportunity cost of 3 years of growing bahiagrass for a farmer that forgoes vegetable production, would be $6,189 for conventional tillage (three years net revenue from CT0) and $4,696 for strip-tillage (three years net revenue from ST0)[1]. This estimation suggests that a farmer would have profit even after accounting for opportunity cost of forgoing vegetable production, such that for CT3, the two years net revenue after subtracting opportunity cost is $5,853, and for ST3 the two years net revenue after subtracting opportunity cost is $3,395.

Higher variation in profit translates into higher risk. Therefore, in this study, we present "Coefficient of Variation (CV)" to illustrate riskiness of each alternative compare to other alternatives. CT1 was the least risky production alternative of all treatments and ST3 is the least risky production alternative among ST treatments (Table 5). Also, CT3 has the highest profitability among all treatments, and ranks third in term of least riskiness of profitability.

---

[1] For this estimation, we assume that any three sequential years of growing bush bean provide approximately the same net revenue for the farmers as in our experiment in CT0 and ST0.

Table 5. A breakdown of all the key economic parameters including net revenue and coefficient of variations for the different sod-based rotation of bush bean, 2011-2013 (Dollar/acre)

Treatment	Pre-harvest variable costs	Total Harvest and marketing	Total Variable costs	Total Fixed costs	Total costs	Total Revenue	Net Revenue	Coefficient of variation
CT0	2,094	1,777	3,871	840	4,711	6,774	2,063	0.41
CT1	2,215	1,706	3,921	858	4,780	6,479	1,699	0.06
CT2	2,094	1,544	3,638	840	4,478	5,884	1,406	1.04
CT3	2,215	3,215	5,430	858	6,288	12,309	6,021	0.28
CT4	2,364	2,892	5,256	881	6,137	11,747	5,610	- *
ST0	2,094	1,560	3,654	805	4,458	6,024	1,565	1.46
ST1	2,215	1,371	3,586	823	4,409	5,323	914	1.92
ST2	2,094	1,515	3,608	805	4,413	5,742	1,329	1.21
ST3	2,215	2,499	4,714	823	5,537	9,583	4,046	0.37
ST4	2,364	2,294	4,658	845	5,504	9,319	3,815	- *

* Since there is only one year of net revenue of bush bean available for CT4 and ST4, measure of Coefficient of variation is not computable for these two treatments.

Estimated economic parameters for broccoli showed that first year of cultivation for zero and two years of bahiagrass and CT treatment have the two highest net revenue among the ten treatments (Table 6). The first year of cultivation in zero and two years of bahiagrass with ST treatments ranks 3rd and 4th in term of profitability. Variation in total variable cost was higher than total fixed cost, which was induced by variation in yield.

Table 6. A breakdown of all the key economic parameters and net revenue for the different sod-based rotation of broccoli, 2011-2013 (Dollar/acre)

Treatment	Year	Pre-harvest variable costs	Total Harvest and marketing	Total Variable costs	Total Fixed costs	Total costs	Total Revenue	Net Revenue
CT0	2011	2,134	2,554	4,688	851	5,539	11,293	5,754
	2012	2,853	385	3,238	958	4,196	1,697	(2,500)
	2013	3,030	1,034	4,065	1,006	5,071	4,608	(463)
CT1	2012	2,903	1,375	4,278	966	5,244	6,057	813
	2013	3,080	1,194	4,274	1,013	5,287	5,320	32
CT2	2011	2,184	2,487	4,671	1,398	6,069	10,996	4,928
	2012	2,903	1,160	4,064	966	5,029	5,111	82
	2013	3,080	1,284	4,364	1,013	5,377	5,719	342
CT3	2012	2,903	2,293	5,197	966	6,162	10,103	3,940
	2013	3,080	1,692	4,772	1,013	5,786	7,539	1,754
CT4	2013	3,080	1,720	4,800	1,013	5,813	7,662	1,848
ST0	2011	2,184	2,187	4,371	823	5,194	9,672	4,478
	2012	2,903	387	3,291	931	4,221	1,707	(2,515)
	2013	3,080	994	4,074	978	5,052	4,428	(624)
ST1	2012	2,903	306	3,210	931	4,140	1,349	(2,791)
	2013	3,080	1,422	4,502	978	5,480	6,335	855
ST2	2011	2,184	2,264	4,447	823	5,270	10,009	4,739
	2012	2,903	868	3,771	931	4,702	3,824	(878)
	2013	3,080	1,254	4,334	978	5,312	5,586	274
ST3	2012	2,903	1,018	3,921	931	4,852	4,485	(367)
	2013	3,030	1,557	4,586	971	5,557	6,935	1,378
ST4	2013	3,080	1,452	4,532	978	5,510	6,469	959

The results of ANOVA matrix for assessing treatment effect of tillage method and bahiagrass rotation on net revenue of broccoli is presented in Table 7. Model specification of effect of both treatments on the net revenue of broccoli has no statistically significant explanatory power. The corresponding P-value of F statistic for treatments' interaction in ANOVA matrix was P<0.18 which illustrated that net revenue of broccoli was not sensitive to years in bahiagrass or tillage treatments. Although difficult to explain why the lack of net revenue sensitivity thereof, we are tempted to partially blame the insensitive broccoli yield/revenue results for both treatments (bahiagrass and tillage treatments) to any of the several unfavorable good agricultural practices (GAP) such as changes in soil quality, insufficient weed, disease and insect control etc., adopted for the crop that could have an impact on the yield during the growing season of the study. The most noticeable feature for broccoli was sever cutworm damage to very young broccoli that occurred at high rates in some blocks but not in others. However, converting to ST had significant main effect on the increase in net revenue of broccoli.

Table 7. Statistical analysis of treatment effect on net revenue of broccoli

Source	Partial sum of square	Degree of freedom	Mean sum of square	F - stat	Prob <F
Model	98 016 122.1	9	10 890 680.2	1.45	0.1785
Years of bahiagrass	34 672 614.2	4	8 668 153.56	1.16	0.3355
Tillage	57 882 739.1	1	57 882 739.1	7.73	0.0067
Bahiagrass*tillage	21 585 286.8	4	5 396 321.7	0.72	0.5804
Residual	644 325 213	86	7 492 153.64		
Total	742 341 335	95	7 814 119.32		

The analysis of riskiness of profitability of broccoli, CT3 is the least risky production alternative among conventional tillage treatments (CT0, CT1, CT2, CT3, CT4) and ST2 and ST3 are the least risky production alternatives among strip tillage treatments (ST0, ST1, ST2, ST3, ST4) (Table 8). ST0, CT0 and ST1 are the riskiest treatments for broccoli.

Table 8. A breakdown of all the key economic parameters including net revenue and coefficient of variations for the different sod-based rotation of broccoli, 2011-2013 (Dollar/acre)

Treatment	Pre-harvest variable costs	Total Harvest and marketing	Total Variable costs	Total Fixed costs	Total costs	Total Revenue	Net Revenue	Coefficient of variation
CT0	2,673	1,324	3,997	938	4,935	5,866	930	4.6
CT1	2,992	1,284	4,276	990	5,266	5,688	422	1.3
CT2	2,722	1,644	4,366	1,126	5,492	7,275	1,784	1.5
CT3	2,992	1,993	4,985	990	5,974	8,821	2,847	0.5
CT4	3,080	1,720	4,800	1,013	5,813	7,662	1,848	-*
ST0	2,722	1,190	3,912	911	4,823	5,269	447	8.1
ST1	2,992	864	3,856	954	4,810	3,842	(968)	2.7
ST2	2,722	1,462	4,184	911	5,095	6,473	1,378	2.2
ST3	2,966	1,287	4,254	951	5,205	5,710	506	2.4
ST4	3,080	1,452	4,532	978	5,510	6,469	959	-*

* Since there is only one year of net revenue of broccoli and bush bean available for C4 and S4, measure of Coefficient of variation is not computable for these two treatments.

Considering CT3 as the most profitable treatment for the broccoli, net revenue after subtracting opportunity cost is $2,904 ($5694 – $2790).

The average net revenue of sod-based rotation and tillage treatment was calculated for both crops (Table 9). Two main sources of variability in the enterprise budget were yield and the variable cost associated with yield. From the table, we can conclude that CT treatments with four and three years of previous bahiagrass rotation (CT3 and CT4) are ranked 1st (two treatments) in term of economic profitability which is induced by highest yield in both cash crops (CT3 provide the highest profitability). The ST3 and ST4 treatments are considered 2nd best in term of profitability.

Analyzing the coefficient of variation for each sod-based rotation and tillage treatments shows that less years of bahiagrass (CT0, ST0, ST1) lead to higher variation in profit, which is stated to be more risky from the perspective of risk-averse farmers. This result is consistent for each of the cash crop separately, while production of bush bean is less risky than broccoli for each treatment.

Table 9. A breakdown of all the key economic parameters including net revenue and coefficient of variations for the different sod-based rotations, 2011-2013 (Dollar/acre)

Treatment	Pre-harvest variable costs	Total Harvest and marketing	Total Variable costs	Total Fixed costs	Total costs	Total Return	Net Rev.	Rank of Net Rev.	Coefficient of variation	Rank of Coeff. of variation
CT0	4,766	3,102	7,868	1,778	9,646	12,640	2,994	6	5.0	7
CT1	5,207	2,991	8,197	1,848	10,045	12,167	2,122	8	1.4	2
CT2	4,816	3,188	8,004	1,966	9,970	13,159	3,190	5	2.6	3
CT3	5,207	5,208	10,415	1,848	12,263	21,130	8,868	1	0.8	1
CT4	5,444	4,612	10,056	1,894	11,950	19,409	7,459	2	-*	-*
ST0	4,816	2,749	7,566	1,715	9,281	11,293	2,012	9	9.6	8
ST1	5,207	2,235	7,442	1,777	9,219	9,165	(54)	10	4.6	6
ST2	4,816	2,976	7,793	1,715	9,508	12,215	2,707	7	3.4	5
ST3	5,181	3,786	8,968	1,774	10,741	15,293	4,551	4	2.8	4
ST4	5,444	3,746	9,191	1,824	11,014	15,788	4,774	3	-*	-*

* Since there is only one year of net revenue of broccoli and bush bean available for C4 and S4, measure of Coefficient of variation is not computable for these two treatments.

The statistical analysis for assessing the effectiveness of CT versus ST and sod-based rotation treatment on farm aggregate net revenue (summation of net revenue of both cash crops) was accomplished by applying two-way ANOVA and utilizing appropriate testing tools (Table 10). Model specification, separate main effectiveness of more years of bahiagrass and tillage treatments and their interaction effect are all proved to be significant in explaining the increase in net revenue ($P < 0.05$ level).

Table 10. Statistical analysis of treatment effect on total net revenue of both bush bean and broccoli

Source	Partial sum of square	Degree of freedom	Mean sum of square	F - stat	Prob > F
Model	481 850 879	9	53 538 986.6	6.83	0.0000
Years of bahiagrass	315 773 097	4	78 943 274.4	10.07	0.0000
Tillage	134 675 857	1	134 675 857	17.18	0.0001
Bahiagrass*tillage	81 171 322.7	4	20 292 830.7	2.59	0.0423
Residual	673 974 150	86	7 836 908.72		
Total	1.1558e+09	95	12 166 579.3		

5. Discussion

The objective of this study was to evaluate sod-based rotation and CT versus ST on profitability of farming with bush bean and broccoli as cash crops. Our economic analyses of data from an experimental crop rotation project conducted over three years, indicated that 3 and 4 years bahiagrass were the most profitable treatments for both bush bean and broccoli, and consequently for whole system containing both cash crops. These results parallel results from Marois and Wright (2003) and Marois et al. (2002) documenting the economic benefits of utilizing sod-based rotations in conventional production.

Economic benefits of tillage treatments are less clear cut.   Close investigation on cost components of CT and ST illustrates that ST was better agricultural practice due to reducing cost of production and providing larger net revenue for the growers of bush bean.   Broader benefits of minimum-till treatments were illustrated in the studies of Ascough et al. (2009) who utilized "Stochastic Dominance" methodology to examine which of three different tillage systems (chisel plow, no-till, and ridge-till) were the most risk-efficient in terms of maximizing economic profitability (net return) across a range of risk aversion preferences. Results showed that net return of corn crop using no-till tillage system was preferred across the entire range of risk aversion regardless of whether the objective of the decision maker was minimizing risk or maximizing net return.

In our studies, however, CT3 had the highest profitability for both bush beans and broccoli induced by large yield for these treatments. Conventional tillage with four years of bahiagrass (CT4), strip tillage with three years of bahiagrass (ST3) and strip tillage with four years of bahiagrass (ST4) were the second, third and fourth most profitable treatments respectively. This study also investigated riskiness of each treatment by examining coefficient of variation. It should be noted that risk analysis based on CV cannot be generalized especially that our result was strictly based on the limited observation years' research data. The least risky treatments for bush bean, broccoli and whole system were conventional tillage with one year of bahiagrass (CT1), conventional tillage with three years of bahiagrass (CT3) and conventional tillage with four years of bahiagrass (CT4) as they performed much better than all ST treatments. These advantages of CT parallel results of Chan et al. (2010) comparing the profitability of four different organic cropping systems for winter squash and cabbage in New York which found and a combination of compost for nitrogen, occasional cover crops and CT had the highest revenues for cabbage.

Statistical analyses based on ANOVA matrix provide growers a prediction of profitability of sod-based rotation and tillage alternatives. Although the model for testing impact of treatments on net revenue of broccoli did not show any significant effect, net revenue of bush bean and system total revenue were sensitive to both treatments. This suggests that a move toward organic vegetable production by incorporating sod-based rotation and CT can lead to higher profit for farmers.

References

Afifi, A. A., & Azen, S. P. (1973). Statistical analysis: a computer oriented approach. *Biometrics, 29*(3), 603. http://dx.doi.org/10.2307/2529182

Archer, D. W., Halvorson, A. D., & Reule, C. A. (2008). Economics of irrigated continuous corn under conventional-till and no-till in northern Colorado. *Agronomy Journal, 100*(4), 1166. http://dx.doi.org/10.2134/agronj2007.0291

Ascough II, J. C., Fathelrahman, E. M., Vandenberg, B. C., Green, T. R., & Hoag, D. L. (2009). Economic risk analysis of agricultural tillage systems using the SMARTstochastic efficiency software package. 18th World IMACS / MODSIM Congress, 13-17 July 2009, Cairns, Australia. Retrieved from http://www.mssanz.org.au/modsim09/B1/ascough.pdf

Balkcom, K., Hartzog, D, Katsvairo, T., & Smith, J. L. (2007). Yield and economics for peanut under two tillage system and perennial grass vs conventional rotation. Proceeding of 29th Southern Conservation Agricultural Systems Conference, Quincy, Florida. pp. 25-27. Retrieved from http://www.ag.auburn.edu/auxiliary/nsdl/scasc/Proceedings/2007/posters/Katsvairo_c.pdf

Bliss, C., Andersen, P., Brodbeck, B., Wright, D., Olson, S., & Marois, J. (2016). The influence of bahiagrass, tillage, and cover crops on organic vegetable production and soil quality in the Southern Coastal Plain. *Sustainable Agriculture Research, 5*(2), 65. http://dx.doi.org/10.5539/sar.v5n2p65

Byrd, M. M., Escalante, C. L., Fonsah, E. G., & Wetzstein, M. E. (2006). Financial efficiency of methyl bromide alternatives for Georgia bell pepper industries. *Journal of the American Society of Farm Managers and Rural Appraisers, 69*(1), 31-39. Retrieved from
http://ageconsearch.umn.edu/bitstream/190689/2/251_Escalante.pdf

Byrd, M. M., Escalante, C. L., Fonsah, E. G. & Wetzstein, M. E. (2007). Feasible fumigant-herbicide system alternatives to methyl bromide for bell pepper producers. *Journal of Agribusiness, 25*(1), 31-45. Retrieved from http://ageconsearch.umn.edu/bitstream/62283/2/JAB%2cSpr07%2c%2303%2cpp31-45.pdf

Chan, S., Caldwell, B., & Rickard, B. J. (2010). An economic examination of alternative organic cropping systems in New York State. College of Agriculture and Life Sciences, Department of Applied Economics and Management. Cornell University. EB 2010-14. Retrieved from http://publications.dyson.cornell.edu/outreach/extensionpdf/2010/Cornell_Dyson_eb1014.pdf

Clark, S., Klonsky, K., Livingston, P., & Temple, S. (1999). Crop yield and economics comparison of organic, low input and conventional farming system in California's Sacramento Valley. *American Journal of Alternative Agriculture, 14*(3), 109-121. http://dx.doi.org/10.1017/s0889189300008225

Dimitri, C., & L. Oberholtzer. (2009). *Marketing US organic foods: Recent trends from farms to consumers*: DIANE Publishing. Retrieved from http://www.ers.usda.gov/media/185272/eib58_1_.pdf.

Drinkwater, L. E., Letourneau, D. K., Workneh, F., Van Bruggen, A. H. C., & Shennan, C. (1995). Fundamental differences between conventional and organic tomato agroecosystems in California. *Ecological Applications*, 1098-1112. http://dx.doi.org/10.2307/2269357

Drost, D. Long, G. Wilson, D., Miller, B. & Campbell, W. (1996). Barriers to adopting sustainable agricultural practices. *Journal of Extension, 34*(6). Retrieved from http://www.joe.org/joe/1996december/a1.php

Ferrer, M. C., Fonsah, E. G., & Escalante, C. (2011). Risk-efficient fumigant-mulching system alternatives for bell pepper production. *Journal of the American Society of Farm Managers and Rural Appraisers*, 162-171. http://ageconsearch.umn.edu/bitstream/118963/2/355_Ferrer.pdf

Fonsah, E. G., & Hudgins, J. (2007). Financial and economic analysis of producing commercial tomatoes in the southeast. *Journal of the American Society of Farm Managers and Rural Appraisers, 70*(1), 141-148. Retrieved from http://portal.asfmra.org/userfiles/file/journal/277_Fonsah_1.pdf

Fonsah, E. G. (2009a). MALTAG Group. 2009 MALTAG organic vegetables. Retrieved from http://www.agecon.uga.edu/extension/budgets/non-beef/documents/2009OrgaVegbudget.pdf. (Viewed: 20 June, 2016)

Fonsah, E. G. (2009b). MALTAG Group. 2009 MALTAG conventional vegetables. Retrieved from http://www.agecon.uga.edu/extension/budgets/non-beef/documents/2009ConvVegbudget.pdf (Viewed: 20 June, 2016)

Fonsah, E. G., & Chidebelu, Angus S. N. D. (2012). Economics of banana production and marketing in the tropics: A case study of Cameroon. African Books Collective. Langaa Research and Publishing CIG, Bamenda.

Fonsah, E. G., Krewer, G., Harrison, K. & Bruorton, M. (2007). Risk rated economic returns analysis for southern highbush blueberries in soil in Georgia. *HortTechnology, 17 (4): 571-579*. Retrieved from http://horttech.ashspublications.org/content/17/4/571.full.pdf+html

Fonsah, E. G., Krewer, G., Harrison, K., & Stanaland, D. (2008). Economic returns using risk rated budget analysis for rabbiteye blueberries in Georgia, *Journal of American Society for Horticultural Science, HortTechnology, 18*, 506-515.
Retrieved from http://horttech.ashspublications.org/content/18/3/506.full.pdf+html

Gallaher, R. N., Ashburn, E. L., Gallaher, K., Baldwin, J. A., Lang, T. A., Oclumpaugh, W. R., Prine, G. M., Robertson, W. K., & Stanley, R.L. (2003). No-till management of agronomic row crops in perennial sod. The Institute of Food and Agricultural Sciences. http://nfrec.ifas.ufl.edu/sodrotation/postconference.htm.

Gebremedhin, B., & Schwab, G. (1998). The economic importance of crop rotation systems: evidence from the literature. Michigan State University, Department of Agricultural, Food, and Resource Economics. No. 11690. Retrieved from http://ageconsearch.umn.edu/bitstream/11690/1/sp98-13.pdf

Garcıa-Préchac, F., Ernst, O., Siri-Prieto, G., & Terra J. A. (2004). Integrating no-till into crop–pasture rotations in Uruguay. *Soil & Tillage Research, 77*, 1-13. http://dx.doi.org/10.1016/j.still.2003.12.002

Greene, C., & Kremen, A. (2003). U.S. organic farming in 2000-2001: adoption of certified Agricultural Information Bulletin No. AIB780, Economic Research Service, United States Department of Agriculture. Retrieved from http://ageconsearch.umn.edu/bitstream/33769/1/ai030780.pdf

Hancock, D. W., Lacy, R. C., Stewart, R. L., Tubbs, R. S., Kichler, J., Green, T. W., & Hicks, R. (2013). The management and use of bahiagrass. The university of Georgia cooperative extension. Retrieved from http://athenaeum.libs.uga.edu/bitstream/handle/10724/12227/B1362.pdf?sequence=1

Hoorman, J. J., Islam, R., & Sundermeier, A. (2009). Sustainable crop rotations with cover crops. The Ohio State University Extension. Retrieved from http://ohioline.osu.edu/factsheet/SAG-9

Ikemura, Y., & Shukla, M. K. (2009). Soil quality in organic and conventional farms of New Mexico, USA, Department of Plant and Environmental Sciences, New Mexico State University. *Journal of Organic Systems*, 4(1), 34-47.
Retrieved from http://www.organic-systems.org/journal/Vol_4(1)/pdf/34-47_Ikemura_Shukla.pdf

Johnson, A. W., Minton, N. A., Brenneman, T. B., Burton, G. W., Culbreath, A. K., Gascho, G. J., & Baker S. H. (1999). Bahiagrass, corn, cotton rotations, and pesticides for managing nematodes, diseases, and insects on peanut. *J. Nematology. 31*, 191-200.

Retrieved from http://www.ncbi.nlm.nih.gov/pmc/articles/PMC2620366/pdf/191.pdf

Katsvairo, T. W., Wright, D. L., Marois, J. J., & Rich, J. R. (2007). Transition from conventional farming to organic farming using bahiagrass. *J. Sci. Food Agric., 87*(15), 2751–2756. http://dx.doi.org/10.1002/jsfa.3002

Klonsky, K. (2012). Comparison of production costs and resource use for organic and conventional production systems. *American Journal of Agricultural Economics, 94*(2), 314-321. http://dx.doi.org/10.1093/ajae/aar102

Marois, J. J., & Wright, D. L. (2003). A working business model for cattle/peanuts/cotton. In: Rhoads, F.M. (Ed.), Proceedings of the Sod-Based Cropping Systems Conference. North Florida Research and Education Center-Quincy, University of Florida, 20-21 February, pp. 180-186.

Marois, J. J., Wright, D. L., Baldwin, J. A., & Hartzog, D. L. (2002). A multi-state project to sustain peanut and cotton yields by incorporating cattle and a sod based rotation. In: van Santen E (ed.) Proceedings of 25th Annual Southern Conservation Tillage Conference for Sustainable Agriculture. Auburn, pp 101-107.

Martin, S. W., Hanks, J. (2009). Economic analysis of no tillage and minimum tillage cotton-corn rotations in the Mississippi Delta. *Soil and Tillage Research. 102*(1), 135-137. http://dx.doi.org/10.1016/j.still.2008.08.009

Siri-Prieto, G., & Ernst, O. (2009). Effect of Perennial Pasture and Tillage system on Soil Organic Carbon and aggregate stability in western in western Uruguay. *Soil and Tillage Research, 105*(2), 260-268. http://dx.doi.org/10.1016/j.still.2009.08.001

Sullivan, P. (2003). Applying the principles of sustainable farming fundamentals of sustainable agriculture. The National Sustainable Agriculture Information Service. Retrieved from https://attra.ncat.org/attra-pub/viewhtml.php?id=295

Sumner, D. R., Minton, N. A., Brenneman, T. B., Burton, G. W., & Johnson, A. W. (1999). Root diseases, and nematodes in bahiagrass-vegetable rotations. *Plant Disease, 83*, 55-59. http://dx.doi.org/10.1094/pdis.1999.83.1.55

Tew, B. V., Rafsnider, G. T., Lybecker, D. W., & Chapman, P. L. (1986). The relative risk efficiencies of alternative tillage practices for irrigated corn production. *North Central Journal of Agricultural Economics, 8*, 83-91. http://dx.doi.org/10.2307/1349084

USDA. (2009). 2007 Census of agriculture. United States summary and state data. Vol. 1. Geographic area series. Part 51. AC-07-A-51.

Waldon, H., Gliessman, S., & Buchanan, M. (1998). Agroecosystem responses to organic and conventional management practices. *Agricultural systems, 57*(1), 65-75.
http://dx.doi.org/10.1016/s0308-521x(97)00078-4

Zwingli, M. E., Hardy Jr, W. E., & Adrian Jr, J. L. (1989). Reduced risk rotations for fresh vegetable crops: an analysis for the Sand Mountain and Tennessee Valley regions of Alabama. *Southern Journal of Agricultural Economics, 21*(2), 155-165. Retrieved from http://ageconsearch.umn.edu/bitstream/30111/1/21020155.pdf

Note

Note 1: 1) four-year rotation: processing tomato, safflower, corn and bean and a winter grain and/or legume double-cropped with bean. 2) two-year rotation: tomato and wheat rotation 3) conventional input, low-input and organic input.

Note 2: Based on the same experiment, Bliss et al. (2016) have studied the influence of bahia grass, tillage, and cover crop on organic production and soil quality in the southern coastal plain. This study focuses on the economic profitability of bahia grass and tillage treatments as information presented before plus the economic analyses would be two unwieldy to present in one paper.

Note 3: Two years of bahia grass has more replication. In the complementary study by Bliss et al., (2016) (Note 1), treatments of "years of bahia grass" are distinguished based on **the number of years in the vegetable** preceding the bahia grass treatment. In this paper, the economic profitability of two treatments of bahiagrass rotation and tillage, plots with two years of bahiagrass but different years of vegetation are assumed as the replication of the same treatment.

Appendix A

U.S. Units Conversion to International System Units

Table A1. U.S. Units Conversion to International System Units

To convert from (U.S. Units)	To (IS Units)	Multiply By
Acre	Hectare (ha)	0.404
Pound (lbs)	Kilogram (kg)	0.453
Pound (lbs)	Ton	0.0005
lbs/acre	Kg/ha	1.1209
$/acre	$/ha	2.471

Appendix B:

Expected Return from Total Acreage

Table B1. Expected Return from Total Acreage (Dollar)

Crop	Treatment	Year	Net Revenue	Chance for profit
Bush bean	CT0	2011	5,754	99%
		2012	(2,500)	1%
		2013	(463)	32%
	CT1	2012	813	73%
		2013	32	51%
	CT2	2011	4,928	98%
		2012	82	53%
		2013	342	61%
	CT3	2012	3,940	96%
		2013	1,754	86%
	CT4	2013	1,848	87%
	ST0	2011	4,478	98%
		2012	(2,515)	1%
		2013	(624)	26%
	ST1	2012	(2,791)	1%
		2013	855	73%
	ST2	2011	4,739	99%
		2012	(878)	14%
		2013	274	59%
	ST3	2012	(367)	35%
		2013	1,378	82%
	ST4	2013	959	75%
Broccoli	CT0	2011	2,849	90%
		2012	1,168	75%
		2013	2,172	84%
	CT1	2012	1,766	82%
		2013	1,633	80%
	CT2	2011	(239)	40%
		2012	1,875	83%
		2013	2,582	86%
	CT3	2012	4,843	93%
		2013	7,199	96%
	CT4	2013	5,610	94%
	ST0	2011	(158)	44%
		2012	704	68%
		2013	4,150	92%
	ST1	2012	(330)	38%
		2013	2,158	84%
	ST2	2011	(525)	27%
		2012	2,404	86%
		2013	2,108	84%
	ST3	2012	2,986	89%
		2013	5,105	94%
	ST4	2013	3,815	91%

17

Factors Affecting the Adoption of Water Harvesting Technologies: A Case Study of Jordanian Arid Area

Samia Akroush[1], Boubaker Dehehibi[2], Bezaiet Dessalegn[3], Omamah Al-Hadidi[1] & Malek Abo-Roman[1]

[1]Socioeconomic Studies Directorate, National Center for Agricultural Research and Extension (NCARE), Baq'a (19381). Amman, Jordan

[2]Sustainable Intensification and Resilient Production Systems Program (SIRPSP), International Center for Agricultural Research in the Dry Areas (ICARDA), Amman, Jordan

3Integrated Water and Land Management Program (IWLMP), International Center for Agricultural Research in the Dry Areas (ICARDA), Amman, Jordan

Correspondence: Samia Akroush, Socioeconomic Studies Directorate, National Center for Agricultural Research and Extension (NCARE), Baq'a (19381). Amman, Jordan. E-mail: samia_akroush@yahoo.com

Abstract

In this article, we investigate the determinants of farmers' decisions to adopt water harvesting technologies (WHT) in the arid agricultural area of Jordan. In particular, we investigate the effect of different socio-demographic, economic, and institutional factors on the adoption of WHT. For doing so, we empirically apply a binary logistic regression model on a micro-dataset (59 farmers) in Jordanian *Badia*. Empirical findings indicate that there is no significant relationship between age and the probability of adoption of WHT. However, our findings show significantly positive relationships at 10% level for farmer education and experience which implies that farmers with higher education and experience level are more likely to adopt WHT. In contrast, it was found that labor and institutional variables such as credit services do not significantly influence adoption of WHT. Results also reveal a significant relationship between land tenure and adoption implying higher adoption rates on communal land as opposed to privately owned land. Based on our empirical results, this research will assist decision makers to prioritize the factors influencing adoption of WHT and provide insights for targeted dissemination, adoption, and diffusion of WHT in the Jordanian arid areas.

Keywords: adoption decisions, water harvesting technologies, arid areas, logit, Jordan

1. Introduction

Water is the primary constraint to agricultural livelihoods in the drylands. Based on prediction of climate change in the region, farmers will be faced with continued and increased water scarcity and rising temperature which will negatively affect agricultural production (Olarinde *et al.*, 2012). Improvement in agricultural water management is thus crucial to sustain current patterns of agricultural production and development over the coming decades (Verner and Breisinger, 2012; IBRD, 2007).

The Water and Livelihood Initiative (WLI) is a regional undertaking (Note 1) that aims to improve the livelihoods of rural households and communities in the MENA region, particularly in semi-arid and arid agricultural areas where water scarcity, land degradation, food security, and water quality deterioration are widespread. Its main objective is to develop and pilot test integrated water, land-use and livelihoods strategies in selected benchmark sites for scaling-out to agro-ecologically and socio-economically suitable areas in the region.

This study was undertaken within the framework of the WLI, to assess factors affecting adoption of the technologies developed and/or improved through the Initiative, with special emphasis on the identification of the constraints to and opportunities for adoption of selected WHT.

Previous research on adoption of water and land conservation technologies provide several insights into the relationship between farmers' demographic status and their decision to adopt or not adopt water and land conservation technologies. For instance, Teshome *et al.,* (2015) examined the drivers of different stages on the adoption of soil and water conservation (SWC) technologies in the north-western highlands of Ethiopia. Their study was based on a detailed farm survey among 298 households in three watersheds. A simple descriptive statistics and an ordered probit model were used to analyze the drivers of different stages of adoption of SWC. The results of the ordered probit model show that some socio-economic and institutional factors affect the three adoption stages of SWC differently. Farm labor, parcel size, ownership of tools, training in SWC, presence of SWC program, social capital (e.g., cooperation with adjacent farm owners), labor sharing scheme, and perception of erosion problem have a significant positive influence on the actual and final adoption phases of SWC. Additionally, the final adoption phase of SWC was positively associated with tenure security, cultivated land sizes, parcels slopes, and perception on SWC profitability. The study recommended that policy makers should take into consideration factors affecting adoption of SWC such as profitability, tenure security, social capital, technical support, and resource endowments (e.g., tools and labor) when designing and implementing SWC policies and development programs.

Ahmed *et al.,* (2013) evaluated the factors influencing adoption of rainwater harvesting technologies among households in *Yatta* district (Kenya). A semi-structured questionnaire was used to gather data from 60 households. Logistic regression was applied to evaluate different factors influencing adoption determinants of rainwater harvesting technologies. Findings from this study indicates that most of the farmers were aware of a variety of WHT, with roof WH (45%) and dams (36.1%) being rated high, and were willing to adopt them within their local context. The empirical regression findings showed that factors such as education level of household head, experience of water shortage, and awareness of WHT and age of farmers were significantly positive influencing WHT adoption. The study concluded that for effective implementation and subsequent adoption of rainwater harvesting technologies, farmers would require technical know-how and skills, capital, raw material and organizational support. In addition, farmers need to be mobilized and trained on the use of rainwater harvesting technologies and sensitized on the potential socioeconomic benefits of adopting them.

This paper offers insight into the constraints and determinants that influence Jordanian farmers' decision on adoption of WHT, and presents policy makers with important considerations for designing effective policies that could stimulate and sustain adoption of these technologies in arid agricultural areas.

The remaining part of this paper is organized as follows: Section 2 presents background information about the project and the technologies tested. The objectives of the study, study area and the methodology used are presented is Section 3. Section 4 presents the results of the binary regression model, the results of the technology characteristics, and the criteria that farmers used to judge the technology. The last section summarizes the main concluding remarks and implications.

2. Background and Research Objective

The WLI in Jordan aims to pilot test strategies for sustainable land and water management in the Jordanian rangeland areas commonly referred to as the *Badia*, focusing on the adaptation of WHT to ensure efficient use of scarce and fragile resources; to restore and reverse degraded and drought-prone areas; and to expand income generation opportunities for improved livelihoods.

Different types of WHT have been tested and applied through the WLI including the *Vallerani, Run-off strips*, and *Marabs*. The *Vallerani* mechanized system is a special tractor-pulled plow that automatically constructs water-harvesting catchments and is ideally suited for large-scale reclamation work (ICARDA Caravan, 2006). The other category of WHT is the Run-off strips, where barley is planted in strips using a seed drill, with unplanted strips in between catchment areas. The catchment areas allows rainfall water to be harvested in the barley planted strips, which will maximize the availability of water for barley, and as a result, the barley crop will give reasonable straw and grain yield. Another type of WHT practiced at the project area are the *Marabs*. These are natural formations found in the *Badia* where water spreads naturally over a relatively wide, slightly shallow "stream" beds and thus allow the use of this area for agricultural production (e.g., barley cultivation for feeding livestock). Streams are temporary due to low rainfall and aridity, and may not, in fact, exist every year. A *Marab* is a WHT that is constructed at the lowest point of the watershed to collect and spread excess runoff water in order to maximize the size of cultivated land.

The aim of this study is twofold: Firstly, to assess farmer's perceptions of and exposure to WHT, and to determine the major factors influencing farmer's adoption decisions and their willingness to adopt these practices. To the best of our knowledge, this is the first paper that offers such insight in the case of Jordan. According to Anderson (1993), willingness to adopt refers to a farmer's motivation to adopt a new innovation, technology, and/or practice. Secondly, the paper aims to estimate the elasticity of adoption of factors that are significant in explaining farmers' decisions in the study area and draw conclusions that might help in developing policy and institutional interventions in order to enhance and accelerate the adoption process and widespread of such technologies.

The study objectives are based on the following hypothesis:

(a) The adoption of the WHT is dependent upon the level of human capital such as age, education, training, membership in farmer's organization and perception of soil degradation;

(b) It is assumed that farmers with large land holdings are more likely to adopt WHT than farmers with smaller land holdings, and

(c) The readiness of farmers to adopt WHT is conditional upon the availability and accessibility of appropriate equipment (case of *Vallerani*).

3. Methodological Framework

3.1 Study Area

The WLI project was implemented in two communities (*Majidyya & Muharib*) at the middle *Badia* of Jordan. *Majidyya* is a rural village 70 km south of Jordan's capital city of Amman (Fig 1). With a small population of 40 household heads and approximately 250 inhabitants, the *Majidyya* community members traditionally use their lands for grazing.

Land degradation in the area is mainly attributed to continuous grazing which affects biodiversity, vegetation composition, and availability of essential biomass to sustain livelihoods. As an adaptation strategy, farmers have switched to barley cultivation because barley is a government-subsidized crop (farmers receive barley with lower prices than the world prices), and because despite its small yield potential due to water scarcity, nonetheless both the seed and the stalk offer farmers a good source of fodder for their livestock.

Muharib is also a small agricultural village with 30 household heads and about 190 inhabitants. Like *Majidyya*, *M*uharib agriculture area has been suffering from water shortages/scarcity and degradation of arable land. Unable to earn enough in their agricultural livelihoods, some families have left *Muharib* for other neighboring villages or cities such as Amman where private and public services and employment provide more prospects for a better life.

The study site (Figure1) was selected as an area that is representative of conditions throughout the Jordanian *Badia*, based on carefully selected criteria including the area layout, runoff, slope, soil, rainfall, and the community. On average, the area receives between 100 and 200 mm of rain per year, and has one of the highest rates of evapotranspiration in Jordan. The selected pilot testing area also suffers from crust formation which reduces water infiltration. The Jordanian *Badia* is severely degraded due to continuous over-grazing, cutting, and ploughing; population growth, water shortages and climate change effects (drought).

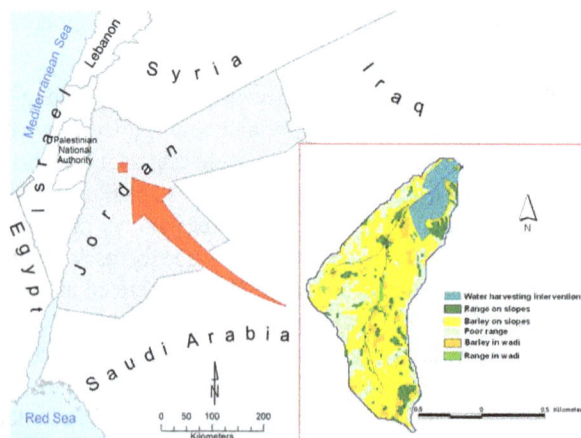

Figure 1. Location, land use and map of the study watershed in Jordan

Source: Akroush *et al.,* (2014).

3.2 Survey Design and Data Collection

For the purpose of achieving the study objectives, sample of 59 farmers were selected using random sampling procedure. The survey was done on a sample of 25 adopters and 34 non-adopters of WHT. The distribution of the sample collected and sample size from the different locations is displayed in Table 1. Farmers were interviewed and survey was administered to the household heads for all selected households to gather the needed information.

Pre-testing of the questionnaire was done in the targeted communities to test its validity with farmer's conditions. Minor modifications were done through feedback from farmer's interviews and with participation of a multidisciplinary team. The information collected using the questionnaires covering several sections included information about farmers socio economic conditions, main crops planted and inputs used, animals owned, marketing livestock products, source of fodder/animal feed, water sources, access to credit, engagement in community based organizations, farmers' knowledge of the new technologies, basic criteria for technology adoption, WHT characteristics and challenges in current agricultural practices.

In addition to this questionnaire, and to facilitate the interpretation of the empirical findings, individual and collective informal discussions with farmers, farmers' associations and local institutions were organized. This was mainly to understand the local farming systems in this area with special emphasis on the WHT and WH management techniques in use by farmers.

Table 1. Distribution of sample by location

Farmers group/location		No. of farmers	%
Adopters	Middle Badia (WLI site)	13	52.0
	Karak Governorate	12	48.0
	Total	25	100.0
Non-adopters	Middle Badia (WLI site)	34	100.0

Source: Own elaboration from survey database (2015).

This study used both qualitative and quantitative methods to analyze the gathered data. The data was collected, coded, entered and checked for consistency before proceeding with its statistical and empirical analysis using IBM SPSS statistics software (v.20). The gathered data was processed and analyzed to generate descriptive statistics (frequencies, mean, minimum, maximum, standard deviations, etc.) of factors expected to influence adoption of WHT and to describe farmers socio-economic characteristics, main crops planted, farmers assets, ownership of animals, marketing of livestock products, sources of fodder, water sources and the community's engagement in organizations in addition to farmer's knowledge to new technologies and finally the main challenges of farmers with respect to the current agricultural practices.

3.3 Econometric Logit Model

In the literature, the two most popular functional form used in adoption modeling are probit (which assumes an underlying normal distribution) and logit (which corresponds to a logarithmic distribution function). These models have got desirable statistical properties as the probabilities are bounded between 0 and 1. Apparently, the choice of which continuous probability distribution to use cannot be justified on theoretical grounds (Amemiya, 1985). In this study, therefore the binary logistic regression is adopted.

A binary logistic regression is used to regress the dependent variable, Y, of weather the farmer had adopted WHT: Prob (event) = Prob (Y, 1 represents i[th] farmer adopted, and 0, otherwise)

$$Y = \begin{cases} 1: adopted \\ 0: otherwise \end{cases}$$

Against the estimated factors affecting adoption of WHT, the below section describes the variable selection and hypothesis of these variables included in the logistic regression model.

3.3.1 Variable Selection and Hypotheses

The description of the following dependent and explanatory variables as well as the main hypotheses are given in Table 2. Table 3 shows the summary statistics of the variables used in the empirical model.

3.3.1.1 Dependent Variable

Given that our conceptual framework is based on the adoption process of WHT and on the concept of dis-adoption of the earlier adopted technologies, in the analytical framework adoption is defined in terms of a binary variable (adoption/no adoption). ADOP (referring to the dependent variable Y_i) is defined and categorized in this study as a binary variable with a value of 1 for those farmers who have adopted WHT and 0 for those who have not adopted it. The explanatory variables and hypotheses are discussed in the following section.

3.3.1.2 Explanatory Variables and Hypotheses

The adoption of WHT is a complicated process similar to any other research on agricultural technology adoption (Adesina and Chianu, 2002) that may be influenced by a set of interrelated biophysical, social, economic, environmental and institutional factors. The seven potential explanatory variables (Table 2) expected to influence the adoption decision of WHT in the study areas include a wide variety of household socio economic, farming, and institutional factors.

Review of major farmer variables and their expected effects

AGE: This variable measures the age of household head. According to the theory of human capital, young farmers are expected to have a greater chance of absorbing and applying new knowledge (Sidibe´, 2005). Therefore, we hypothesized that successful use of WHT will be a challenge for older people i.e. older farmers will be less likely to adopt WHT than young people.

EDUC: Measures the level of education attained by a farmer. In most adoption studies (*Ramji et al.,* 2002), farmers with higher levels of educational attainment are expected to adopt new technologies or practices than less educated farmers. Hence, it is expected that EDUC has a positive impact on adoption.

LABE: Refers to the size of the active-labor force. The unavailability of labor often limit farmers' use of WHT. Indeed, with the presence of a larger active-labor, a positive influence on the adoption of WHT is expected.

FEXP: Refers to farmers experience in farming. This consists of the knowledge and experience gained in farming. A positive influence on the adoption of WHT is expected with the presence of a larger experience in farming.

Institutional and policy variables and their expected effects

CRED: The accessibility of the households to cash credit (CRED) can help rural households in poor areas increase their production and consumption. Credit is expected to affect adoption of WHT positively. The CRED dummy variable would be expected to exert a positive influence on adoption of WHT.

Farming variables and their expected effects

TENUR: measures status of land ownership. There are three types of land tenure presented at the project site, owned, rented and shared lands. Land ownership, considered with a positive expected profit determinant, is widely believed to speed and encourage the adoption of new technologies but still it is difficult to forecast the sign of TENUR WH adoption.

Table 2. Variables used in the empirical Binary Logistic Model

Acronym	Description	Type of measure	Expected Sign
Dependent variable			
ADOP	Whether a farmer has adopted or not	Dummy (1 if yes, 0 if no)	
Explanatory variables			
AGE	Household head's age	Years (1, 2, 3, 4)	-
EDUC	Educational background of the household head	1, 2, 3, 4,5, 6	+
FEXP	Household head's farming experience	Years	+
LABE	Labor force size	Active labor force numbers	+
TENUR	Status of land ownership	1, fully owned; 2, rented; 3, shared	?
CRED	Obtained credit	Dummy (1 if yes, 0 if no)	+

Source: Own elaboration from survey database (2015).

3.3.2 Modeling WH Harvesting Adoption Decision

Binary logistic regression is a popular statistical technique in which the probability of a dichotomous outcome (such as adoption or non-adoption) is related to a set of explanatory variables (Table 3) and has been widely applied in adoption studies (Adesina and Chianu, 2002; Asfawa and Admassie, 2004; Chianu and Tsujii, 2004). In this development framework, farmers' adoption of WHT is based on an assumed basic utility function. According to this theory, WHT will be adopted by the farmer i, if the utility obtained from WHT exceeds that of non-adoption. The farmer's behavior towards WHT is described by Equations (1)-(4).

$$\text{Prob(event)=Prob(Y, 1 represents i}^{\text{th}}\text{ farmer adopted, and 0, otherwise)} \qquad (1)$$

Let's define X_i a vector representing the set of parameters including the factors which influence the adoption decisions of the ith farmer on WHT. For the farmer i, Z_i is an indirect utility derived from the adoption decision, which is presented as a linear function of K explanatory variables (X), and is expressed as:

$$Z_i = \beta_0 + \sum_{i=1}^{n} \beta i X_{ki} \qquad (2)$$

where β_0 is the intercept term, and $\beta_1, \beta_2, \beta_3,.. ., \beta_i$ are the coefficients associated with each explanatory variable $X_1, X_2, X_3, ..., ., X_{ki}$. Grouped in a vector X, these factors explain the WHT adoption decision, or the probability that the ith farmer adopts WHT:

$$P_i = \frac{e^{Zi}}{1+e^{Zi}} \qquad (3)$$

Where; P^i denotes the probability that the ith farmer's adoption decision and $(1-P_i)$ is the probability that Y_i is 0. The odds (Y =1 versus Y = 0) to be used can be defined as the ratio of the probability that a farmer adopts (P_i) to the probability of non- adoption $(1-P_i)$, namely odds = $P_i /(1-P_i)$. By taking the natural log, we get the prediction equation for an individual farmer:

$$\ln\left(\left(\frac{Pi}{1-Pi}\right)\right) = \ln \text{ odds} = \beta_0 + \sum_{i=1}^{n} \beta i X_{ki} = Z_i \qquad (4)$$

Where Z_i is also referred to as the log of the odds ratio in favor of adoption.

Table 3. Statistics of the variables used in the empirical logit model

Variable	Minimum	Maximum	Mean	Std. Deviation
ADOP	0	1	0.42	0.498
AGE	25	75	49.61	13.618
EDUC	1	5	3.10	1.213
FEXP	5	60	20.47	12.338
LABE	1	7	1.98	0.777
TENUR	1	6	1.69	1.290
CRED	0	1	0.32	0.471

Source: Own elaboration from survey database (2015).

3. Results and Discussion

4.1 Basic Criteria for WHT Adoption Decision

Adoption and diffusion theory has been widely used to identify the factors that influence an individual's decision to adopt or reject an innovation. The perceived originality of the idea for the individual (farmer) determines his or her response to it. Rogers (1995: p11) identified five characteristics of an innovation that affect an individual's adoption decision:

Relative advantage: Is a determinant on the degree to which the innovation is perceived as better than the idea it supersedes. Relative advantage refers to the extent to which the innovation is more productive, efficient, costs less, or improves in some other manner upon existing practices.

Compatibility: Is the characteristic measuring the degree to which the innovation is perceived as being consistent with existing values, past experiences, and needs of potential adopters. An innovation must be considered socially acceptable to be implemented. And some innovations require much time and discussion before they become socially acceptable.

Complexity: Explain accurately how difficult the innovation is to understand and use.

Trialability: Indicates the degree to which the innovation may be experimented with on a limited basis. Innovations are easier to adopt if they can be tried out in part, on a temporary basis, or easily dispensed with after trial.

Observability: This is an indicator on the degree to which the results of the innovation are visible to others. The chances of adoption are greater if folks can easily observe relative advantages of the new technology. In fact, after some adopt, observability can improve the diffusion effect, a critical component of technology transfer.

A Likert scale of five, strongly agree (5) and strongly disagree (1) was used to assess the above mentioned characteristics of adopters of WHT at the WLI project site. This Likert scale is an ordered scale from which respondents choose one option that best aligns with their view. It was used to measure farmer's attitudes by asking the extent to which they agree or disagree with a particular question or statement on WHT characteristics. Results from this assessment are displayed in Table 5 and Figure 1.

The promising results indicates that farmers agreed to adopt WHT because it reduces risk in agriculture production (Coefficient 9.56), the technology is compatible (coefficient 9.48), is triable, easy to follow up and easy to implement (coefficient 9.33), and has environmental benefits (coefficient 9.4). However, the majority do not agree that this technology is affordable especially the *Vallernai* WHT (Coefficient 6.96), it needs skills and knowledge (coefficient 5.35), and is complex (coefficient 4.43). Please refer to Figure 2.

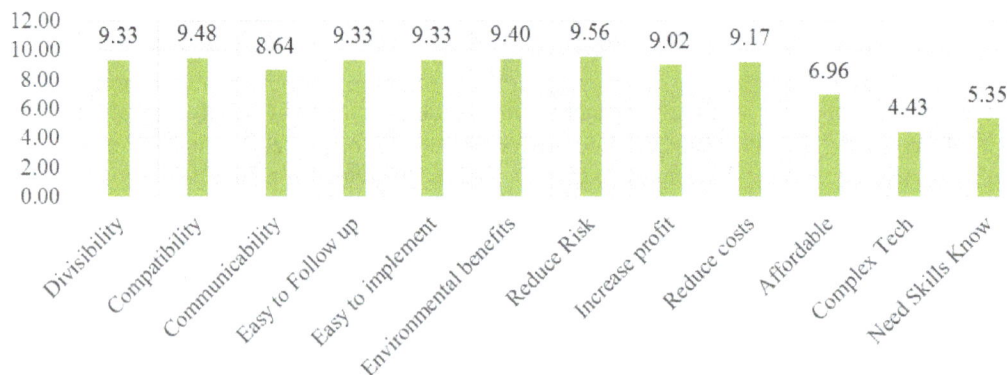

Figure 2. Criteria of WHT characteristics

Source: Own elaboration from survey database (2015).

4.2 Regression Model-Binary Logistic Model (Logit)

The following section presents the results of the binary logistic regression model. The validity of the model is initially discussed before presenting the empirical findings.

4.2.1 Validity of the Model

The maximum likelihood methods (MLE) was used to estimate the coefficients of the binary logistic regression model. The Hosmer and Lemeshow statistic is one of the most consistent goodness of fit test to assess if the model fits for the binary regression (Sidibe´, 2005). The results of the model are given in Table 4. The overall percentage of correct predictions is about 69.5 %. The p-value 0.868 uses the Hosmer and Lemeshow Goodness-of-Fit Test, which is computed from the Chi-square distribution with 8 degrees of freedom ($d.f$), confirms the model's estimates very well fit the data. This implies that we fail to reject the null hypothesis that there is no difference between the observed and predicted values of the dependent variable. This implies that, at an acceptable level, the model's estimates are very consistent with the data. However, another strand of the econometric literature (Sidibe´, 2005) argues that a p-value less than 0.05 indicate a poor fit for a binary logistic regression model.

The corresponding odds ratio (Column, Exp (β)), in Table 5, gives the exponential of expected value of β raised to the value of the logistic regression coefficient, which is the predicted change in odds for a unit increase in the corresponding explanatory variable.

Table 4 displayed that only two (2) explanatory variables in the model were statistically significant at the 5% level in explaining WH adoption behavior in the Jordan *Badia* expressed in terms of the variables used in this case, and the logistic regression equation is:

$$\ln\left(\left(\frac{Pi}{1-Pi}\right)\right) = \ln \text{ odds} = -2.44 + 0.015 \text{ AGE}(0,03) + 0.457 \text{ EDU}(0,298) + 0.067 \text{ FEXP}(0.036) + 0.324$$

$$\text{LABE}(0.501) - 1.67 \text{ TENURE}(0.695) + 0.567 \text{ CRED}(0.719) \tag{5}$$

Information on the relationship between the explanatory variables and the dependent variable is provided in these estimates, where the dependent variable is on the Logit scale.

4.2.2 Empirical Results of the Model and Discussion

Farmers' resource endowment and socioeconomic characteristics

Results for the logistic model show that two characteristics (age and credit) of farmers are non-significant in the decision to adopt WHT and two characteristics (education and farm experience) are significant (Table 5). The empirical findings showed that there is no significant relationship between age and the probability of adoption of WHT so farmer age is not a determinant factor for adopting WHT. However, farmer education (EDU) has a positive impact in the adoption of WHT demonstrated by a significant relationship at 10% level of significance.

We hypothesized that farmers with higher levels of education are more likely to adopt WHT. In line with our predictions, our results showed significant relationship between adopting WHT and higher educational level at 10% level of significance. As hypothesized, the farmer experience (FEXP) coefficient was also found to be significant and positively correlated with adoption decision at the 10% level of significance. The variable labor (LABE) was found to be non-significant in its relationship with adopters. This finding is consistent with our expectations which indicate that such technologies do not need much labor force especially the mechanized WHT.

Exp (β) shows that holding all other explanatory variables constant, for every one unit increase in EDUC score, we expect a 1.58 times increase in the log-odds of adoption (the probability of adoption). There is a tendency for increasing adoption of WHT with increasing farmer's education level. The Exp (β) also shows that holding all other explanatory variables constant, for every one -unit increase in FEXP score, we expect a 1.07 times increase in the log-odds of adoption (the probability of adoption). The same trend was found for increasing adoption of WHT with increasing farmer's experience in agriculture. The Exp (β) of land tenure showed also that holding all other explanatory variables constant, for every one-unit increase in TENURE score, we expect a 0.188 times decrease in the log-odds of adoption (the probability of adoption). Finally, the results show that there is a propensity for decreasing adoption of WHT with increasing farmer's private ownership of land. While this sounds contrary to adoption literature, the reason lies in the high investment costs associated with adoption of these technologies at a large-scale level. Farmers thus tend to adopt such large scale WHT interventions as a group.

Institutions and policies factors

The results of our study show that institutional variables such credit services (CRED) have no impact on the decision to adopt WHT. This implies that farmers do not use credit for the purpose of establishing WHT which is very high (the establishment cost of the technologies). We found a significant relationship between land tenure (TENUR) and adoption. This implies that adoption decreases when land is privately owned, and giving the fact that the upfront cost is too big, farmers are more interested to invest as a group or on communal lands and are more willing to share the cost of adoption. (Table 5).

Table 5. Parameter estimates of the binary logistic regression model for factors influencing adoption of WHT in Jordan *Badia*

Variable	β	S.E.	Wald	Sig.	Exp(β)
AGE	0.015	0.030	0.254	0.615	1.015
EDUC**	0.457	0.298	2.358	0.125	1.580
FEXP*	0.067	0.036	3.448	0.063	1.070
LABE	0.324	0.501	0.420	0.517	1.383
TENUR**	-1.670	0.695	5.777	0.016	0.188
CRED	0.567	0.719	0.622	0.430	1.763
Constant	-2.440	2.350	1.078	0.299	0.087

Hosmer and Lemeshow Test: Chi-square, 3.881; df., 8; Sig., 0.868
-2 Log likelihood 57.613a ; Cox & Snell R Square, .321; Nagelkerke R Square, .431
The overall percentage of correct predictions, 69.5 %.
*Significance at 5%.
**Significance at 10%.
*** Significance at 1%.

Source: Own elaboration from model results (2015).

3. Concluding Remarks

The basic objective of the WLI in Jordan is to improve livelihoods of smallholder farmers in the *Badia*. This study was conducted to enhance our understanding of factors influencing adoption of proven WHTs in an effort to provide insights on pathways to increase the adoption of WHT in Jordanian *Badia*.

Based on the results of the study, we recommend that institutions involved in the development and dissemination of WHT in the Jordanian *Badia*, take into consideration farmers' age, educational level, and availability of active labor force in their efforts to promote adoption of WHT. The study offers a different perspective to earlier efforts to promote the use of WHT which were based on purely technical dimensions such as agronomic and biophysical characteristics. The lessons from this study are that non-consideration of household socio demographic, economic and institutional characteristics may lead to inappropriate targeting of WHTs and would delay the dissemination and widespread of the technology at larger scale. Farmers at the study area facing many challenges related to applying WHT such as weather conditions, high prices of agricultural inputs in addition to high investment costs in using mechanized WH equipment. This emphasized that technologies should not be imposed, but should be driven by the willingness of farmers. Creating strong awareness amongst farmers' communities, relevant stakeholders and decision makers before trying to scaling-up this WH technology is crucial.

In arid areas, increasing farmers' knowledge and perception of the merits of WHT through better access to technical information, extension, and training is often cited as central to helping them on developing a positive assessment of WHT. This highlighted the need for policy support to encourage farmers to adopt this technology at their farms by providing loans with low interest rates targeted only for investment with such technologies, since the mechanized WH is costly and not affordable by farmers, and consequently accelerate the adoption process of these promising technologies. Therefore, any development intervention intended to enhance agricultural productivity through promoting WHT in the study area need to take into account the most important variables with respect to the type of innovation and farmers' preference.

Finally, along these lines, our research findings revealed the need for greater political and institutional input into WHT projects. In precise manner, there is a need to design and develop alternative effective policy instruments and mechanisms, strong institutional options for extension services, technical assistance, training and capacity building that will facilitate adoption the technology through participatory practices to ensure better fit to the needs of farmers. Creation of strong networking among different institutions related to applying WHT and involvement of civil societies, public and private financial institutions and support services could be an example of mechanisms to enhance WHT adoption in Jordan *Badia*. More specifically, linking mechanisms between research and extension and extension education on WHT would further push the adoption of such water management technologies at farm level. Yet, only technologies with a high financial feasibility should be promoted and therefore farmers should be encouraged to join established and strengthened associations through which training, technical assistance and help with access to extension information can be provided. This can only be achieved through planned and designed programs in partnerships with all concerned organizations and targeting the right beneficiaries.

Acknowledgment

The authors thank the Water Livelihood Initiative (WLI) initiative (http://wli.icarda.org/) led by ICARDA (http://www.icarda.org) and sponsored by the USAID (https://www.usaid.gov/) for funding this research.

References

Ahmed, I., Onwonga, R., Mburu, M., & Elhadi, D. (2013). Evaluation of Types and Factors Influencing Adoption of Rainwater Harvesting Techniques in Yatta district , Kenya. *International Journal of Education and Research* Vol. 1 No. 6 June 2013Adesina, A.A., & Chianu, J. (2002). Determinants of farmers' adoption and adaptation of alley farming technology in Nigeria. *Agroforestry Systems, 55*, 99-112.

Akroush, S., Shideed, K., & Bruggeman, A. (2014). Economic analysis and environmental impacts of water harvesting techniques in the low rainfall areas of Jordan. *International Journal of Agricultural Resources Governance and Ecology, 10*(1), 34-49. https://doi.org/10.1504/IJARGE.2014.061040

Amemiya, T. (1985). Advanced Econometrics, Harvard Univ. Press.

Anderson, J. R. (1993). The economics of new technology adaptation and adoption. *Revue of Marketing and Agricultural Economics, 61*, 301-309.

Asfawa, A., & Admassie, A. (2004). The role of education on the adoption of chemical fertiliser under different socioeconomic environments in Ethiopia. *Agricultural Economics, 30*, 215-228. https://doi.org/10.1111/j.1574-0862.2004.tb00190.x

Chianu, J. N., & Tsujii, H. (2004). Determinants of farmers' decision to adopt or not adopt inorganic fertilizer in the savannas of northern Nigeria. *Nutrient Cycling in Agroecosystems, 70*, 293-301. https://doi.org/10.1007/s10705-004-0715-z

ICARDA., (2006). The Vallerani Water Harvesting System, *ICARDA Caravan* No. 23, December 2006.

International Bank for Reconstruction and Development (IBRD) (2007). Making the most of scarcity : accountability for better water management results in the Middle East and North Africa. In MENA development report on water, ed. W. Bank, 270. Washington D.C.: World Bank.p. cm. https://doi.org/10.1596/978-0-8213-6925-8

Olarinde, L., Oduo, L., Binam, J., Diagne, A, Njuki, J., & Adekunle, A. A. (2012). Impact of the Adoption of Soil and Water Conservation Practices on Crop Production: Baseline Evidence of the Sub Saharan Africa Challenge Programme. *American-Eurasian Journal of Agricultural & Environmental Sciences, 12*(3), 293-305, 2012 ISSN 1818-6769 © IDOSI Publications, 2012.

Ramji, P., Neupane, K. R., & Sharma, G. B. T. (2002). Adoption of agroforesty in the hills of Nepal: a logistic regression analysis. *Agricultural Systems, 72*, 177-196. https://doi.org/10.1016/S0308-521X(01)00066-X

Rogers, E. M. (1995). Diffusion of innovations. Everett M. Rogers, 4th ed.

Sidibe´, M. (2005). Farm-level adoption of soil and water conservation techniques in northern Burkina Faso. *Agricultural Water Management, 71*, 211-224. https://doi.org/10.1016/j.agwat.2004.09.002

Teshome, A, Graaff., J., & Kassie, M. (2015). Household-Level Determinants of Soil and Water Conservation Adoption Phases: Evidence from North-Western Ethiopian Highlands. *Environmental Management, 57*(3), 620-636. https://doi.org/10.1007/s00267-015-0635-5

Verner, D., & Breisinger, C. (2012). Adaptation to a Changing Climate in the Arab Countries-A Case for Adaptation Governance and Leadership in Building Climate Resilience. 441. Washington D.C.: World Bank. https://doi.org/10.1596/978-0-8213-9459-5

18

Farmers Perception of Low Soil Fertility and Hybrid Maize and the Implications in Plant Breeding

Priscilla F. Ribeiro[1], Baffour Badu-Apraku[2], Vernon E. Gracen[3], Eric Y. Danquah[3], Manfred B. Ewool[1], Charles Afriyie-Debrah[1] & Benedicta N. Frimpong[1]

[1]CSIR-Crops Research Institute, Ghana

[2]International Institute Tropical Agriculture, Ibadan, Nigeria

[3]West Africa Centre for Crop Improvement, University of Ghana, Legon, Ghana

Correspondence: Priscilla F. Ribeiro, CSIR-Crops Research Institute, P.O. Box 3785, Kumasi, Ghana. E-mail: prisboat@yahoo.com

Abstract

In spite of efforts by national and international scientists to improve crop productivity, varieties of crops grown in Africa have low productivity. Varieties improved for yield have had low adoption rates among small scale farmers. Productivity of maize remains low in the smallholder sector because the crop continues to be grown under stress-prone environments and with limited resources. Participatory Rural Appraisal (PRA) tools, including two focus group discussions and interviews with 120 individual farmers were conducted in Wenchi and Ejura-Sekyedumase districts in Ghana to determine maize production constraints, assess farmers' perceptions of low soil fertility in maize production and their coping strategies for the control of low soil fertility. Opportunities for breeding new maize varieties with tolerance to low soil fertility and improving farmers' perception on adoption and utilisation of maize hybrids were also examined. Results from interviews revealed that low soil fertility, drought, diseases and insect pests are the dominant constraints in maize productivity in the two districts. Farmers in the study area also have preference for low soil nitrogen (low N) tolerant, drought tolerant, disease and pest resistant varieties that require lower inputs. They prefer maize varieties which produce slender cobs that are light in weight with lots of grain. The farmers lack knowledge about hybrids but are willing to adopt maize hybrids that are tolerant to low N.

Keywords: drought, hybrid maize, low soil nitrogen, participatory rural appraisal

1. Introduction

The Ghanaian economy and livelihood of its population heavily depend on agriculture. For this reason, efforts to sustain and improve the sector's productivity are extremely crucial to the country's economic development and also to the welfare of her people. Securing food and livelihood is inextricably linked to the exploitation of the natural resource base (land, water and forest). Soil degradation, mainly due to soil erosion and nutrient depletion, has become one of the most important environmental and economic problems in Ghana (Adu, 2012).

Soil fertility decline is a major biophysical factor challenging crop production in Ghana (Logah et al., 2010). The estimated annual loss of maize grain yield due to low N stress alone varies from 10 to 50% (Wolfe et al., 1988; Logrono and Lothrop, 1997). Reducing losses caused by low soil N could increase maize productivity in Ghana and significantly contribute towards national food security and poverty alleviation. Furthermore, understanding the production constraints that farmers face, low soil fertility in particular could greatly assist in the design of an effective breeding programme that could incorporate tolerance to low soil N and also improve other agronomic traits as well. Hence, the inclusion of farmers' perceptions of low soil N varieties in the national maize breeding programs is essential.

More than 30 improved maize varieties have been released in Ghana between 1942 and 2014 (CRI library, 2015). Though these have been made available to farmers through agricultural extension personnel, maize productivity remains low. This may be attributed in part to low adoption of productivity-enhancing technologies including improved varieties and management practices, low use of purchased inputs, especially fertilizer, and lack of appropriate government policies (Ragasa et al., 2013).

Participatory rural appraisals (PRAs) have been used worldwide to solicit farmers' views on various agricultural resource management options necessary to ensure household food security and improvement in their welfare (Odendo et al., 2002). Formal plant breeding approaches in the public sector have not been seeking input from farmers, as is evident in both the slow adoption rate of improved varieties by farmers and the poor performance of adopted varieties under low input conditions (Bänziger and Cooper, 2001). To improve the adoption rate and increase performance of adopted varieties, an assessment of attributes of maize varieties preferred by farmers and the socio-economic environment under which the farmers operate is critical. By integrating farmers' concerns and circumstances into agricultural research, researchers can develop technologies that would be more widely adopted, resulting in more productive, stable, equitable and sustainable agricultural systems. The objectives of this study were to:

i) Examine maize production constraints of farmers in Ghana.

ii) Assess farmers' perceptions of the problem of low soil fertility in maize production and their strategies for coping with low soil fertility.

iii) Evaluate their potential interest in new maize varieties with enhanced tolerance to low soil fertility and other important traits.

iv) Evaluate farmers' perceptions on adoption of hybrids.

2. Methodology

2.1 Brief Description of Study Areas

The PRA was carried out in two districts, Wenchi in the Brong Ahafo Region and Ejura-Sekyedumase in the Ashanti Region of Ghana. The EjuraSekyedumase district is located in the Northern part of Ashanti Region and the Wenchi district in the western part of Brong Ahafo Region. Both districts are located in the Forest-Savanna transition agro-ecological zone. The two districts are characterized by a bimodal rainfall pattern where the major season is between April and July and the minor season is between September and November. Hence there are two cropping seasons. The average annual rainfall is about 1,140 - 1,270 mm in Wenchi and 1,200 - 1,500 mm in Ejura-Sekyedumase. The temperatures in the districts range from 21^0 C – 300^C. The major occupation in these districts is farming and maize is one of the main staple crops grown. Assisted by extension officers, three communities (villages) were chosen in each district based on the maize production volumes and accessibility and presence of maize related research activities. Akrobi (07.743N, 082.129W), Awisa (07.81N, 0.02.11W) and Amposahkro (07.86N, 002.08W) were the three villages in the Wenchi district while those in the Ejura-Sekyedumase district were Adiembra (07.43N, 001.49W), Aframso (07.31N, 001.39W) and Teacherkrom (07.33N, 001.43W).

2.2 Sampling Technique

One hundred and twenty small-scale farmers were involved in the PRA study. They were identified by local agricultural officials and randomly selected without any bias towards age, gender, experience in farming or status in the community.

2.3 Data Collection Procedure and Analysis

The study used both qualitative and quantitative approaches to elicit data from maize farmers. The qualitative approaches employed key informant interviews and focus group discussions to draw collective responses from farmers. This was followed a formal survey using a structured interview schedule to collect information on farmers' perception on low soil nitrogen and coping strategies, awareness of hybrid maize varieties, constraints in maize production and preferred maize traits for breeding purposes. Data was analyzed using Statistical Package for Social Sciences (SPSS) version 16. Descriptive statistics such as frequency counts, percentages, and charts were used to describe the attributes of the variables collected.

3. Results

3.1 Constraints to Maize Production

Low soil fertility was identified as the most important constraint to maize production in all the villages (Figure. 1). This was followed by drought, pests and diseases. All the farmers interviewed described the symptoms of low soil fertility as yellowing of leaves, stunted growth and reduced yield, which is related to symptoms of low N or low P or both. In response to questions about the factors responsible for low soil fertility, ninety-eight percent of the farmers from all the villages identified continuous cropping on the same land as the main cause. They also identified lack of fertilizer application and lack of crop rotation activities as the second and third major causes (Figure. 2). In their opinion, excessive rainfall did not contribute significantly to low soil fertility in the locations.

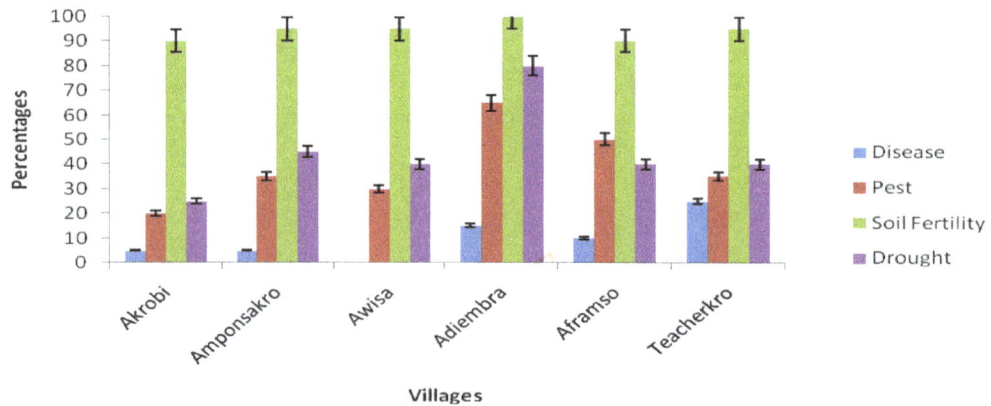

Figure 1. Constraints to maize production in the various villages

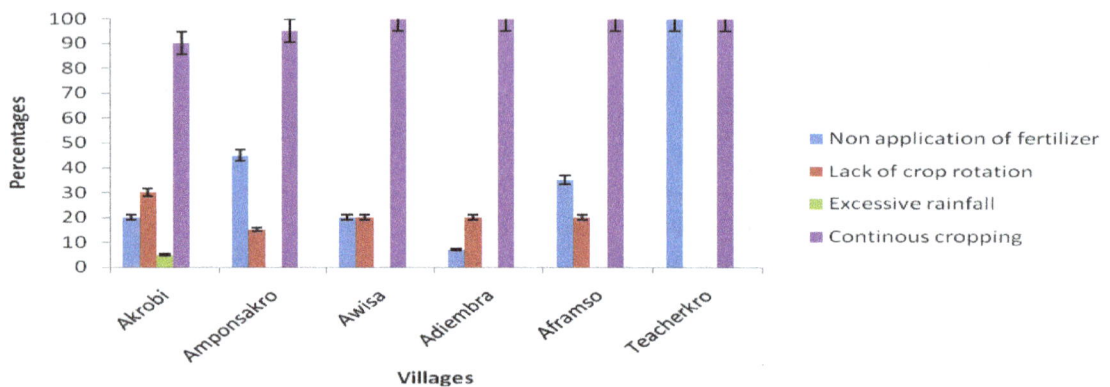

Figure 2. Causes of low soil fertility

3.2 Management of Low Soil Fertility (Coping Strategies)

The majority of the farmers (94%) mentioned application of fertilizer as the most important method to replenish soil of low fertility (Fig. 4). Twenty-three percent of the farmers ranked crop rotation as an effective method for replenishing low soil fertility. Only 5% of farmers mentioned fallowing as useful for managing fertility. Land rotation was not mentioned, as land is generally very scarce in the area.

3.3 Fertilizer Use

The majority of the farmers (90%) purchase unsubsidized fertilizers, which they claim was very expensive. They indicated that, while the distance to agro-chemical stores is not a problem, availability of fertilizer during the period when it is needed most is a major challenge.

3.4 Perception on Hybrids

Ninety-five percent (95%) of the farmers interviewed showed preference for local and improved open-pollinated variety (OPV) seeds over hybrid seeds because they believed that the former was high yielding and responded better to drought. Only 4.1% of the farmers interviewed had ever planted hybrid seed, although 25.8% had heard about it. The few farmers who grew hybrids testified to the high yielding characteristics and appreciable low soil fertility tolerance levels but 95.9% of the farmers still prefer to grow local OPVs because the seeds can be saved and recycled. Their major concern with hybrids was with the high cost of seeds. The 25.8% who had heard about hybrids confused them with improved OPV seeds.

3.5 Willingness to Use Low N Tolerant Maize Varieties

All the farmers showed interest in maize hybrids tolerant to low soil N but expressed reservations about the seed price. Most of the respondents were not willing to pay anything more than the existing market seed price of GHC4.40 per kilo (as at 2015). However, they indicated that they would only purchase the new hybrids if they had superior performance in on-farm trials.

4. Discussion

Farmers identified low soil fertility, drought, pests and diseases as major production constraints impeding maize productivity in the smallholder sector. Of these low soil fertilities was the outstanding constraint. These constraints are common to the majority of small-scale farmers in sub-Saharan Africa (De Vries and Toenniessen, 2001; Odendo et al., 2002; Wekesa et al., 2003). Farmers identified leaf yellowing, stunted growth and reduced yield as symptoms characteristic of low soil fertility. Though farmers were not able to directly link to low soil fertility to vital soil nutrients including nitrogen, there is ample evidence in literature which links the two. For instance, Lafitte and Edmeades, (1994) posit that high yields for maize production is largely linked to utilization and uptake efficiency of Nitrogen, Phosphorus and Potassium, hence the encouragement for farmer to apply inorganic fertilizer in the absence of the organic alternative. The farmers have adopted a number of coping strategies to mitigate the problem of low soil fertility. Fertilizer application is the major method employed to replenish soil fertility. Lack of fertilizer application is a major cause of low soil fertility. This is consistent with the findings of Ragasa et al. (2013).

Apart from the high price of fertilizer, it is sometimes not readily available when needed. This means that low levels of fertilizer are applied which results in low yield. Farmers indicated that occasionally the government subsidy on fertilizers arrives quite late after planting activities were completed, making it not useful to them. Efforts at enhancing timely availability of subsidized fertilizers to farmers at the appropriate time are highly encouraged. These findings disagree with those of Ragasa et al. (2013) who observed that the fertilizer subsidy program may have encouraged farmers to use more fertilizer in their maize plots.

Some farmers in this study practice crop rotation with cowpea, cassava and groundnut. The farmers were aware that the crops used for the rotation could help fix nitrogen in the soil and hence improve soil fertility. This finding is consistent with those of Morris et al. (1999). Land rotation, however, is scarcely employed because land is scarce and very expensive to acquire.

Many of the participants of this study could not afford to purchase fertilizer and land. It is obvious, therefore, that cultivation of varieties tolerant to low soil fertility is the most practical and cost-effective means to manage low soil fertility.

Farmers demonstrated a lot of interest in new maize varieties. They are willing to adopt them if they have assurance that farmers' preferences have been incorporated and they are adapted to local farming conditions. This finding is consistent with those of Nkonya and Featherstone (2001), who reported that varieties with farmer preferred traits are easily adopted and confirmed that farmers' personal experiences influence what varieties they grow.

Farmer perception of hybrid seeds released in Ghana is that they do not possess their preferred characteristics compared to OPVs. Hence, there is generally low adoption of hybrid seeds for maize production in Ghana. This phenomenon may, in part, be attributed to lack of interest and / or knowledge of these varieties by the farmers. Lack of attention and promotion of hybrid varieties in Ghana are the most likely reasons for low uptake by farmers. The effect of low opinion of farmers about hybrid seeds could also be one of the reasons why Ghana remains far behind other African countries in hybrid adoption. For instance, more than 90 percent adoption of hybrid maize is observed in Zambia, Kenya, and Zimbabwe compared with only 3 percent in Ghana. (Tripp and Mensah-Bonsu 2013).

Though hybrid seeds offer many benefits, including greater yield, improved pest resistance, and better stress tolerance than OP varieties, to farmers who buy them year after year (Dogor, 2013), this research indicates that farmers may not invest in hybrid seeds for reasons included high prices of hybrid seed, non-availability of hybrid seed at local shops, belief that there is a high requirement of fertilizer for cultivation, small or no differences in yield when compared to local or improved varieties, poor storability and poor processing quality.

Hybrid seeds with potential to thrive and increase maize production in conditions such as low soil nitrogen is even more critical under the prevailing circumstances. It is a major game-changer with a potential to transform maize productivity in Africa in general and specifically in Ghana.

Farmers interviewed in this study area lacked knowledge about hybrids. For this reason, the need for education on the benefits of hybrids cannot be overemphasized. Farmers generally obtain information on fertilizers and other production matters from extension workers of MOFA. It is important that the extension and other agencies educate farmers on the importance and potential benefits of hybrids. Breeders need to collaborate with farmers at the initial stages of varietal development as suggested by Sperling et al. (2001). This would allow breeders to identify farmer-preferences in order to incorporate them into improved varieties developed for their specific areas.

5. Conclusion and Implication

The survey identified low soil fertility, drought, diseases and insect pests as major constraints to maize production in the study area. It also confirmed low soil fertility as the main constraint. As iterated, for maize production, low soil nitrogen is one of the major causes for low soil fertility. The research confirmed that maize farmers in the districts preferred maize with specific traits including low N tolerance, drought tolerance, good storability, disease and insect resistance and with little need for inputs. Additionally, they prefer slender cobs, light in weight with lots of grains. Additionally, farmers are willing to adopt low N tolerant hybrid maize varieties if the price of seeds is affordable and after observing their performance on-farm.

It is imperative to breed for maize hybrids tolerant to low soil nitrogen, as this will mitigate challenges farmers face with unavailability and high cost of fertilizer. There is urgent need to encourage the adoption of hybrid varieties which are more tolerant to low soil fertility and are responsive to applications of inputs to raise maize productivity. This will require varieties significantly superior to those currently in use, an effective seed production and delivery system, and mechanisms to overcome some of the market failures that discourage technology adoption. There will also be the need to integrate targeted farmers as well as farmer– preferences and concerns in low fertility / low – N – related maize breeding programmes to ensure their success and also smooth and sustainable adoption of the hybrid seeds for increased production.

Acknowlegdement

The authors are grateful to Alliance for a Green Revolution in Africa (AGRA) for sponsoring this study. Thanks to the Director (Professor Eric Y. Danquah) and staff of West Africa Centre for Crop Improvement (WACCI), University of Ghana for your support. Staff of CSIR-Crops Research Institute and the extension agents at Ministry of Food and Agriculture.

References

Adu, R. (2012). Assessment of The Potential of Reclaimed Mined Land for Agricultural Production. Masters dissertation from Kwame Nkrumah University of Science and Technology, pp. 1. Retrieved from the University library, http://hdl.handle.net/123456789/5877

Banziger, M., & M, Cooper. (2001). Breeding for low input conditions and consequences for participatory plant breeding: Examples from tropical maize and wheat. *Euphytica, 122*, 503-519. https://doi.org/10.1023/A:1017510928038

De Vries, J., & Toenniessen, G. H. (2001). Securing the harvest: Biotechnology, breeding and seed systems for African crops. CAB International, Wallingford, UK.

Dogor, M. M. K. (2013) The effect of fertilizer formulation on yield components and yield of hybrid maize (*Zea mays* L.) varieties in the Guinea savannah zone of Ghana. (Masters dissertation from University of Ghana Pp 11) Retrieved from http://ugspace.ug.edu.gh/

Lafitte, H. R., & Edmeades, G. O. (1994). Improvement for tolerance to low soil nitrogen in tropical maize II. Grain yield, Biomass production, and N accumulation, *Field Crops Research, 39*, 15-25, https://doi.org/10.1016/0378-4290(94)90067-1

Logah, V., Sarfo, E. Y., Quansah, C., & Danso, I. (2010). Soil microbial biomass carbon, nitrogen, and phosphorus dynamics under different amendments and cropping systems in the semi-deciduous zones of Ghana. *West African Journal of Applied Ecology, 17,* 121- 133.

Morris, M. L., Tripp, R., & Dankyi, A. A. (1999). *Adoption and Impacts of Improved Maize Production Technology: A Case Study of the Ghana Grains Development Project.* Economics Program Paper 99-01. Mexico, D.F.:CIMMYT

Nkonya, E. M., & Featherstone, A. M. (2001). Cross-pollinated crop variety adoption studies and seed recycling: the case of maize in Tanzania. *Eastern African Journal of Rural Development, 17*, 25-34.

Odendo, M., De Groote, H., Odongo, O., & Oucho, P. (2002). Participatory Rural Appraisal of Farmers' Maize Selection Criteria and Perceived Production Constraints in the Moist Mid altitude Zone of Kenya. IRMA Socio-Economic Working Paper No. 02-01. Nairobi, Kenya: CIMMYT and KARI.

Ragasa, C., Dankyi, A., Acheampong, P., Wiredu, A. N., Chapo-To, A., Asamoah, M., & Tripp, R. (2013). Patterns of adoption of improved maize technologies in Ghana. *Ghana strategy support programme.* International Food Policy Research Institute.

Sperling, L., Ashby, J. A., Smith, M. E., Weltzien-R. E., & Mcguire, S. (2001). A framework for analyzing participatory plant breeding approaches and results. *Euphytica, 122*, 439-450. https://doi.org/10.1023/A:1017505323730

Tripp, R., & Mensah-Bonsu, A. (2013). Ghana's Commercial Seed Sector: New Incentives or Continued Complacency? Ghana Strategy Support Program Working Paper 32. Washington, DC: International Food Policy Research Institute. WABS Consulting Ltd. 2008. "Draft Report: Maize Value Chain Study in Ghana, Enhancing Efficiency and Competitiveness," WABS Consulting Ltd. December 2008.

Wekesa, E., De Groote, H., Ndungu, J., Chivatsi, W., Mbutha, P., Hadullo, L., Odhiambo, J., & Ambajo, G. (2003). Participatory rural appraisals for insect resistant varieties in the Coastal lowlands, Kenya. IRMA socio-economic working paper 03-01.

Wolfe, D. W., Henderson, D. W., Hsiao, T. C., & Alvio, A. (1988). Interactive water and nitrogen effects on maize. II. Photosynthetic decline and longevity of individual leaves. *Agronomy Journal, 8*, 865- 870. https://doi.org/10.2134/agronj1988.00021962008000060005x

19

Evaluation of Low Heat Unit Corn Hybrids Compared to Barley for Forage Yield and Quality on the Canadian Prairies

Herbert Andrew Lardner[1, 2], Leah Pearce[2] & Daalkhaijav Damiran[1, 2]

[1]Department of Animal and Poultry Science, University of Saskatchewan, 51 Campus Drive, Saskatoon, SK S7N 5A8, Canada

[2]Western Beef Development Centre, P.O. 1150, Humboldt, SK S0K 2A0 Saskatchewan, Canada

Correspondence: H.A. Lardner, Department of Animal and Poultry Science, University of Saskatchewan, 51 Campus Drive, Saskatoon, SK, S7N 5A8. Canada. E-mail: blardner.wbdc@pami.ca

Abstract

Corn (*Zea mays* L.) production is expanding in the prairie region of western Canada. The objectives of this study were to compare three new low heat unit corn hybrids to barley (*Hordeum vulgare* L.) for forage yield, nutrient profile, and total nutrient production. The study was conducted at 4 sites (Evansburg and Fairview, Alberta; Melfort and Scott, Saskatchewan) with different soil characteristics (Gray Luvisolic, Grey Wooded, Dark Brown, and Black soil zones) over three consecutive years (2012-2014). At each site, annually, 16 plots (2.4×7.2 m) were randomly assigned to one of four forage crops (corn: Monsanto DKC26-25, Hyland 2D093, Pioneer P7443R; barley: cv. AC Ranger) in a replicated ($n = 4$) trial. Number of cobs per plant was not different ($p = 0.23$) between corn hybrids averaging 1.22 ± 0.32/plant (mean ± sd). Forage yield among the corn hybrids was negligible ($p > 0.05$), but the corn hybrids exhibited 40% higher yield ($p < 0.05$; avg. 11.3 ± 3.6 t/ha on DM basis) compared to barley (avg. 6.7 ± 1.7 t/ha). Corn hybrids were lower ($p < 0.05$) in CP content [7.6 ± 1.4% versus (vs.) 12.4 ± 0.1%] than barley. No difference was observed between the 4 forage crops in TDN content (68.2 ± 2.8% DM). Study results suggest that new cool-season corn hybrids can produce high quality forage to meet the nutrient requirements of grazing beef cows in mid- and late-stage pregnancy. New corn hybrids may be suitable alternatives for winter grazing strategies since forage harvest costs would be eliminated.

Keywords: barley, corn, forage, nutritive value, warm-season crop, growing degree days

1. Introduction

Winter feeding costs are a major contributor to the overall cost of production for cow-calf producers (Krause et al., 2013). Traditionally, these costs are due to feeding cows in drylot pens over the winter period, which includes costs for harvesting, handling, and transporting feed and removal of manure (Kelln et al., 2011; Krause et al., 2013). Grazing beef cows on annual stockpiled forages or swath grazing during the winter months are options to potentially reduce the costs of wintering beef cows (Van De Kerckhove et al., 2011; Krause, et al., 2013). Cool season annual forages such as barley (*Hordeum vulgare L.)* are well suited to western Canadian growing conditions and provide acceptable forage yield and quality (McCartney et al., 2008; Kelln et al., 2011; Kumar et al., 2012). Corn (*Zea mays* L.) as a winter grazing crop for beef cattle has been more limited to the southern areas of eastern Canada, or southern areas of the prairies. Corn is a warm season annual, usually seeded late with variation in the date of maturity depending on the geographic location and available corn heat units (CHU) (May et al., 2007). However, early seeding of late maturing hybrids of corn has immense potential for use in extensive winter grazing systems (Saskatchewan Ministry of Agriculture [SMA], 2008). Previous grazing trials at Lanigan, Saskatchewan suggested that corn can serve as an excellent winter foraging crop, left either standing or swathed (Lardner, 2002). Standing corn can also serve as an effective windbreak for grazing cattle during winter and snow depth will not limit animal access to the crop as corn stands well above the ground (Baron et al., 2003). The corn grown in the Canadian prairies is different from the corn hybrid varieties grown in warmer climates (Lassiter et al., 1958). The main differences are due to the shorter growing season and lower growing temperatures in the Canadian prairies compared with the areas of warm-season corn production, such as the United States (Lauer et al., 2001; Abeysekara et al., 2013). Many corn hybrids for western Canada require ≥2,300 crop heat units (CHU) to reach the silage harvest stage, with a kernel maturity of 45% dry matter (DM).

Recently, with the introduction of low heat unit corn hybrids [i.e., Pioneer P7443R (Pioneer Hi-Bred International Inc., Johnston, IA), Hyland 2D093 (Hyland Seeds, Blenheim, ON, Canada), Monsanto DKC26-25 (Monsanto Creve Coeur, Greater St. Louis, Missouri] suited to western Canadian weather, there is an increased interest in the use of warm season annuals in extensive grazing systems (Lardner et al., 2012; Jose, 2015). In order to establish these hybrids as a forage crop, these must be compared against other commonly used (conventional) forage crops such as whole plant barley. In addition, differences in forage quantity and quality of these newly developed corn hybrids may occur due to growing environment and soil quality. The objectives of this study were to compare new low-heat-unit corn hybrids to forage barley in terms of forage yield and nutrient value in different soil zones on the western Canadian prairies.

2. Materials and Methods

2.1 Site Selection, Pre-Planting Soil Analysis and Experimental Design

The study was conducted at 4 different sites located in Alberta and Saskatchewan (western Canada) over 3 consecutive years (2012, yr 1; 2013, yr 2; 2014, yr 3). The research sites were established on a Gray Luvisolic soil at Fairview, Alberta, a Grey Wooded soil at Evansburg, Alberta, a Dark Brown soil at Scott, Saskatchewan and a Black soil at Melfort, Saskatchewan. All sites were selected to be typical of areas for expansion of growing whole plant corn for grazing management. Each year, certified seed of 3 corn hybrids, Monsanto *DKC 26-25*, Pioneer *P7443R*, Hyland *2D093*, and 1 forage barley hybrid, *AC Ranger* were established at all 4 sites. At each site, 16 plots were randomly assigned to 1 of 4 annual crops (3 corn hybrids [DKC26-25; 2D093; P7443R]; 1 barley variety [AC Ranger]) in a replicated ($n = 4$) forage crop production plot trial. Individual plot size was 3.8×14 m with a 0.5 m gap between treatment plots and a 3 m gap between forage replicates. Each year, in order to determine nutrient requirements, prior to seeding, soil samples were collected from 10 random locations at each site from the top 0-15 cm and analysed by a commercial laboratory for nitrate-N, ammonium-N, phosphorus (P), potassium (K), and sulphur (S). The P7443R, 2D093, and DKC26-25 corn hybrids were rated at 2150, 2000 and 2250 CHU maturity, respectively.

2.1.1 Fairview Alberta

The study site was located at the Peace Country Beef and Forage Association Research Farm, Fairview, Alberta, Canada (56°08′N 118°44′W). The soil type was classified as a Gray Luvisolic soil, loamy argileux (Perkins et al., 1986) and long-term average CHU's were 1800 to 2000 [Alberta Agriculture Food and Rural Development (AAFRD), 2001]. The site history was fallow in summer of 2011, and canola grown in 2010. A soil test was conducted each year prior to seeding for determination of N, P, K and S, followed by harrowing of the site. The soil had a pH of 5.2 and 8.1% organic matter. Soil N, P, and K levels were maintained as recommended by the Alberta Fertilizer Guide (AAFRD, 2004). Each year, before seeding, corn plots were fertilized with 112 kg actual N/ha + 45 kg actual P/ha, while the barley plots received 45 kg actual N/ha + 26 kg actual P/ha. The fertilizer was drilled into the plots using a small plot drill. Corn and barley crops were seeded on May 28, May 23, and May 30 in yr 1, yr 2, and yr 3, respectively. Corn was seeded with a 6-row John Deere 7200 corn planter with a row spacing of 75 cm at a rate of 74100 seeds/ha at a depth of 4 cm, while the barley was seeded at the rate of 112 kg/ha using a 7-row plot drill (spaced at 17.5 cm between rows) at a seeding depth of 4 cm. Each yr, the barley was sprayed with 2-4 D amine at 1.7 L/ha at the 4-leaf stage (mid-June), and corn was sprayed with glyphosate [N-(phosphonomethyl) glycine] at 1.7 L/ha application rate at the 4-leaf stage (late-June). In addition during yr 2, plots were hand-weeded so that plots were weed-free. When barley reached soft dough stage on August 4, August 1, and July 29 for yr 1, yr 2, and yr 3, respectively, barley plots were harvested for estimation of forage biomass yield. Corn forage yield was determined (see *Section 2.2* for details) at kernel half milk line stage on October 4, September 26, and September 29, for yr 1, yr 2, and yr 3, respectively. All forage samples were clipped, weighed and dried for DM determination and laboratory analyses.

2.1.2 Evansburg Alberta

Field trials were conducted at the West Central Forage Association Research Farm (53°57′N 115°12′W), located at Evansburg, Alberta, Canada. Soils were classified as imperfectly drained Gleyed Orthic Gray Wooded soils developed on weakly to moderately calcareous stratified lacustrine material and long-term average CHU's were 1800 to 2000 (AAFRD, 2001). Corn was seeded June 4, June 11, and May 13 in yr 1, yr 2, and yr 3, respectively with a Versatile 2.5 cm hoe seeder at a row spacing of 75 cm at 74100 seeds/ha at a depth of 4 cm and barley was seeded on June 20, June 7, and May 13 in yr 1, yr 2, and yr 3, respectively at a depth of 4 cm at 108 kg/ha and a row spacing of 25 cm. At time of seeding, all plots were side-banded with 113.1 kg/ha N, 44.8 kg/ha P_2O_5, 33.6 kg/ha K_2O, and 16.8 kg/ha sulfur. Corn plots were sprayed each yr with a pre-seeding burn and at 4-leaf stage (early July) with 0.84 L/ha glyphosate. Barley plots were sprayed with 2-4 D amine at 0.67 L/ha at the 4-leaf stage (late June). Corn and barley forage yield was determined (see *Section 2.2* for details) on October 16, October 28, and September 10 in yr 1, yr 2, and yr 3, respectively.

2.1.3 Scott Saskatchewan

Field trials were conducted at the Agriculture and Agri-Food Canada Research Farm, Scott, Saskatchewan, Canada, located in the Dark Brown Chernozemic soil zone (52°21'N 108°49'W). Soil type was classified as an Elstow loam (Saskatchewan Soil Survey, 1992). Long-term average CHU's were 2001 to 2100 (SMA, 2010). The site grew canola in 2010 and was left to fallow in the summer of 2011 A soil test was conducted each year prior to seeding for determination of N, P, K and S, followed by harrowing of the site. Corn plots were fertilized with 52 kg actual N/ha, 22 kg actual P_2O_5/ha, 20 kg/ha actual K_2O broadcast pre-seeding and at seeding with 76 kg/ha N side banded. Then, 6.7, 11.2, 5.6, and 5.6 kg/ha actual N, P_2O_5, K_2O, and S, respectively was placed adjacent to seed at planting. Each yr, barley plots received 67 kg actual N/ha through side banding at seeding. Corn plots were seeded May 18, May 30, and May 28 in yr 1, yr 2, and yr 3, respectively. Corn was seeded with a Versatile 2.54 cm hoe seeder at a row spacing of 75 cm at 74100 kernels/ha at a 4 cm depth. Barley was seeded on June 22, June 12, and May 28 in yr 1, yr 2, and yr 3, respectively at 108 kg/ha and a depth of 4 cm with a row spacing of 25 cm. All plots received a mid-May pre-seeding burn off with glyphosate at 1 L/ha. Each yr, corn was sprayed with 0.67 L/ha glyphosate at the 4-leaf and 8-leaf stages and barley plots were sprayed with 2-4 D amine at 0.67 L/ha at the 4-leaf stage. Corn forage yield was determined (see *Section 2.2* for details) at half milk line stage on October 4, September 26, and July 29, in yr 1, yr 2, and yr 3, respectively, while barley plots were harvested at soft dough on August 4, August 1, and July 29 in yr 1, yr 2, and yr 3, respectively.

2.1.4 Melfort Saskatchewan

The site was located at the Agriculture and Agri-Food Canada Research Farm, Melfort, Saskatchewan, Canada (52°44'N 104°47' W). The soil was classified as thick Black Chernozem (Udic Boroll) silty clay soil and long-term average CHU's were 2100 to 2200 (SMA, 2010). In early June, N fertilizer was broadcasted at 84 kg N/ha prior to seeding. Corn was seeded on May 15 (yr 1), May 23 (yr 2), and June 9 (yr 3), at 71400 kernels/ha using a Hege plot drill (Hege Equipment Inc., Colwich, KS) at 4 cm depth with a 75 cm row spacing, while barley was seeded at 108 kg/ha on June 12 (yr 1), July 2 (yr2, and July 3 (yr 3) with a 7-row plot drill (spaced at 17.5 cm between rows) at 4 cm depth. Corn plots were sprayed with glyphosate at 1.7 L/ha at the 4-leaf stage (June 12, June 24, and July 2 for yr 1, yr 2, and yr 3, respectively), while barley plots were hand weeded each yr eliminating any weeds or volunteer canola. Barley forage yield was determined on August 20, September 4, and August 25 in yr 1, yr 2, and yr 3, respectively. Corn forage yield was determined (see *Section 2.2* for details) on September 14, September 20, and October 8 in yr 1, yr 2, and yr 3, respectively.

2.2 Estimation of Whole Plant Dry Matter Yield and Number of Cobs Per Plant

Each yr, corn plots were harvested following the first killing frost, with most hybrids at the 2/3 milk line (R5 stage) stage of maturity. Prior to harvesting, the number of cobs per plant was also determined. At each site, corn forage yield was determined by randomly harvesting 3 entire rows (5.25 m) of corn followed by weighing material to obtain a wet weight. From each corn plot, 3 random corn stalks including cobs were selected as sample (> 1 kg) and chopped to 1 cm in size. Forage yield of barley was determined at the soft dough stage by harvesting full plants above 7 cm stubble height followed by determining wet weight. A 10 random grab samples (>1 kg) of swathed forage were taken from each barley plot. All selected samples were placed in a forced air oven at 55°C for 72 h to determine DM percentage and saved for subsequent forage quality analysis. Plot DM yield was determined by multiplying the DM concentration by the plot fresh weight and was expressed in tonne (t)/ha.

2.3 Forage Quality Analysis

All forage samples were ground to pass through a 1-mm screen using a Thomas-Wiley Laboratory Mill (Model 4, Thomas Scientific, Swedesboro, NJ, USA). Crude protein (CP) was determined by nitrogen combustion (method 990.03, AOAC, 2000) using a Leco FP-528 Nitrogen Combustion Analyzer (Leco, MI, USA). Neutral detergent fiber (NDF) was determined as described in Van Soest et al. (1991) and acid detergent fiber (ADF) according to method 973.18 (AOAC, 2000). Calcium (Ca) and phosphorus (P) were analyzed using the dry ashing procedure (methods 927.02 and 965.17; AOAC, 2000, respectively). The Ca was determined using an atomic absorption spectrophotometer (Perkin-Elmer, Model 2380, CN, USA), while P concentration was read at 410 nm on a spectrophotometer (Pharmacia, LKB-Ultraspec® III, Stockholm, Sweden). Total digestible nutrients (TDN; % DM) and metabolizable energy (ME; Mcal/kg DM) were calculated using the grass-legume Penn-State equation based on ADF as described Adams (1995). All data on forage yield and nutritive quality are reported on a DM basis.

2.4 Calculation of Nutrient Yield

Nutrient yield per hectare was calculated by multiplying crop forage yield (t/ha) by nutrient concentration to allow a comparison of nutrient yield potential for animal feed production among the forage crops.

2.5 Climate Data

Monthly average precipitation and temperature (Table 1) were obtained from Environment Canada (www.climate.weatheroffice.ec.gc.ca) for each site. The difference in average temperature between trial years and long-term averages (30 year avg.) was negligible for all four sites (ranged 98.7 to 104.2% of long-term average). Major difference was in rainfall, with 11 and 15% above the long-term average observed for April to October at Melfort and Scott, Saskatchewan, respectively. However, trial years were drier (precipitation lower by 35%) compared to long-term average at Fairview, Alberta. Overall, these data suggested that the current study was conducted in an environment with comparable temperatures and with higher (Melfort and Scott, SK) or lower (Fairview, AB) precipitation relative to the long-term average. Total precipitation from April to October (growing season) was 235, 302, 326, and 347 mm for Fairview, Evansburg, Scott, and Melfort, respectively, averaged over the three trial years. Therefore, soil moisture was assumed to be not limiting at any location.

Table 1. Precipitation and temperature over 3 growing seasons (2012-2014) and long-term average at 4 sites in Canadian prairies*

	Fairview AB		Evansburg AB		Scott SK		Melfort SK	
Month	3yr-avg.	Avg.†	3yr-avg.	Avg.	3yr-avg.	Avg.	3yr-avg.	Avg.
Precipitation (mm)								
April	23.7	23.0	38.5	23.0	26.1	23.8	25.6	22.0
May	33.1	36.0	49.4	36.0	41.5	35.5	32.9	47.0
June	57.5	60.0	44.9	60.0	124.6	67.3	127.4	70.0
July	29.9	59.0	93.1	59.0	64.9	65.6	79.7	77.0
August	25.9	45.0	35.9	45.0	46.1	43.7	45.6	33.0
September	31.9	32.0	26.5	32.0	15.5	31.8	13.0	34.0
October	32.9	31.0	13.7	31.0	7.3	13.9	22.6	29.0
Temperature (°C)								
April	0.9	4.1	2.6	3.2	1.0	3.3	-0.8	3.1
May	10.1	9.1	10.5	10.2	10.5	10.7	11.6	9.6
June	13.8	13.0	14.2	14.5	14.6	15.2	14.8	14.9
July	16.6	14.9	17.3	17.3	17.5	17.2	17.8	17.3
August	15.9	13.8	16.5	16.1	17.0	16.4	17.4	17.3
September	11.5	9.8	12.0	10.4	12.5	10.3	13.0	11.2
October	2.9	4.0	3.8	4.6	3.4	4.3	3.0	3.5

*Source = Environment Canada (www.climate.weatheroffice.ec.gc.ca); farmzone Canada (www.farmzone.com), The Weather Network (www.theweathernetwork.com).

†30 yr-average

For each experimental site, crop heat units (CHU) were calculated on a daily basis, using the maximum (Tmax) and minimum (Tmin) daily air temperatures, measured from midnight to midnight, in °C. The following equation was used to calculate daily CHU: Daily CHU = (Ymax + Ymin)/2; where Ymax = [3.33 × (Tmax-10)] - [0.084 × (Tmax-10)^2] (if Ymax < 0, set Ymax = 0) and Ymin = [1.8 × (Tmin - 4.4)] (if Ymin <0, set Ymax = 0). Many corn hybrids for western Canada require ≥2300 CHU to reach the silage harvest stage, with a kernel maturity of 45% DM. In corn cultivation, CHU accumulation from planting until first killing frost (-2°C) was not different ($p = 0.69$; data not shown) due to trial site and was 1749, 1850, 1767 and 1894 for Fairview, Evansburg, Scott, and Melfort, respectively (Table 2). In barley cultivation, growing degree days (GDD) are calculated using the following equation: GDD = [(Tmax, °C + Tmin, °C)/2 - 5]. In the current study, 3-yr average CHU was not different [1219 ± 87 (mean ± sd); $p = 0.60$] among the sites. Days to harvest were 92, 118, 121, and 100 for Evansburg, Fairview, Melfort, and Scott, respectively.

Table 2. Summary of planting, harvesting date, and crop heat unit (CHU) at four sites in Canadian prairies over the 3 yr study

Year (YR)	Seeding date	Harvesting date	Days to harvest, d	GDD*	CHU†
Fairview, AB					
2012	May 28	October 03	130	1251	1997
2013	May 23	September 26	127	1185	1954
2014	May 30	September 25	118	1124	1749
Average	-	-	**125**	**1187**	**1900**
Evansburg, AB					
2012	May 14	October 07	147	1304	2175
2013	June 11	October 28	140	1132	1759
2014	May 13	September 10	121	1168	1850
Average			**136**	**1202**	**1928**
Scott, SK					
2012	May 18	September 12	118	1239	2056
2013	May 30	September 27	120	1311	2131
2014	May 28	September 5	100	1074	1767
Average	-	-	**113**	**1208**	**1984**
Melfort, SK					
2012	May 15	September 14	123	1312	2138
2013	May 23	September 21	121	1346	2171
2014	June 9	October 08	121	1190	1894
Average	-	-	**122**	**1283**	**2068**

*Data calculated from seeding until harvest; Data calculated from seeding until swathing date; GDD = [(Tmax + Tmin)/2 - 5], where Tmax = Maximum daily temperature (°C), Tmin = Minimum daily temperature (°C).

†CHU = [1.8 (Tmin - 4.4) + 3.3 (Tmax - 10) - 0.084 (Tmax - 10)²] / 2.

2.6 Statistical Analysis

Due to non-homogeneity of variances (Steel et al., 1997) between site-yrs, and yr x site, yr x treatment interactions ($p < 0.05$), the experiment was analyzed separately for each site-yr and no attempt was made to explain these interactions.

Numerous significant crop × site interactions were detected for many of the measured variables and thus data were subsequently analyzed by site. The experimental design for this study was a completely randomized design with subsampling and with a 4 × 4 factorial arrangement of treatments. For each location, crop production data (number of cobs per plant, DM, and yield) and nutrient density data (CP, TDN, Ca, P, and ME) and nutrient yield obtainable from each hectare field (CPY, TDNY, CaY, PY, and MEY), CHU, and GDD were analyzed using the Proc Mixed Model procedure of SAS (2003); each plot was considered an experimental unit (in total 192 plots for 3-yr trial) using the Satterthwaite method. The model used for the analysis was: $Y_{ij} = \mu + T_i + e_{ij}$; where Y_{ij} was an observation of the dependent variable ij; μ was the population mean for the variable; T_i was the fixed effect of treatment (crops: either DKC26-25 or Monsanto DKC26-25; 2D093 or Hyland 2D093; P7443R or Pioneer P7443R; AC Ranger); and e_{ij} was the random error associated with the observation ij. Means were determined using the least squares means statement of SAS and were separated using Tukey's multi-treatment comparison method (Saxton, 1998). Year was included as a random (block) variable in all analyses. In order to examine the relationship between CHU and forage dependent variables (yield and cob number per plant), correlation analysis was performed using the CORR procedure of SAS (2003). For all statistical analyses, significance was declared at $p < 0.05$.

3. Results

3.1 Fairview Site

3.1.1 Forage Nutrient Profile

Effect of treatment on nutritive value, number of cobs per corn plant, DM, and forage yield at the Fairview site over the 3-yr study is presented in Table 3. All four forage crops were similar ($p > 0.05$) in DM (32.8 ± 6.0%; mean ± sd), TDN (69.8 ± 2.0% DM), ME (2.5 ± 0.07 Mcal/kg DM), and P (0.24 ± 0.03% DM) contents. The three corn hybrids were similar with each other ($p > 0.05$) in CP, Ca, as well as NDF, which was however, lower in CP (8.9 ± 1.5 vs. $14.3 \pm 2.5\%$ DM) and in Ca (0.29 ± 0.07 vs. $0.46 \pm 0.04\%$ DM), but greater ($p < 0.05$) in NDF (51.4 vs. 4.2% DM) than barley. When pooled, corn hybrids had higher values for ADF (~28.6% DM) compared with the barley (~25.3%). The DKC26-25 was greater ($p < 0.05$), but 2D093 and P7443R were similar ($p > 0.05$) to barley in ADF.

Table 3. Forage yield and nutrient profile of corn and barley hybrids at Fairview, Alberta CANADA over 3 yr (2012, 2013, and 2014)

Item*	Corn			Barley	
	DKC26-25	2D093	P7443R	Ranger	SEM†
Nutrient Profile					
DM, %	32.2	30.4	32.9	35.5	0.86
CP, % DM	9.3[b]	8.6[b]	8.8[b]	14.3[a]	0.52
ADF, % DM	29.3[a]	29.0[ab]	27.5[ab]	25.3[b]	1.06
NDF, % DM	51.9[a]	51.9[a]	50.4[a]	43.2[b]	1.39
TDN, % DM	68.8	70.1	70.5	69.8	0.56
ME, Mcal/kg	2.49	2.53	2.55	2.52	0.021
Ca, % DM	0.30[b]	0.28[b]	0.29[b]	0.46[a]	0.018
P, % DM	0.24	0.24	0.23	0.24	0.007
Yield					
Cobs/plant	1.2	1.3	1.2	-	0.04
Forage Yield t/ha	10.3[ab]	9.4[b]	12.0[a]	6.8[c]	0.40
CPY, t/ha	0.93[ab]	0.81[b]	1.04[a]	0.97[ab]	0.030
TDNY, t/ha	7.09[b]	6.58[b]	8.40[a]	4.74[c]	0.279
CaY, t/ha	0.030	0.027	0.035	0.031	0.002
PY, t/ha	0.025[a]	0.023[a]	0.027[a]	0.016[b]	0.001
MEY, Mcal/ha	25624[b]	23752[b]	30363[a]	17140[c]	1010.2

*Means within a row with different superscripts differ ($p < 0.05$).

†SEM, Standard error of mean.

3.1.2 Forage and Nutrient Yield

Corn hybrids did not differ ($p > 0.05$) in number of cobs per plant, averaging 1.2 ± 0.25. Corn hybrids had higher forage yield (from 38 to 76% greater; $p < 0.05$) than barley. Differences ($p < 0.01$) were also detected in the forage yield among corn hybrids: 2D093 was lower ($p < 0.05$), while DKC26-25 was similar ($p > 0.05$) to P7443R. There were no differences among crop hybrids for forage derived CaY (kg/ha) ($p > 0.05$). As expected based on the observed Forage Yield results, all three corn hybrids were 39 to 77% greater ($p < 0.05$) than barley in TDNY and consequently, 38 to 77% greater than barley in MEY. The DKC26-25 and 2D093 were similar with each other ($p > 0.05$), but were 19% lower than P7443R in TDNY and consequently, 19% lower in MEY.

3.2 Evansburg Site

3.2.1 Forage Nutrient Profile

Effect of corn hybrids on forage nutrient profile, number of cobs per plant, DM, and forage yield at the Evansburg site over the study years is presented in Table 4. All three corn hybrids had similar CP ($7.6 \pm 1.4\%$; $p > 0.05$) content, but that was lower ($12.4 \pm 0.09\%$; $p < 0.05$) than CP of the barley. Barley exhibited 17% lower ADF and 22% lower NDF concentrations than corn resulting in 4% higher TDN and ME for barley. Barley exhibited 46% lower Ca concentration and 16% lower P concentration than corn. There were nutrient differences among corn hybrids at Evansburg, Alberta. Hyland 2D093 exhibited 29 % lower ADF, 13% lower NDF, 43% lower Ca and 17% lower P concentrations than DKC26-25 and P7443R hybrids.

Table 4. Forage yield and nutrient profile of corn and barley hybrids at Evansburg, Alberta CANADA over 3 yr (2012, 2013, and 2014)

Item*	Corn			Barley	SEM†
	DKC26-25	2D093	P7443R	Ranger	
Nutrient Profile					
DM, %	31.8	31.2	33.2	33.7	1.66
CP, % DM	8.0^b	7.2^b	7.6^b	12.4^a	0.35
ADF, % DM	30.7^a	21.7^b	30.8^a	23.1^b	1.79
NDF, % DM	54.2^a	47.4^b	55.0^a	40.7^c	1.73
TDN, % DM	66.7^b	67.0^{ab}	68.5^{ab}	70.0^a	0.85
ME, Mcal/kg	2.41^b	2.42^{ab}	2.47^{ab}	2.53^a	0.030
Ca, % DM	0.33^{ab}	0.20^{bc}	0.36^a	0.16^c	0.041
P, % DM	0.24^a	0.20^b	0.24^a	0.19^b	0.006
Yield					
Cobs/plant	0.9	1.0	1.0	-	0.06
Forage Yield t/ha	13.0^a	10.9^a	11.9^a	6.3^b	1.13
CPY, t/ha	1.01	0.82	0.89	0.78	0.049
TDNY, t/ha	8.68^a	7.17^{ab}	7.93^a	4.44^b	0.737
CaY, t/ha	0.045^a	0.024^{ab}	0.051^a	0.011^b	0.004
PY, t/ha	0.030^a	0.022^b	0.027^{ab}	0.012^c	0.001
MEY, Mcal/ha	31374^a	25956^{ab}	28657^a	16052^b	2663

*Means within a row with different superscripts differ ($p < 0.05$).

†SEM, Standard error of mean.

3.2.2 Forage and Nutrient Yield

Corn hybrids did not differ ($p > 0.05$) in number of cobs (0.97 ± 0.33/plant) and forage DM content (32.5 ± 0.56). Barley forage yield was 47% lower than corn yield at Evansburg, Alberta. No differences ($p > 0.05$) were detected between all crops in forage derived CPY (avg. 8.7 ± 0.27 t/ha) per hectare. However, barley TDNY, CaY, PY and MEY were lower than those of corn. Forage yield difference between corn and barley had greater impact on forage nutrient yield parameters than nutrient concentration differences. Monsanto DKC26-25 was lower ($p < 0.05$), although 2D093 and P7443R were similar ($p > 0.05$) to barley in TDNY and in MEY.

3.3 Scott Site

3.3.1 Forage Nutrient Profile

Effect of hybrids on number of cobs per corn plant, forage nutrient profile, and yield at the Scott site over the 3-yr study is presented in Table 5. Corn hybrids did not differ ($p > 0.05$) in number of cobs, averaging 1.3 ± 0.25 cobs per plant. All 3 corns had similar forage yield ($p > 0.05$; 10.0 ± 2.73 t/ha), that was, nevertheless, higher (6.9 ± 1.3 t/ha; $p < 0.05$) than that of barley. However, all four forage crops were similar ($p < 0.05$) in DM content ($28.7 \pm 7.7\%$) at the time of harvesting. Corn vs. All 3 corn hybrids were lower ($p < 0.05$) than barley in CP and Ca content. Pioneer P7443R was similar ($p > 0.05$) with both DKC26-25 and 2D093 in CP, ADF, NDF, TDN, as well as ME content. However, compared to 2D093, DKC26-25 was greater ($p < 0.05$) in CP, TDN, and ME, but it was lower ($p < 0.05$) in ADF and NDF. All 3 corns had similar P content ($p > 0.05$) with barley.

Table 5. Forage yield and nutrient profile of corn and barley hybrids at Scott, Saskatchewan CANADA over 3 yr (2012, 2013, and 2014)*

| Item* | Corn | | | Barley | |
	DKC26-25	2D093	P7443R	Ranger	SEM†
Nutrient Profile					
DM, %	27.5	27.0	28.5	32.0	1.12
CP, % DM	9.1[b]	7.8[c]	8.0[bc]	10.3[a]	0.32
ADF, % DM	26.5[b]	33.1[a]	30.5[ab]	28.4[b]	1.12
NDF, % DM	49.4[b]	55.8[a]	52.4[ab]	49.0[b]	0.97
TDN, % DM	69.8[a]	66.7[b]	67.6[ab]	66.3[b]	0.79
ME, Mcal/kg	2.52[a]	2.41[b]	2.44[ab]	2.40[b]	0.028
Ca, % DM	0.25[b]	0.21[bc]	0.19[c]	0.31[a]	0.016
P, % DM	0.28[a]	0.20[b]	0.21[b]	0.24[ab]	0.018
Yield					
Cobs/plant	1.2	1.3	1.3	-	0.04
Forage Yield, t/ha	9.0[a]	10.7[a]	10.2[a]	6.9[b]	0.41
CPY, t/ha	0.82	0.84	0.81	0.72	0.033
TDNY, t/ha	6.28[a]	7.09[a]	6.89[a]	4.62[b]	0.268
CaY, t/ha	0.023	0.023	0.019	0.021	0.001
PY, t/ha	0.026[a]	0.022[ab]	0.021[ab]	0.016[b]	0.001
MEY, Mcal/ha	22707[a]	25628[a]	24923[a]	16698[b]	969.6

*Means within a row with different superscripts differ ($p < 0.05$).

†SEM, Standard error of mean.

3.3.2 Forage and Nutrient Yield

Forage yield of barley was 31% lower than corn at Scott, Saskatchewan. The average values of all four crops were similar for CPY (8.8 ± 1.5 t/ha) and CaY (0.02 ± 0.01 t/ha). All 3 corns were similar ($p > 0.05$) with each other, but they were greater ($p < 0.05$) than barley in TDNY (6.8 ± 1.75 vs. 4.6 ± 0.94 t/ha) and MEY (24419 ± 6357 vs. 16697 ± 3399 Mcal/ha). Forage yield differences were more important when determining nutrient yield differences than crop nutrient concentration at Scott, Saskatchewan.

3.4 Melfort Site

3.4.1 Forage Nutrient Profile

All 3 corn hybrids had similar DM concentration ($p > 0.05$) with each other, but greater ($p < 0.05$) than barley DM (32.1 ± 4.1 vs. 24.8 ± 4.1%) and TDN (68.5 ± 2.1 vs. 64.5 ± 2.1%) (Table 6). The corn hybrids were lower ($p < 0.05$) than barley in CP (7.2 ± 1.2 vs. 12.6 ± 2.3%), Ca (0.21 ± 0.05 vs. 0.43 ± 0.16%), as well as in P content (0.20 ± 0.02 vs. 0.26 ± 0.07%). However, barley exhibited 5% lower ME concentration than corn.

Table 6. Forage yield and nutrient profile of corn and barley hybrids at Melfort, Saskatchewan CANADA over 3 yr (2012, 2013, and 2014)

Item*	Corn			Barley	
	DKC26-25	2D093	P7443R	Ranger	SEM†
Nutrient Profile					
DM, %	32.5[a]	30.8[a]	33.2[a]	24.8[b]	0.78
CP, % DM	7.3[b]	7.6[b]	6.8[b]	12.6[a]	0.44
ADF, % DM	29.0[b]	33.0[a]	32.6[a]	33.4[a]	0.79
NDF, % DM	50.7[b]	55.8[a]	54.2[ab]	54.1[ab]	1.25
TDN, % DM	69.1[a]	67.2[a]	69.0[a]	64.5[b]	0.52
ME, Mcal/kg	2.49[a]	2.39[b]	2.51[a]	2.33[b]	0.017
Ca, % DM	0.21[b]	0.24[b]	0.18[b]	0.44[a]	0.026
P, % DM	0.20[b]	0.20[b]	0.20[b]	0.26[a]	0.011
Yield					
Cobs/plant	1.4	1.5	1.5	-	0.04
Forage Yield t/ha	13.2[a]	12.0[a]	12.7[a]	6.7[b]	0.64
CPY, t/ha	0.94	0.89	0.85	0.84	0.036
TDNY, t/ha	9.03[a]	8.07[a]	8.74[a]	4.31[b]	0.437
CaY, t/ha	0.026	0.027	0.022	0.029	0.001
PY, t/ha	0.026[a]	0.024[a]	0.025[a]	0.017[b]	0.001
MEY, Mcal/ha	32628[a]	28874[a]	31875[a]	15570[b]	1577.7

*Means within a row with different superscripts differ ($p < 0.05$).

†SEM, Standard error of mean.

3.4.2 Forage and Nutrient Yield

Corn hybrids did not differ ($p > 0.05$) in number of cobs, averaging 1.4 ± 0.27 cobs per plant (Table 6). In the same fashion with the other 3 sites, all 3 corn hybrids had similar forage yield (12.6 ± 4.12 t/ha; $p > 0.05$), which was 47% higher ($p < 0.05$) than barley (6.70 ± 1.1 t/ha). No variation in forage derived CPY (avg. 8.6 ± 2.8 t/ha) and CaY (0.27 ± 0.13 t/ha) occurred among the four crops. A hybrid effect on the CPY or CaY was not found ($p > 0.05$) between the new corn hybrids. Corn was greater ($p < 0.05$) than barley in TDNY (8.6 ± 2.7 vs. 4.3 ± 0.68 t/ha) and MEY (31125 ± 9813 vs. 15570 ± 2465 Mcal/ha) and PY (0.025 ± 0.008 vs. 0.017 ± 0.017 t/ha). In contrast to the other sites, the higher barley CP is compensated by the higher Forage Yield of corn at Melfort so that CPY was not significantly ($p > 0.05$) different. For the other calculated forage nutrients however, Forage Yield differences contributed to significantly higher TDNY, CaY, PY and MEY for corn compared to barley at Melfort, Saskatchewan.

3.5 The Relationship between CHU and Forage Yield and Nutrient Profile

Summarizing all four sites, corn hybrids had good cob development by the end of the growing season over the trial years. When sites were compared (data not shown), Melfort site (24.8%) was similar ($p > 0.05$) with Scott (32.0%), and Evansburg (33.7%), but was lower ($p < 0.05$) than Fairview (35.5%) site on barley DM content by the time of harvesting. Whereas, DM content at harvesting was not different ($p > 0.05$) across the corn hybrids as well as across the trial sites and averaged $30.9 \pm 6.3\%$ (mean ± sd). When all 4 trial sites were pooled (data not shown), forage yield did not vary ($p > 0.05$) among the 3 corn hybrids. However, on average all corn hybrids had approximately 40% higher forage yield (avg. 11.3 ± 3.6 t/ha) than the AC Ranger barley (avg. 6.7 ± 1.7 t/ha). Summarizing over all four sites, the number of cobs per corn plant had a moderate ($r = 0.57$, $p < 0.01$) correlation, but corn forage yield had a weak correlation ($r = 0.31$, $p = 0.06$) with CHU's, the latter suggesting that low CHU's access leads to a relatively low number of cobs per plant and a minor effect on forage yield. However, corn TDNY and CPY had a moderate positive ($r = 0.51$, $p < 0.01$) and a weak positive correlation ($r = 0.34$, $p < 0.04$) with CHU, respectively. The relationships between CHU and TDNY, and between CHU and CPY were a moderately positive ($r = 0.51$, $p < 0.01$) and a weakly positive ($r = 0.34$, $p < 0.04$), respectively.

4. Discussion

Forage yield and nutrient profile are influenced by soil nutrients, plant species, genotype within species, stage of harvest, weather and other environmental conditions (Baron et al., 2012). Specially, successful growth of corn depends on the availability of CHU, and corn is considered more suitable to areas receiving a minimum of 2000 to 2100 CHU's (McCartney et al., 2009). On average, CHU's in the current trial were closer with the McCartney et al. (2009) bench mark. Comparable forage yield and nutritive value among corn hybrids and barley at different trial sites could be explained in part by the similar accumulated CHU and GDD across sites during the study years (Table 2). Additionally, plant-available soil nutrients probably were adequate at all 4 sites since soil N, P, and K were maintained at the levels recommended despite each site representing different soil associations. These observations may help explain the apparent lack of difference that was found among the corn hybrids at these trial sites.

The DM content of corn forage has previously been reported as 32% to 40% (Bal et al., 2000; Jurjanz & Monteils, 2005), a value comparable with the current study findings. Forage yield of barley was in agreement with previous (Aasen et al., 2004; Baron et al., 2012) studies (i.e., 4.2-13.6 t/ha) conducted in Alberta, Canada. Corn forage yield shown in the current study fell within the range of that observed previously in the Canadian prairies (Baron et al., 2006; Baron et al., 2014; Lardner et al., 2012; 8.3-15.9 t/ha).

The NRC (2000) model predicts that a dry beef cow in early to mid-gestation requires 7 to 8% of CP in the diet for maintenance which increases to 11 to 13% CP in young (first parity) growing or lactating cows. Protein content was higher for barley (10.3 to 14.3% CP) than the 3 corn hybrids (6.8 to 9.3% CP). The latter CP values were in agreement with those of previous work (Baron et al., 2003; Abeysekara et al., 2013) for corn forage. Taking into consideration the protein requirements of beef cattle, in the current study, only AC Ranger barley had sufficient amounts of CP required by gestating and lactating cows. The 3 corn hybrids were only adequate in the CP for cows in the mid-gestation stage. Therefore, cows in the late pregnancy stage grazing these corn hybrids (regardless of grown site and of fertility sources) would consequently need some form of protein supplementation (Van De Kerckhove et al., 2011; Jose, 2015; Damiran et al., 2016), or good legume hay (Krause et al., 2013) with high CP content. Nutrient values for low CHU region corn plants (NDF 49, CP 7.1, ADF 28.3, and TDN 67.3% of DM, ME 2.2 Mcal/kg DM) have been reported previously (Abeysekara et al., 2013) and are in close agreement with our current results.

Energy is probably the most important nutritional consideration in beef cattle production in cold climates. Findings of the present study were in agreement with published values for regular corn forage (TDN = 68.8%; NRC, 2001). Using TDN as the energy source for beef cow, the rule of thumb is 55-60-65 (%, DM) (Yurchak & Okine, 2004). This rule says that for a mature beef cow to maintain her body condition score through the winter, the ration must have a TDN energy reading of 55% in mid pregnancy, 60% in late pregnancy and 65% after calving (Yurchak & Okine, 2004). Therefore, even accounting for weathering effects on plant material for winter grazing (Baron et al., 2003) , the result of the current study suggested that all three new corn hybrids had adequate energy needed by dry gestating cows, mid pregnancy cows, as well as late pregnancy cows at all sites. This further confirms that similar with barley, corn is a high energy feed that will normally match the nutritional needs of a dry cow in mid and late pregnancy and after calving. Likewise, a recent Saskatchewan winter field study (Jose, 2015) also suggests that the corn grazing cows were more efficient compared to swathed barley or bale grazing cows, in converting each unit of feed DM to net energy for maintenance and production.

Animal mineral requirements can also vary with the stage and level of production (NRC, 2000). Forage mineral content can be affected due to changes in soil, season and stage of maturity (Kappel et al., 1985). All crops tested in the current study had sufficient amounts of Ca needed for dry gestating cows. None of the crops was however adequate to meet the 0.58% Ca required by lactating cows (NRC, 2000). Only DKC26-25 and P7443R corn hybrids at Evansburg, Alberta were short of meeting the 0.31% Ca requirements of growing and finishing beef cattle. Forage P content was between 0.19 and 0.28% for corn hybrids and AC Ranger barley. These values were within the ranges suggested for growing and finishing beef cattle (0.21% P) and dry gestating cows (0.16% P) (NRC, 2000). But for lactating cows, which require 0.26% P (NRC, 2000), the crops fell short in meeting their P requirements. This, therefore, indicates that for cows in the late pregnancy stage, some form of mineral supplementation to address the short fall of both forage Ca and P contents is needed. In addition, based on the nutrient composition of these new corn hybrids, fresh forage (green chop) would have adequate levels of NDF, ADF, and protein (Weiss et al., 1992; Taylor & Allen, 2005) for ensiling.

Sustainable management of beef cattle involves balancing the nutrient needs of the animal with the nutritional opportunities of the forage resource. For example, it may be possible to reduce the daily feed cost of swath grazing below that of swathed barley forage by using crop species with potential for higher forage yield and quality, assuming similar costs of production (Baron et al., 2014). As stated in previous section (*Section 3*), on average, each corn field produced a relatively similar amount of CPY (0.89 vs. 0.86 t/ha), but were approximately 65% greater for TDNY (7.7 vs. 4.7 t/ha) than what the barley field produced, indicating greater carrying capacities for wintering beef cow grazing as this suggests a substantial savings in feed energy costs (Kelln et al., 2011). Thus, the new corn hybrids evaluated in the current study, would fit very well where higher yield from a limited land base is desirable.

5. Conclusions & Implications

Study results confirmed that the new low heat unit corn hybrids had higher biomass and lower CP content than barley across Canadian prairie environments. Very little differences were observed in the yield and nutrient profile among the new corn forage hybrids. The Monsanto DKC 26-25, Pioneer P7443R, and Hyland 2D093, all produced good forage yield and were of suitable quality to meet nutrient requirements of grazing beef cows in the mid and late pregnancy stages. In addition, grazing these corn hybrids whole plant may fit very well where winter grazing is practiced since forage harvest costs would be eliminated. Overall, as the present study demonstrated, new bred low heat unit forage corn is comparable to conventional barley forage in major nutrients content (excluding CP) and availability to animals in major soil zones of Canadian prairies. A long term field grazing research program (with multiple years) needs to be conducted to make a more definitive conclusion.

Acknowledgments

The present study was supported by the Alberta Livestock and Meat Agency, Saskatchewan Agriculture Development Fund, Hyland Seeds, Monsanto Canada, DuPont Pioneer, West Central Forage Association, Peace Country Beef & Forage Association, Northeast Agricultural Research Foundation, Western Applied Research Corporation, and Western Beef Development Centre. The authors acknowledge the technical assistance of Dr Paul Jefferson and Enkhjargal Darambazar (University of Saskatchewan).

References

Aasen, A., Baron, V. S., Clayton, G. W., Dick, A. C., & McCartney, D. H. (2004). Swath grazing potential of spring cereals, field pea and mixtures with other species. *Canadian Journal of Plant Science, 84*, 1051-1058. http://dx.doi.org/10.4141/P03-143

Abeysekara, S., Christensen, D. A., Niu, Z., Theodoridou, K., & Yu, P. (2013). Molecular structure, chemical and nutrient profiles, and metabolic characteristics of the proteins and energy in new cool-season corn varieties harvested as fresh forage for dairy cattle. *Journal of Dairy Science, 96,* 6631-6643. https://doi.org/10.3168/jds.2013-6841

Adams, R. S. (1995). Dairy nutrition. In: C. Walker (Ed.), *Dairy Reference Manual.* (pp. 108-109). Northeast Regional. Agricultural Engineering Service, NY: Ithaca.

Alberta Agriculture Food and Rural Development [AAFRD]. (2004). *Alberta fertilizer guide.* Government of Alberta. Accessed April 2016.
http://www1.agric.gov.ab.ca/$department/deptdocs.nsf/all/agdex3894/$file/541-1.pdf?OpenElement.

Alberta Agriculture Food and Rural Development [AAFRD]. (2001). *Annual total corn heats units: 1971-2000.* http://www1.agric.gov.ab.ca/$department/deptdocs.nsf/all/sag6442/$FILE/onl_s_9_twp_annual_normals_1 9712000.gif. Accessed 18 April 2016.

AOAC (Association of Official Analytical Chemists). (2000). *Official methods of analysis* (17th ed.). AOAC International, Arlington, VA. USA.

Bal, M. A., Shaver, R. D., Shinners, K. J., Coors, J. G., Lauer, J. G., Straub, R. J., & Koegel, R. G. (2000). Stage of maturity, processing, and hybrid effects on ruminal *in situ* disappearance of whole-plant corn silage. *Animal Feed Science and Technology, 86,* 83-94. https://doi.org/10.1016/S0377-8401(00)00163-2

Baron, V. S., Aasen, A., Oba, M., Dick, A. C., Salmon, D. F., Basarab, J. A., & Stevenson, C. F. (2012). Swath-grazing potential for small-grain species with a delayed planting date. *Agronomy Journal, 104,* 393-404. https://doi:10.2134/agronj2011.0234

Baron, V. S., Dick, A. C., McCartney, D., Basarab, J. A., & Okine, E. K. (2006). Carrying capacity, utilization and weathering of swathed whole plant barley. *Agronomy Journal, 98,* 714-721. http://dx.doi.org/10.2134/agronj2005.0171

Baron, V. S., Doce, R. R., Basarab, J., & Dick, C. (2014). Swath-grazing triticale and corn compared to barley and a traditional winter feeding method in central Alberta. *Canadian Journal of Plant Science, 94,* 1125-1137. https://doi.org/10.4141/cjps2013-412

Baron, V. S., Najda, H. G., McCartney, D. H., Bjorge, M., & Lastiwka, G. W. (2003). Winter weathering effects on corn grown for grazing in a short-season area. *Canadian Journal of Plant Science, 83,* 333-341.

https://doi.org/10.4141/P01-202

Damiran, D., Lardner, H. A., Larson, K., & McKinnon, J. J. (2016). Effects of supplementing spring-calving beef cows grazing barley crop residue with canola meal and wheat-based dry distillers' grains with solubles on performance, reproductive efficiency, and system cost. *Professional Animal Scientist, 32,* 400-410. https://doi.org/10.15232/pas.2015-01479

Jose, D. (2015). *Evaluation of winter feeding systems for crop yield and agronomy, beef cow performance, metabolism and economics.* (Unpublished MSc. Thesis). University of Saskatchewan, Saskatchewan, Saskatoon, Canada.

Jurjanz, S., & Monteils, V. (2005). Ruminal degradability of corn forages depending on the processing method employed. *Animal Resources, 54,* 3-15. https://doi.org/10.1051/animres:2004041

Kappel, L. C., Morgan, E. B., Kilgore, L., Ingraham, R. H., & Babcock. D. K. (1985). Seasonal changes of mineral content of southern forages. *Journal of Dairy Science, 68,* 1822-1827. https://doi.org/10.3168/jds.S0022-0302(85)81033-X

Kelln, B. M., Lardner, H. A., McKinnon, J. J., Campbell, J. R., Larson, K. & Damiran, D. (2011). Effect of winter feeding system on beef cow performance, reproductive efficiency and system cost. *Professional Animal Scientist, 27,* 410-421. https://dx.doi.org/10.15232/S1080-7446(15)30513-1

Krause, A. D., Lardner, H. A., McKinnon, J. J., Hendrick, S., Larson, K., & Damiran, D. (2013). Comparison of grazing oat and pea crop residue versus feeding grass-legume hay on beef-cow performance, reproductive efficiency, and system cost. *Professional Animal Scientist, 29,* 535-545. https://doi.org/10.15232/S1080-7446(15)30275-8

Kumar, R., Lardner, H. A., Christensen, D. A., McKinnon, J. J., Damiran, D., & Larson, K. (2012). Comparison of alternative backgrounding systems on beef calf performance, feedlot finishing performance, carcass traits and system cost of gain. *Professional Animal Scientist, 28,* 541-551. http://dx.doi.org/10.15232/S1080-7446(15)30403-4

Lardner, H. A. (2002). *Comparison of grazing corn varieties.* Fact sheet. No. 2002-02. Lanigan, Saskatchewan, Canada: Western Beef Development Centre. Saskatoon, Canada. http://www.wbdc.sk.ca/pdfs/fact_sheets/2002/comparison_of_grazing_corn_varieties.pdf

Lardner, H. A., Larson K., & Pearce, L. (2012). *Winter grazing beef cows with standing corn.* Fact sheet No. 2012-03. Lanigan, Saskatchewan, Canada: Western Beef Development Centre. Saskatoon, Canada. http://www.wbdc.sk.ca/pdfs/fact_sheets/2012/2012.03_WBDC_winter_grazing_beef_cows_with_standing_corn.pdf

Lassiter, C. A., Huffman, C. F., Dexter, S. T., & Duncan, C. W. (1958). Corn versus oat silage as a roughage for dairy cattle. *Journal of Dairy Science, 41,* 1282-1285. https://doi.org/10.3168/jds.S0022-0302(58)91085-3

Lauer, J. G., Coors, J. G., & Flannery, P. J. (2001). Forage yield and quality of corn cultivars developed in different eras. *Crop Science, 41,* 1449-1455. https://doi.org/10.2135/cropsci2001.4151449x

May, W. E., Klein, L. H., Lafond, G. P., McConnell, J. T., & Phelps, S. M. (2007). The suitability of cool- and warm-season annual cereal species for winter grazing in Saskatchewan. *Canadian Journal of Plant Science, 87,* 739-752. https://www.nrcresearchpress.com/doi/pdf/10.4141/P06-026

McCartney, D., Fraser, J., & Ohama, A. (2008). Annual cool season crops for grazing by beef cattle. A Canadian review. *Canadian Journal of Animal Science, 88,* 517-533. https://doi.org/10.4141/CJAS08052

McCartney, D., Fraser, J., & Ohama, A. (2009). Potential of warm-season annual forages and brassica crops for grazing: A canadian review. *Canadian Journal of Animal Science. 89,* 431-440. https://doi.org/10.4141/CJAS09002

NRC. 2000. *Nutrient requirements of beef cattle.* 7th rev. ed. National Academy Press, Washington, DC.

NRC. 2001. *Nutrient requirements of dairy cattle.* 7th rev. ed. National Academy Press, Washington, DC.

Perkins, D. N., Alward, S. S., Davis, B. A., Gallien, O. J., Labelle, J. A. G., & Monette, B. (1986). *Soil landscapes of Canada: Alberta.* Land Resource Research Centre, Research Branch, Agriculture Canada, Ottawa.

SAS Institute. (2003). SAS/STAT User's Guide, Version 8.2. SAS Institute, Inc. p. 707. Cary, NC, USA:

Saskatchewan Ministry of Agriculture [SMA]. 2010. *Saskatchewan accumulated corn heat units average CHUs*

for silage production.
http://irrigationsaskatchewan.com/icdc/wp-content/uploads/2014/11/CHU_average.pdf. Accessed 18 April 2016.

Saskatchewan Soil Survey. (1992). *The soils of Prairie Rose rural municipality No. 309*, Saskatchewan. Saskatchewan Institute of Pedology. University of Saskatchewan., Saskatoon, Canada.

Saxton, A. M. (1998). A macro for converting mean separation output to letter groupings in Proc. Mixed Model. Pages 1243-1246. *In: Proceedings of* 23rd SAS Users Group Int., Cary, NC.

Saskatchewan Minstry of Agriculture (2008). *Annual crops for greenfeed and grazing.* [online]. Available: http://www.agriculture.gov.sk.ca/Default.aspx?DN=b45a5f86-4eaf-41eb-a731-4a7bb90c646d.

Steel, R. G. D., Torrie, J. H., & Dickey, D. A. (1997). Principles and procedures of statistics. A biometric approach. McGraw-Hill Co., Inc. 3rd Ed. New York, NY.

Taylor, C. C., & Allen, M. S. (2005). Corn grain endosperm type and brown midrib 3 corn silage: Ruminal fermentation and N partitioning in lactating cows. *Journal of Dairy Science, 88,* 1434-1442. https://doi.org/10.3168/jds.S0022-0302(05)72811-3

Van De Kerckhove, A. Y., Lardner, H. A., Walburger, K., McKinnon, J. J., & Yu, P. (2011). Effects of supplementing spring-calving beef cows grazing barley crop residue with a wheat-corn blend dried distillers grains with solubles on animal performance and estimated dry matter intake. *Professional Animal Scientist, 27,* 219-227. https://dx.doi.org/10.15232/S1080-7446(15)30477-0.

Van Soest, P. J., Robertson, J. B., & Lewis, B. A. (1991). Symposium: Carbohydrate methodology, metabolism, and nutritional implications in dairy cattle. *Journal of Dairy Science, 74,* 3583-3597. https://doi.org/10.3168/jds.S0022-0302(91)78551-2

Weiss, W. P., Conrad, H. R., & Pierre, N. R. St. (1992). A theoretically-based model for predicting total digestible nutrient values of forages and concentrates. *Animal Feed Science and Technology, 39,* 95-110. http://dx.doi.org/10.1016/0377-8401(92)90034-4

Yurchuk, T. & Okine, E. (2004). *Agri-facts: Beef Ration Rules of Thumb.* Agdex 420/52-4. http://www1.agric.gov.ab.ca/$department/deptdocs.nsf/all/agdex9146/$file/420_52-4.pdf?OpenElement. Accessed April 2016. Alberta Agriculture Food and Rural Development.

20

Occurrence of the Common Amber Snail Succinea Putris (L.) (Gastropoda: Styllomatophora) on Japanese Knotweed (*Fallopia Japonica* [Houtt.] Ronse Decraene) in Slovenia - Possible Weed Biocontrol Agent?

Žiga Laznik[1] & Stanislav Trdan[1]

[1]Department of Agronomy, Biotechnical Faculty, University of Ljubljana, Jamnikarjeva 101, SI-1000 Ljubljana, Slovenia

Correspondence: Žiga Laznik, Department of Agronomy, Biotechnical Faculty, University of Ljubljana, Jamnikarjeva 101, SI-1000 Ljubljana, Slovenia. E-mail: ziga.laznik@bf.uni-lj.si

Abstract

Japanese knotweed, *Fallopia japonica* is one of the most troublesome invasive alien plant in Europe and North America. In 2012 we started monitoring for possible indigenous natural enemies of Japanese knotweed in Slovenia. In Zgornji Log, near Litija ($46°4'38.09''$N, $14°49'31.65''$E; 245 m) on the river Sava, we recorded several plants of *F. japonica* damaged by the common amber snail *Succinea putris*. In this paper we discuss the possibility of the biological control of Japanese knotweed with *S. putris*.

Keywords: *Fallopia japonica*, *Succinea putris*, Slovenia, biological control

1. Introduction

Japanese knotweed, *Fallopia japonica* (Houtt.) Ronse Decraene var. *japonica* (Polygonaceae) is a perennial herb, native to Asia including China, Japan and Korea. Japanese knotweed was introduced into Europe in the first half of the nineteenth century (Bailey, 1994) and into US (Patterson, 1976) as an ornamental plant, and has become one of the most troublesome invasive alien plant in Europe and North America in recent years (Gerber et al., 2008). It can propagate by both sexual reproduction and clonal growth from rhizomes and stem fragments (Wang et al., 2007). The species thrives in riparian ecosystems, where water is readily available, but it can also be found in relatively dry areas. Once a ramet has established, it can spread clonally by rhizomes, forming dense stands, and reducing the available habitat for other organisms (Gerber et al., 2008). The rapid spread of this invasive plant in its adventive range is likely to be a result of decreased regulation by specialist herbivorous natural enemies. In addition to a lack of co-evolved specialist natural enemies, increased growth and invasion of non-native weed could be the result of growth adaptations (phenotypic plasticity) or selection for fast-growing genotypes in the invasive range (evolution of increased competitive ability) (Keane & Crawley, 2002).

Strgar (1982) reported a massive expansion of Japanese knotweed in Slovenia expecially near the rivers of Drava, Meža, Sotla, Savinja and Sava. Most recently the research of Jogan (2006) reported of a major range expansion of *F. japonica* to all parts of the country (except the submediteranean), as a result of disturbance, such as flood events which move rhizome fragments along river systems, and human activity, in particular the illegal dumping of waste and top soil contaminated with rhizome fragments. Control is extremly difficult since the plant has an extensive rhizome system which survives despite mechanical and chemical control (Laznik & Trdan, 2012).

Considering the scale of classical biocontrol efforts around the globe (Cowie, 2001) and the world-leading biological pest management strategies implemented in European covered crops (Eilenberg et al., 2000), it is surprising how little effort has been targeted at exotic weeds in Europe. Until recently (Shaw et al., 2009) no full classical weed-biocontrol programme has been carried out in any European Union Member State (Shaw et al., 2003). For a diverse array of other reasons (e.i. strict plant protection regulations in several other countries, including Slovenia) the use of a exotic biological control agent will still be prohibited. Therefore the aim of our current research was to search for an indigenous biocontrol agent of *F. japonica*, which could be systematically used for controlling this increasingly harmful invasive plant in Slovenia.

The taxonomy and ecology of different species of Succineidae has been mostly studied in the United States (Patterson, 1971; Hubricht, 1985). Data on ecology and life history of the Succineidae in Europe are scarce (Cejka, 1999). Rigby (1965) reported some data on the histology, functional morphology and some aspects of life history of *Succinea putris* (Linnaeus 1758). Nine species of the family Succineidae are found in Europe and four in Slovenia. These are *S. putris*, *S. oblonga* (Draparnaud 1801), *Oxyloma elegans elegans* (Risso 1826) and *Oxyloma elegans* (Risso 1826) (Slapnik, 2005; Fauna Europaea, 2013).

S. putris is one of the most abundant Succineidae species in Slovenia. Despite its wide distribution and importance as an intermediate host of *Leucochloridium* spp., there are no published studies on its biology in Slovenia (Velkovrh, 2003). Several other authors (Hubricht, 1985; Đatkauskienë, 2005) report that the common amber snail lives on reed and other plants standing near the water, on constantly humid meadows, in alluvial forests and in forests on swampy ground. The snail often crawls up plant stems and can hibernate at some distance from water (Cejka, 1999).

2. Material and Methods

In 2012 we started monitoring for possible indigenous natural enemies of Japanese knotweed in Slovenia. We surveyed 38 places in Slovenia, mainly alongside rivers Sava, Savinja and Drava. In each place we collected 5 plants of Japanese knotweed. In Zgornji Log, near Litija (46°4′38.09″N, 14°49′31.65″E; 245 m) on the river Sava, we recorded several plants of *F. japonica* damaged by snails (Figure 1). Snails (20) were collected and dissected and determination of the species was made analysing the reproductive system of the snails.

Figure 1. The common amber snail *S. putris* feeding on the plant of Japanese knotweed (*F. japonica*) near river Sava (Slovenia)

3. Results and Discussion

Analysis confirmed the presence of the common amber snail *S. putris*. Results of our observation showed that *S. putris* was feeding only on Japanese knotweed, on other related species such as *F. convolvulus* (L.) Löve, *F. dumetorum* (L.) Holub, *F. baldschuanica* (Regel) Holub, *F. sachalinensis* (F. Schmidt) Ronse Decraene, *F.* x *convolvuloides* (Brügger) (Holub) and *F.* x *bohemica* (Chrtek & Chrtkova) J. P. Bailey was not present.

The goal of weed biological control is not to eradicate the weed but rather to reduce its vigor so that desirable plants can coexist (van Driesche et al., 2008). Compared with the extensive use of insects for biological control, snails have been used in relatively few biocontrol efforts (Cowie, 2001). Most snail species, both herbivorous and predatory species, are generalists. This fact, combined with the fact that most biocontrol introductions using snails have been and continue to be undertaken by people with entomological, but not malacological, experience, as noted for example by Mead (1955) and van der Schalie (1969), has meant that adequate understanding of snail biology has not been incorporated into the biocontrol efforts. Many classical biocontrol introductions had major impacts on non-target organisms, in some cases leading to the extinction of native species, and others have the clear potential for such impacts (Cowie, 2001). Any putative biocontrol agents used to control plants would have to be shown, prior to release, to be specific to the target weed it is intended to control, as well as demonstrating capable to reduce populations of that weed (Cowie, 2001). It is unlikely that many gastropod will be found that can filfill these criteria. An extensive review of attempts to control pest molluscs using predatory gastropods can be found in Barker & Efford (2004).

Nearly all efforts at weed biological control have involved classical biological control based on the introduction of insects or plant pathogens from the host plant's native range (van Driesche et al., 2008). In a few cases, generalists fish, such as the grass carp, have been used to graze down the biomass of macrophytic water plants in a non-specific way. In the review of Cowie (2001) the author presents data on several attempts to control aquatic weeds with snails (Robins, 1971; Thomas, 1975; Horne et al., 1992; Okuma et al., 1994; Perera & Walls, 1996). Some of the biocontrol efforts mentioned above probably can be considered successful from the perspective of controlling the target organisms. For many, however, success has not been demonstrated.

A survey by Shaw et al. (2009) revealed that 186 species of phytophagous arthropods are associated with *F. japonica* in its native range. Four years of research, funded by a consortium of UK and North American sponsors, demonstrated that a psyllid species, *Aphalara itadori* Shinji, had the highest potential as a classical biocontrol agent (Shaw et al., 2011).

Here the snail *S. putris* is in suggested, for the first time, as a possible native biocontrol agent which we recommend should be tested for its ability to regulate *F. japonica*. As a native species none of the dangers inherent to classical introductions are relevant. In a related research, Podroužková (2011) investigated food preferences of two snail species, *Succinea putris* and *Urticicola umbrosus* C. Pfeiffer, among five widespread invasive plant species from river floodplains: *Impatiens glandulifera* Royle, *Helianthus tuberosus* L., *F. japonica*, *F. sachalinensis* F. Schmidt and *F.* x *bohemica* (J. Chrtek & A. Chrtkova) J.P. Bailey and one native species, *Urtica dioica* L. Laboratory experiments showed that the most important factors affecting snails food preferences were plant species and the condition of plant material. *U. umbrosus* consumed much less fresh material than *S. putris*. In their research *Fallopia* spp. were rejected, whether fresh or previously frozen, which does not correspond with our field observations, where several plants of *F. japonica* were severe damaged by *S. putris*. In future we would like to test *S. putris* to determine their food preference to different *Fallopia* species in Slovenia and possible impacts on non-target plants.

Acknowledgements

This study was carried out within Professional Tasks from the Field of Plant Protection, a programme funded by the Ministry of Agriculture and Environemnt - Phytosanitary Administration of the Republic of Slovenia.

References

Bailey, J. P. (1994). Reproductive biology and fertility of *Fallopia japonica* (Japanese knotweed) and its hybrids in the British Isles. Pages 141-158. In: de Waal, LC, LE Child, PM Wade, & JH Brock (eds.): Ecology and Management of Invasive Riverside Plants. New York: J. Wiley.

Barker, M. B., & Efford, M. G. (2004). Predatory gastropods as natural enemies of terrestrial gastropods and other invertebrates. In Barker GM (ed.), Natural Enemies of Terrestrial Molluscs, pp. 279-403. CABI Publishing, Wallingford, UK. https://doi.org/10.1079/9780851993195.0279

Cejka, T. (1999). The terrestrial molluscan fauna of the Danubian floodplain (Slovakia). *Biologia, 54*, 489-500.

Cowie, R. H. (2001). Can snails ever be effective and safe biocontrol agents? *International Journal of Pest Management, 41*, 23-40. https://doi.org/10.1080/09670870150215577

Đatkauskienė, I. (2005). Characteristic of lifespan and reproduction period of *Succinea putris* (L.) (Gastropoda: Styllomatophora). *Ekologija, 3*, 28-33.

Eilenberg, J., Enkegaard, A., Vestergaard, S., & Jensen, B. (2000). Biocontrol of pests on plant crops in Denmark: present and future potential. International symposium on biological control agents in crop and animal protection, Swansea, UK, 24-28 August, 1999. *Biocontrol Science and Technology, 10*, 703-710. https://doi.org/10.1080/09583150020011681

Fauna Europaea. (2013). www.faunaeur.org. *24*(8). 2013

Gerber, E., Krebs, C., Murrell, C., Moretti, M., Rocklin, R., & Schaffner, U. (2008). Exotic invasive knotweeds (*Fallopia* spp.) negatively affect native plant and invertebrate assemblages in European riparian habitats. *Biological Conservation, 141*, 646-654. https://doi.org/10.1016/j.biocon.2007.12.009

Horne, F. R., Arsuffi, T. L., & Neck, R. W. (1992). Recent introduction and potential botanical impact of the giant rams-horn snail, *Marisa cornuarietis* (Pilidae), in the Comal Springs ecosystem of central Texas. *Southwestern Naturalist, 37*, 194-196. https://doi.org/10.2307/3671668

Hubricht, L. (1985). The distribution of the native land mollusks of the eastern United States. *Zoology, 24*, 191.

Jogan, N. (2006). Japonski dresnik (*Fallopia japonica*) rastlina leta 2006. *Proteus, 68*, 437-440. [Slovenian]

Jordaens, K., Pinceel, J., & Backeljau, T. (2005). Mate choice in the hermaphroditic land snail *Succinea putris* (Stylommatophora: Succineidae). *Animal Behaviour, 70*, 329-337. https://doi.org/10.1016/j.anbehav.2004.10.021

Keane, R. M., & Crawley, M. J. (2002). Exotic plant invasions and the enemy release hypothesis. *Trends in Ecology and Evolution, 17*, 164-170. https://doi.org/10.1016/S0169-5347(02)02499-0

Laznik, Ž., & Trdan, S. (2012). Damage potential of Japanese knotweed (*Fallopia japonica*) and its biological control with psyllid *Aphalara itadori* Shinji. *Acta Agriculturae Slovenica, 99*, 93-98. [Slovenian]

Mead, A.R. 1955. The proposed introduction of predatory snails into California. *The Nautilus, 69*, 37-40.

Okuma, M., Tanaka, K., & Sudo, S. (1994). Weed control method using apple snail (*Pomacea canaliculata*) in paddy fields. *Weed Research, 39*, 114-119. https://doi.org/10.3719/weed.39.114

Patterson, C. M. (1971). Taxonomic studies of the land snail family Succineidae. *Malacological Review, 4*, 131-202.

Patterson, D. T. (1976). The history and distribution of five exotic weeds in North Carolina. *Castanea, 41*, 177-180.

Pemberton, C. E. (1956). Defense of a predator. *The Nautilus, 69*, 142-144.

Perera, G., & Walls, J. G. (1996). Apple snails in the aquarium (Neptune City, New Jersey: TFH Publications, Inc.), 121 pp.

Podroužková, Š. (2011). Food preferences of land snails in a river flood-plain invoved with invasive plants. Diploma thesis.

Rigby, J. E. (1965). *Succinea putris*: a terrestrial opisthobranck mollusk. *Proceedings of the Zoological Society of London, 144*, 445-486. https://doi.org/10.1111/j.1469-7998.1965.tb05194.x

Robins, C. H. (1971). Ecology of the introduced snail, *Marisa cornuarietis* (Ampullariidae) in Dade County, Florida. *The Biologist, 53*, 136-152.

Shaw, R. H. (2003). Biological control of invasive weeds in the UK: opportunities and challenges. In: Child, L., Brock, J.H., Brundu, G. (Eds.), 6th International conference on the cology and management of alien plant invasions (EMAPi), 12 to 15 September 2001, Loughborough, UK. Plant Invasions: Ecological threats and management solutions, 337-354.

Shaw, R. H., Bryner, S., & Tanner, R. (2009). The life history and host range of the Japanese knotweed psyllid, *Aphalara itadori* Shinji: Potentially the first classical biological weed control agent for the European Union. *Biological control, 49*, 105-113. https://doi.org/10.1016/j.biocontrol.2009.01.016

Shaw, R. H., Tanner, R., Djeddour, D., & Cortat, G. (2011). Classical biological control of *Fallopia japonica* in the United Kingdom - lessons for Europe. *Weed Research, 51*, 552-558. https://doi.org/10.1111/j.1365-3180.2011.00880.x

Slapnik, R. (2005). Mollusks in park Škocjanske jame. *Annal Ser Hist Nat, 15*, 265-276. [Slovenian]

Strgar, V. (1982). Genus *Reynoutria* v adventivni flori Slovenije, II. *Biološki vestnik, 30*, 151-154. [Slovenian]

Thomas, K. J. (1975). Biological control of *Salvinia* by the snail *Pila globosa* Swainson. *Biological Journal of the Linnean Society, 7*, 243-247. https://doi.org/10.1111/j.1095-8312.1975.tb00227.x

Van der Schalie, H. (1969). Man meddles with nature - Hawaiian style. *The Biologist* 51: 136-146.

Van Driesche, R., Hoddle, M., & Center, T. (2008). Control of pests and weeds by natural enemies. An introduction to biological control. Blackwell publishing, 473 pp.

Velkovrh, F. (2003). Mollusks- Mollusca. In: Sket, B., Gogala, M., Kuštor, V. (Eds.). Animals of Slovenia. 109-132 [Slovenian].

Wang, Q., Wang, S., & Zhang, S. (2007). The artificial multiplication method of *Fallopia japonica*. *Journal of Chinese Medicinal Materials, 30*, 1209-1210.

21

Field Testing of Efficacy of Three Environmentally Friendly Insecticides Against Colorado Potato Beetle (*Leptinotarsa Decemlineata* [Say], Coleoptera, Chrysomelidae) on Potato-Evaluation of the Effect on Yield

Žiga Laznik[1] & Stanislav Trdan[1]

[1]Department of Agronomy, Biotechnical Faculty, University of Ljubljana, Jamnikarjeva 101, SI-1000 Ljubljana, Slovenia

Correspondence: Žiga Laznik, Department of Agronomy, Biotechnical Faculty, University of Ljubljana, Jamnikarjeva 101, SI-1000 Ljubljana, Slovenia. E-mail: ziga.laznik@bf.uni-lj.si

Abstract

In 2007 and 2008 the field experiment was conducted to test the efficacy of three environmentally friendly insecticides against the Colorado potato beetle (*Leptinotarsa decemlineata*), with the aim of evaluating their effect on the yield of potato. 0.25 % water emulsion of Neem-Azal (active ingredient azadirachtin) was applied twice, while 3 % water emulsion of Aktiv (a.i. potassium salt of fatty acids) and 1 % water emulsion of Prima (a.i. refined rape oil) were applied eight times. In 2007, the potato yield was higher (25.3 ± 3.2 t ha^{-1}) than in 2008 (8.2 ± 0.8 t ha^{-1}). In 2007 there were no significant differences in potato yield at different control measures and the yield ranged from 22.5 ± 2.5 t ha^{-1} (Aktiv) to 28.2 ± 2.8 t ha^{-1} (Prima). In 2008, the highest potato yield was recorded in Neem-Azal treatment (10.5 ± 1.7 t ha^{-1}), while in two other insecticide treatments the potato yield did not differ significantly with control treatment neither with the Neem-Azal treatment. Potato tubers were classified into three fractions: fraction 1 (tubers <4 cm), fraction 2 (tubers between 4 and 5 cm), and fraction 3 (tubers > 5 cm). On average we produce 2.11 ± 0.06 t ha^{-1}, 9.93 ± 0.53 t ha^{-1}, and 13.17 ± 0.70 t ha^{-1} of potato in 2007, and 2.11 ± 0.20 t ha^{-1}, 4.68 ± 0.37 t ha^{-1}, 0.84 ± 0.29 t ha^{-1} of potato in 2008, respectivelly.

Keywords: insecticides, azadirachtin, refined rape oil, potassium soap, IPM, colorado potato beetle, yield.

1. Introduction

The Colorado potato beetle [CLB] (*Leptinotarsa decemlineata* [Say]) (Coleoptera: Chrysomelidae) is one of the most economically significant pests in Europe, launching year after year increasingly stronger attacks on potato fields (Ozturk & Yildrim, 2012). The harmful stages consist of larvae and adults. Both are foliar pests and can eat more than 100 cm^2 of leaf surface during their lifetime (Ozturk & Yildrim, 2013). This pest has two to three generations per year (Berry et al., 1997). The first and second generations can cause 100% defoliation of potato foliage if the potato is not protected. By full defoliation the yield of potato can be reduced by more than 50 % (Ozturk & Yildrim, 2013). Laznik et al. (2010) recentlly reported that total defoliation of potato plants influenced the tuber developments in such a way that potato plants produced more small tubers which were unable to grow in size.

Although it was brought to Europe nearly 100 years ago, an effective and suitable control of this potato pest is still lacking. A too intensive use of synthetic insecticides resulted in resistance of this pest in more than one case (Stanković et al., 2004), while the use of environmentally more acceptable substances - though some of them do show considerable efficacy in controlling this insect (Scott et al., 2003) - has not been generally applied in Europe until now. The reason for this situation can be attributed also to the slow mode of action of these substances, as many growers judge the efficacy of a pesticide according to its immediate effect. Our current overall aim is to reduce, or to stop using, highly toxic pesticides. Already, some environmentally friendly methods of pest control are being studied, including the use of entomopathogenic nematodes for the control of CPB (Laznik et al., 2010).

The Indian Neem tree, *Azadirachta indica* A. Juss, has long been recognized as a multipurpose plant species, with uses varying from medicinal to plant protection (Schmutterer, 1990). Hence, the insecticidal properties of azadirachtin, the most active molecule extracted from its kernels, have been exhaustively studied over the past three decades. These properties include antifeedancy, repellency, ovipositional deterrence, inhibition of fertility, reduction of fitness, and disruption of insect growth. The mechanisms involved and the potential of neem-based insecticides to control a wide range of insects, including CPB, have been reviewed by Schmutterer (1990) and Mordue and Blackwell (1993). In addition, to contact toxicity, systemic activity has been documented (Gill & Lewis, 1971; Osman & Port, 1990). The efficacy of azadirachtin to CPB has been demonstrated under laboratory conditions by Trdan et al. (2007). Several neem extracts have been commercialized in the United States and Europe (Immaraju, 1998).

Fatty acids and their derived soap products were first regarded as contact insecticides over 80 years ago (Fulton, 1930). However, research on pesticidal soaps was discontinued because of the popularity of synthetic insecticides, until the resurgence of interest in the 1970s. Since the early 1980s, commercial formulations of soap have been available for the control of soft-bodied pests, such as mites, aphids, and psyllids, on ornamentals as well as on selected fruit and vegetable crops (Koehler et al., 1983). Trdan et al. (2006) also confirmed moderate efficacy of potassium soap against the cabbage stink bugs in white cabbage. Soaps induce rapid mortality by disrupting the permeability of insect cuticles and by causing asphyxiation through obstruction of the spiracles (Koehler et al., 1983).

Insecticidal oils, including those of botanical or mineral origin, are favorable biorational pesticides for management of numerous pest insects, especially soft-bodied insects (Yang et al., 2010). Oils have different effects on pest insects. The most important is that they block the air holes (spiracles) through which insects breath, causing them to die from asphyxiation (Butler et al., 1989). In some cases, oils also may act as poisons, interacting with the fatty acids of the insect and interfering with normal metabolism (Liang & Liu, 2002). Oils also may disrupt how an insect feed, a feature that is particularly important in the transmission of some plant viruses by aphids (Fournier & Brodeur, 2000).

The aim of our research was to investigate the field efficacy of three natural substances (azadirachtin, refined rape oil and potassium salt of fatty acids known as potassium soap [produced by adding potassium hydroxide to the fatty acids present in animal fats and plant oils]) to control CPB and to evaluate the effect of treatment on potato yield. Trdan et al. (2007) confirmed in a preliminary laboratory investigation a high susceptibility of CPB adults and larvae to refined rape oil (high mortality rate of adults five days after treatment), while azadirachtin was less effective in their research.

2. Material and Methods

2.1 Experimental Field

We planted the Kondor potato variety in a plot measuring 40 x 5.9 m on the experimental field of the Biotechnical Faculty in Ljubljana, Slovenia (46°04'N, 14°31'E, 299 m alt.) on 11 April 2007 and on 28 April 2008. The preparation of the field began in autumn 2006, when stable manure (30 t ha^{-1}) was spread and then the field was ploughed. In spring time the field received the mineral fertiliser, NPK (15:15:15). In both years after the potato harvest we seeded the oil radish variety Raula (20 kg ha^{-1}), which later served as a green manure. The potato was planted with a two-row automatic planting machine with shedding discs. The planting speed was 3 km h^{-1}, planting depth was around 5 cm with row distance of 75 cm. The planting density was 45,000 tubers ha^{-1}, and distance between tubers in a row was 29.7 cm. We divided the field into four blocks, and in each there were four treatments: control (unsprayed), Neem-Azal (0.25 % water emulsion; a.i. azadirachtin; manufacturer and supplier Metrob d.o.o., Ljubečna, Slovenia), Prima (1 % water emulsion; a.i. refined rape oil; manufacturer and supplier Unichem d.o.o., Sinja Gorica, Slovenia) and Aktiv (3 % water emulsion; a.i. potassium salt of fatty acids; manufacturer and supplier Unichem d.o.o., Sinja Gorica, Slovenia). The size of each treatment parcel was 14.75 m^2 (2.5 x 5.9 m).

2.2 Agri-technical Measures

The herbicide Sencor WG 70 (0.75 kg ha^{-1}, a.i. metribuzin; manufacturer: Bayer CropScience, Leverkusen, Germany; supplier: Bayer d.o.o., Ljubljana, Slovenia) was used (29 May 2007) after the potato plants grew to 5 cm. On 14 May 2008 we used the herbicide Plateen (2.5 kg ha^{-1}) (a.i. flufenacet and metribuzin; manufacturer: Bayer CropScience; supplier: Bayer d.o.o.). When the economic treshhold (25 adults counted on 50 potato plants) [4] was exceeded, application of insecticides was conducted. The economic treshold was in 2007 exceeded on 18 May and in 2008 on 15 May. In 2007 the insecticide Neem-Azal was applied on 25 May and on 13 July, the application with Aktiv and Prima was conducted on 18 May, 25 May, 5 June, 15 June, 26 June, 5 July, 13 July and 25 July. In 2008 the insecticide Neem-Azal was applied on 21 May and on 9 July, the application with Aktiv and Prima was conducted on 17 May, 23 May, 1 June, 9 June, 17 June, 1 July, 9 July and 17 July. We applied insecticides using the backpack sprayer Solo 425. The jet stream nozzle number 04F110 with a pressure of 2 bars was used.

Due to the large possibility in appearance of late blight (*Phytophthora infestans* [Mont.] de Bary) we sprayed the potatoes with the fungicide Acrobat MZ (2.5 kg ha^{-1}) (a.i. dimetomorf-9 % and mancozeb-60 %; manufacturer: BASF SE, Ludwigshafen, Germany; supplier: BASF Slovenija, d.o.o., Ljubljana) on 29 May and 13 June 2007. In 2008 the fungicide Melody duo (2.5 kg ha^{-1}) (a.i. iprovalicarb-5.5 % and propineb-61.3 %; manufacturer: Bayer CropScience; supplier: Bayer d.o.o.) was applied for the same purpose on 26 June, and on 24 July. The potatoes were harvested with a back output machine with two rolling plates on 8 August 2007 and on 12 August 2008. On the day of harvest the tubers were classified with a special shaking device into three fractions: fraction 1 (tubers <4 cm), fraction 2 (tubers between 4 and 5 cm), and fraction 3 (tubers > 5 cm) and weighed them separately as well as together. Later we calculated this to the t ha^{-1}.

2.3 Statistical Analysis

Differences in yield were analysed with the use of ANOVA. Prior to analysis, each variable was tested for homogeneity of variance, and the data found to be non-homogenous was transformed to log(Y) before ANOVA. Significant differences ($P \leq 0.05$) between mean values were identified using Student-Newman-Keuls's multiple range test. All statistical analyses were done using Statgraphics Plus for Windows 4.0 (Statistical Graphics Corp., Manugistics, Inc., Rockville, MD, USA). The data is presented as untransformed means ± SE (Laznik et al., 2012).

3. Results

A group analysis demonstrated that in 2007 (F = 4.50; df = 3, 47; P = 0.0559) the potato yield was not affected by measures of pest control, while in 2008 (F = 7.95; df = 3, 47; P = 0.0114) the mentioned measures had significant differences in potato yield. In 2007 the highest yield was obtained in the Prima treatment (28.05 ± 1.81 t ha^{-1}) and lowest in the Aktiv treatment (22.53 ± 1.26 t ha^{-1}) (Figure 1). In 2008 the highest yield was obtained in the Neem-Azal treatment (10.41 ± 1.68 t ha^{-1}) and lowest in the control treatment (6.1 ± 1.5 t ha^{-1}). A group analysis demonstrated that in 2007 (F = 216.63; df = 2, 47; P < 0.0001) and 2008 (F = 87.07; df = 2, 47; P < 0.0001) the potato yield was affected by different size fractions. In 2007 the highest yield was obtained with fraction 3 (tubers > 5 cm) [13.17 ± 0.71 t ha^{-1}] and the lowest with fraction 1 (tubers <4 cm) [2.11 ± 0.06 t ha^{-1}]. In 2008 the highest yield was obtained with fraction 2 (tubers between 4 and 5 cm) [4.40 ± 0.43 t ha^{-1}] and the lowest with fraction 3 (tubers > 5 cm) [0.84 ± 0.29 t ha^{-1}].

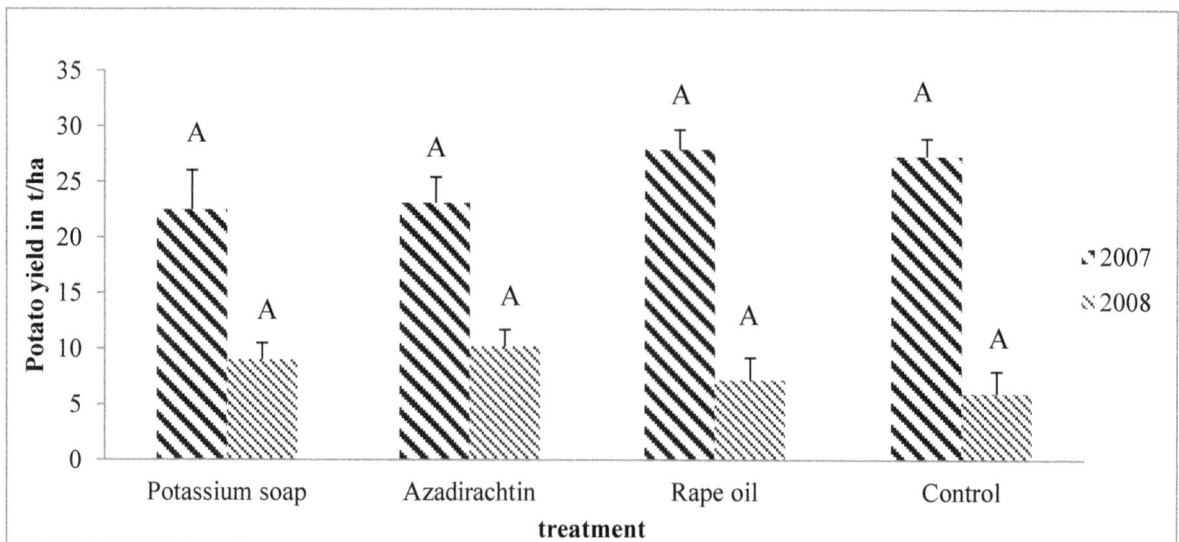

Figure 1. Potato yield [t/ha] at different control treatments in 2007 and 2008. Mean values followed by the same letter do not differ (P≤0.05) according to Student-Newman-Keuls's multiple range test. Bars represent SE of mean potato yield.

No statistically significant differences between the treatments were determined in 2007 (F = 0.75; df = 3, 15; P = 0.5493, while in 2008 (F = 1.03; df = 3, 15; P = 0.0252) such differences were established in fraction 1. On average we produced per treatment 2.11 ± 0.06 t ha^{-1} of potato in 2007 and 2.11 ± 0.20 t ha^{-1} of potato in 2008

(Figures 2-3). In 2007 the yield ranged from 1.94 ± 0.12 t ha^{-1} (Aktiv) to 2.20 ± 0.16 t ha^{-1} (control treatment). In 2008 the yield ranged from 2.11 ± 0.29 t ha^{-1} (Prima) to 2.88 ± 0.48 t ha^{-1} (Aktiv).

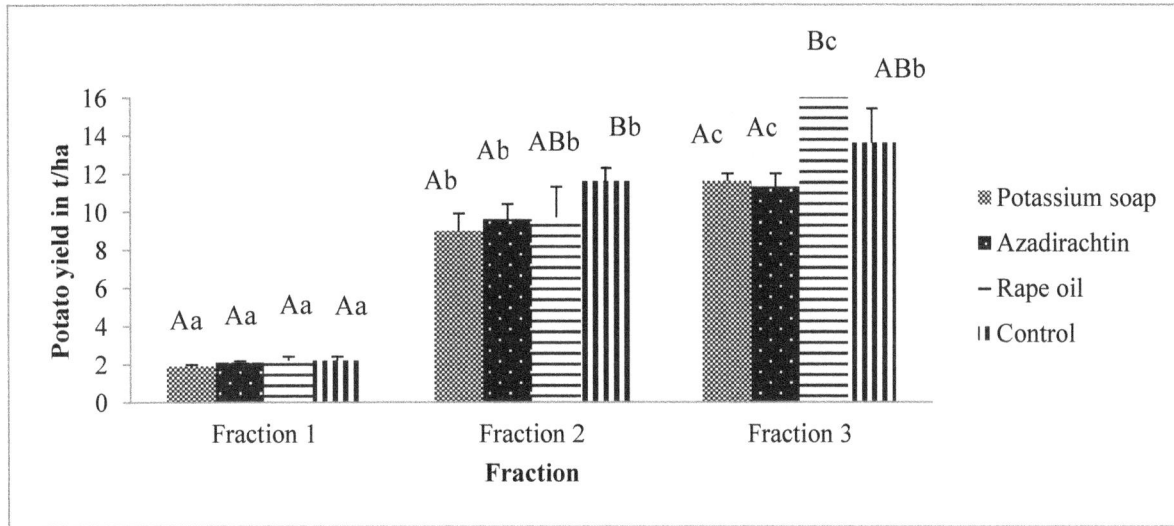

Figure 2. Potato yield [t/ha] at different fractions and treatments in 2007. Mean values followed by the same letter do not differ (P≤0.05) according to Student-Newman-Keuls's multiple range test. Small letters indicates statistical significant differences between different fractions, big letters indicates statistical significant differences between different control treatments. Bars represent SE of mean potato yield. Fraction 1 (tubers < 4 cm); fraction 2 (tubers between 4 and 5 cm); fraction 3 (tubers > 5 cm)

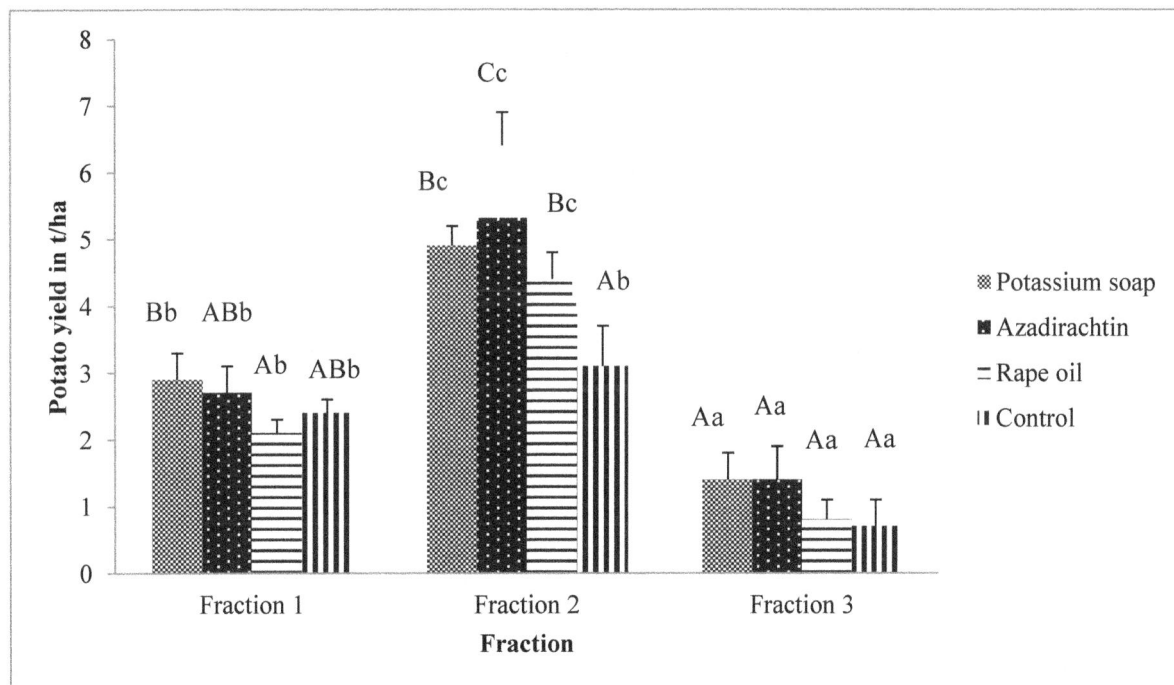

Figure 3. Potato yield [t/ha] at different fractions and treatments in 2008. Mean values followed by the same letter do not differ (P≤0.05) according to Student-Newman-Keuls's multiple range test. Small letters indicates statistical significant differences between different fractions, big letters indicates statistical significant differences between different control treatments. Bars represent SE of mean potato yield. Fraction 1 (tubers < 4 cm); fraction 2 (tubers between 4 and 5 cm); fraction 3 (tubers > 5 cm)

In 2007 at fraction 2 we determined statistically significant differences between different treatments (F = 0.97; df = 3, 15; P = 0.0477) and the values ranged from 8.98 ± 0.89 t ha^{-1} (Aktiv) to 11.52 ± 0.92 t ha^{-1} (control treatment). The harvested potato as classified in fraction 2, yielded statistically significant differences between the treatments in 2008 (F = 6.45; df = 3, 15; P = 0.0127) (Figures 2-3). The highest yield was obtained at treatment Neem Azal (6.35 ± 0.52 t ha^{-1}) and the lowest at the control treatment (3.05 ± 0.63 t ha^{-1}). On average we produced per treatment 9.93 ± 0.53 t ha^{-1} of potato in 2007 and 4.68 ± 0.37 t ha^{-1} of potato in 2008 (Figure 4).

The harvested potato as classified in fraction 3, yielded statistically significant differences between the treatments in 2007 (F = 5.31; df = 3, 15; P = 0.0221). The highest yield was obtained at treatment Prima (16.18 ± 1.00 t ha^{-1}) and the lowest at the Neem-Azal (11.35 ± 0.73 t ha^{-1}). In 2008 at fraction 3 we did not determined statistically significant differences between different treatments (F = 0.96; df = 3, 15; P = 0.4516) and the values ranged from 0.67 ± 0.36 t ha^{-1} (control treatment) to 1.35 ± 0.42 t ha^{-1} (Aktiv) (Figures 2-3). On average we produced per treatment 13.17 ± 0.70 t ha^{-1} of potato in 2007 and 0.84 ± 0.29 t ha^{-1} of potato in 2008 (Figure 4).

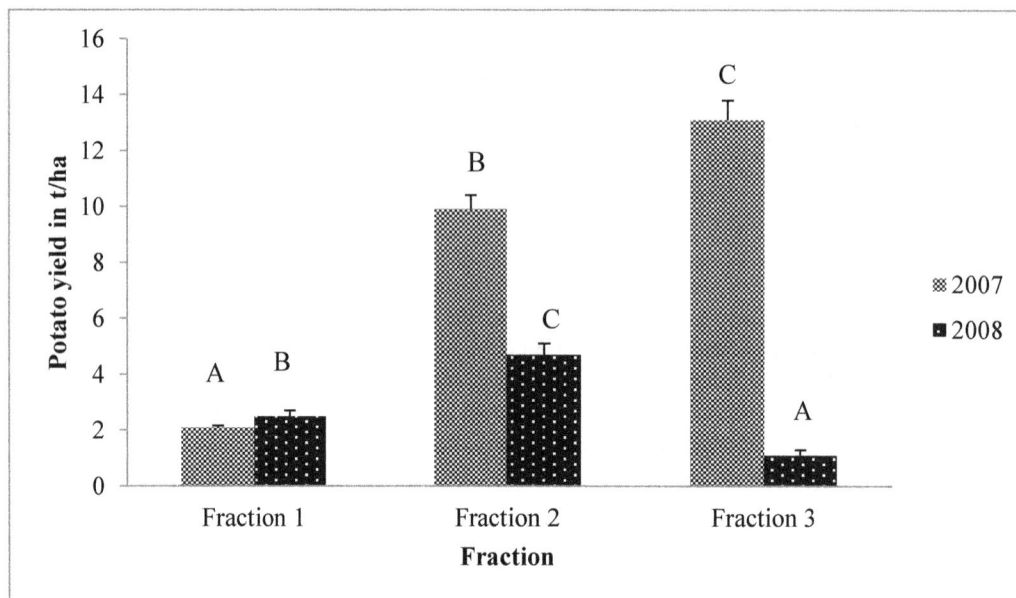

Figure 4. Potato yield [t/ha] at different fractions in 2007 and 2008. Mean values followed by the same letter do not differ (P≤0.05) according to Student-Newman-Keuls's multiple range test. Letters indicates statistical significant differences between different fractions in the same year. Bars represent SE of mean potato yield. Fraction 1 (tubers < 4 cm); fraction 2 (tubers between 4 and 5 cm); fraction 3 (tubers > 5 cm)

4. Discussion

Results of our research showed that in 2007 the potato yield was higher than in 2008. The reason for this is that in 2008 we used as planting material the potato from the previous year (Milošević et al., 2008). The degeneration of potato is appeared to be either due to physiological causes or due to infection of tuber-borne viruses (Chandla et al., 2001). Physiological degeneration can be recovered through proper crop management, but degeneration due to virus can hardly be overcome. Thus the degeneration problem of potato seed tubers due to PVY and PLRV is considered to be the most severe constraint to potato cultivation. It is the most tenacious problem of potato seed tubers resulting spontaneous yield deterioration of the crop (Chandla et al., 2001). When compared to some related research, where the yield of the Kondor potato variety was studied (Ábrahám et al., 2006; Musa et al., 2009) we attained poorer results, also due to the use of small amounts of fertilizers and to high population pressure of CPB in our experiment. We gained similar conclusions also in our related previous research (Laznik et al., 2010).

Although reports of earlier studies indicate that several environmentally friendly insecticides could offer an alterantive to chemical control of CPB (Trdan et al., 2007; Hiiesaar et al., 2009; Amiri-Besheli, 2010) our current research confirm such results only partly. Namely in 2008, the potato sprayed twice with azadirachtin, gave the highest yield. Multiple potato treatments with potassium soap and refined rape oil were less efficient and did not gave a satisfactory results. However, it is also important to note that results from laboratory tests are not always comparable to field testing (Cantelo & Nickle, 1992) as the functioning of bioinsecticides in the open is influenced by an extensive list of factors. Hossain et al. (2008) exposed in their research several factors that influence on the efficacy of azadirachtin in the open. Most often, neem products are applied as foliar sprays to control different foliar pests including CPB. Despite the efficacy of foliar application of neem, major drawbacks are addressed: the fast biodegradability of neem is simultaneously an attractive advantage in domestic areas but a hindering drawback where the short persistence lowers efficacy in field applications. In particular UV light, rainfall and perhaps high acidity on treated surfaces of plants cause a fast degradation or the loss of active material sprayed on foliage (Schmutterer, 1990; Johnson et al., 2003). Moreover, it has been addressed in a number of studies that topical applications of azadirachtin solutions with direct contamination of plant dwelling organism can pose a risk to non-taget beneficials such as predators and parasitoids (Krishnaya & Grewal, 2001). Soil applications such as seed dressing or plant substrate treatments could reduce this risk providing that systemic translocation of the active ingredient is possible (Otto, 1996; Thoeming et al., 2003; Kumar et al., 2005).

Our results demonstrate a poor insecticidal activity of potassium soap and refined rape oil, comparable to some previous research (Liu & Stansly, 2000; Amiri-Besheli, 2010). Nevertheless, previous results were based on excellent to perfect coverage obtained with the spray tower or by leaf dips. The contact activity of these materials requires thorough and complete coverage on the leaf surface (Liu & Stansly, 2000). In addition, insecticidal oils and soaps are only active while wet and become ineffective under drying conditions (Butler et al., 1989).

The population density of CPB in our experiment increased with time and majority of damages appeared in the middle of cultivation period of the potato (end of May, in June, and the beginning of July) (Laznik et al., 2010), which resulted in an expected loss of potato. In our research we found that more damaged plants produced a larger number of small tubers and a small number of large tubers.

Total defoliation of potato plants influenced the tuber developments in such a way that potato plants produced more small tubers which were unable to grow in size. We confirmed previously known facts that the size of tubers at the end of the cultivation period depends on the efficiency of controlling larvae and adults of CPB, because of the influence of defoliation on the poorer development of tubers in the soil (Mannan et al., 1992; Kakaty et al., 1992).

Foliar application of environmentally safe insecticides studied in our research did not show satisfying efficacy of controling CPB and their influence on potato yield although the results from a laboratory test (Trdan et al., 2007) were promising. The low efficacy of tested insecticides could be explained with the rapid degradation under high temperature and UV light (Johnson et al., 2003; Barrek et al., 2004). Therefore, successful use of these materials requires appropriate application methods and environmental conditions. Further research in the field is required to assess more precisely the most effective rate, timing of application and concentration of tested environmentally safe insecticides used under these conditions against CPB.

5. Conclusions

Results of our research showed that in 2007 the potato yield was higher than in 2008. The reason for this is that in 2008 we used as planting material the potato from the previous year. Our results demonstrate a poor insecticidal activity of potassium soap and refined rape oil, comparable to some previous research. Foliar application of environmentally safe insecticides studied in our research did not show satisfying efficacy of controling CPB and their influence on potato yield although the results from a laboratory test (Trdan et al., 2007) were promising.

Acknowledgements

The present work was performed within the programme P4-0013, funded by the Slovenian Research Agency. We thank Assist. Prof. Dragan Žnidarčič, Assist. Prof. Matej Vidrih, Assist. Prof. Filip Vučajnk, Aleksander Bobnar, Janko Rebernik, Boštjan Medved Karničar and Assist. Prof. Ludvik Rozman for technical assistance.

References

Ábrahám, E. B., Zsom, E., & Sárvári, M. (2006). The effect of year and irrigation on the yield quantity and quality of the potato. *Journal of Agricultural Science, 24*, 12-16.

Amiri-Besheli, B. (2010). Efficacy of chlorpyrifos-methyl, metoxyfenozide, spinosad, insecticidal gel, insecticidal soap and mineral oil on citrus leafminer. *Journal of Food, Agriculture & Environment, 8*, 668-671.

Barrek, S., Paisse, O., & Grenier-Loustalot, M. F. (2004). Analysis of neem oils by LC-MS and degradation kinetics of azadirachtin-A in a controlled environment – characterization of degradation products by HPLC-MS-MS. *Analytical and Boanalytical Chemistry, 378*, 753-763. https://doi.org/10.1007/s00216-003-2377-0

Berry, R. E., Liu, J., & Reed, G. (1997). Comparison of endemic and exotic entomopathogenic nematode species for control of Colorado potato beetle (Coleoptera: Chrysomelidae). *Journal of Economic Entomology, 90,* 1528-1533. https://doi.org/10.1093/jee/90.6.1528

Butler, Jr. G. D., Coudriet, D. L., & Henneberry, T. J. (1989). Sweetpotato to whitefly: host plant preference and repellent effect of plant-derived oils on cotton, squash, lettuce and cantaloupe. *Southwestern Entomologist, 14,* 9-16.

Butler, Jr. G. D., Puri, S. N., & Henneberry, T. J. (1991). Plant-derived oil and detergent solutions as control agents for *Bemisia tabaci* and *Aphis gossypii* on cotton. *Southwestern Entomologist, 16,* 331-337.

Cantelo, W. W., & Nickle, W. R. (1992). Susceptibility of prepupae of the Colorado potato beetle (Coleoptera: Chrysomelidae) to entomopathogenic nematodes (Rhabditida: Steinernematidae, Heterorhabditidae). *Journal of Entomological Science, 27,* 37-43.

Chandla, V. K., Shiv, K., Singh, M. N., Verma, K. D., Khurana, S. M. P., Kumar, S., & Pandey, S. K. (2001). Role of aphids in degeneration of seed stocks in the higher hills. *Journal of Indian Potato Association, 28,* 117-118.

Fournier, V., & Brodeur, J. (2000). Dose-Response Susceptibility of Pest Aphids (Homoptera: Aphididae) and their Control on Hydroponically Grown Lettuce with the Entomopathogenic Fungus *Verticillium lecanii,* Azadirachtin, and Insecticidal Soap. *Environmental Entomology, 29,* 568-578. https://doi.org/10.1603/0046-225X-29.3.568

Fulton, B. B. (1930). The relation of evaporation to killing efficiency of soap solutions on the harlequin bug and other insects. *Journal of Economic Entomology, 23,* 625-630. https://doi.org/10.1093/jee/23.3.625

Gill, J. S., & Lewis, C. T. (1971). Systemic action of an insect feeding deterrent. Nature 232:625-630. https://doi.org/10.1038/232402a0

Hiiesaar, K., Švilponis, E., Metspalu, L., Jõgar, K., Mänd, M., Luik, A., & Karise, R. (2009). Influence of Neem-Azal T/S on feeding activity of Colorado Potato Beetles (*Leptinotarsa decemlineata* Say). *Agronomy Research, 7,* 251-256.

Hossain, M. B., Poehling, H.-M., Thöming, G., & Borgemeister, C. (2008). Effects of soil application of neem (NeemAzal®-U) on different life stages of *Liriomyza sativae* (Diptera: Agromyzidae) on tomato in the humid tropics. *Journal of Plant Diseases and Protection, 115,* 80-87. https://doi.org/10.1007/BF03356243

Immaraju, J. A. (1998). The commercial use of azadirachtin and its integration into viable pest control programmes. *Pesticide Science, 54,* 285-289. https://doi.org/10.1002/(SICI)1096-9063(1998110)54:3<285::AID-PS802>3.0.CO;2-E

Johnson, S., Dureja, P., & Dhingra, S. (2003). Photostabilizers for azadirachtin-A (a neem-based pesticide). *Journal of Environmental Science and Health, 38,* 451-462. https://doi.org/10.1081/PFC-120021665

Kakaty, B. M., Thamburaj, S., & Stalin, P. (1992). Study on the growth and yield of sweet potato at different stages of harvest. *Journal of Root Crops, 18,* 73-76.

Koehler, C. S., Barclay, L. W., & Kretchun, T. M. (1983). Soaps as insecticides. *California Agriculture, 37,* 11-12.

Krishnaya, P. V., & Grewal, P. S. (2001). Effects of neem and selected fungicides on viability and virulence of the entomopathogenic nematode *Steinernema feltiae. Biocontrol Science and Technology, 12,* 259-266. https://doi.org/10.1080/09583150210388

Kumar, P., Poehling, H.-M., & Borgemeister, C. (2005). Effects of different application methods of azadirachtin against sweetpotato whitefly *Bemisia tabaci* Gennadius (Hom.: Aleyrodidae) on tomato plants. *Journal of Applied Entomology, 129,* 489-497. https://doi.org/10.1111/j.1439-0418.2005.01009.x

Laznik, Ž., Bohinc, T., Vidrih, M., & Trdan, S. (2012). Testing the suitability of three herbs as intercrops against the *Allium* leaf miner (*Phytomyza gymnostoma* Loew, Diptera, Agromyzidae) in onion production. *Journal of Food, Agriculture and Environment, 10,* 751-755.

Laznik, Ž., Tóth, T., Lakatos, T., Vidrih, M., & Trdan, S. (2010). Control of the Colorado potato beetle (*Leptinotarsa decemlineata* [Say]) on potato under field conditions: a comparison of the efficacy of foliar application of two strains of *Steinernema feltiae* (Filipjev) and spraying with thiametoxam. *Journal of Plant Diseases and Protection, 117,* 129-135. https://doi.org/10.1007/BF03356348

Liang, G., & Liu, T.-X. (2002). Reppelency of a kaolin particle film, Surround, and a mineral oil, SunSpray oil, to silverleaf whitefly (Homoptera: Aleyrodidae) on melon in the laboratory. *Journal of Economic Entomology, 95*, 317-324. https://doi.org/10.1603/0022-0493-95.2.317

Liu, T.-X., & Stansly, P. A. (2000). Insecticidal activity of surfactants and oils against silverleaf whitefly *(Bemisia argentifolii)* nymphs (Homoptera: Aleyrodidae) on collards and tomato. *Pest Management Science, 56*, 861-866. https://doi.org/10.1002/1526-4998(200010)56:10<861::AID-PS217>3.0.CO;2-Y

Mannan, M. A., Bhuiyan, M. K. R., Quasem, A., Rashid, M. M., & Siddique, M. A. (1992). Study on the growth and partitioning of dry matter in sweet potato. *Journal of Root Crops, 18*, 1-5.

Milošević, D., Djalović, I., & Bugarčić, Z. (2008). The importance of a healthy planting material for potato production increases. *Agro-Knowledge Journal, 9*, 5-17.

Mordue, A. J., & Blackwell, A. (1993). Azadirachtin: an update. *Journal of Insect Physiology, 39*, 903-924. https://doi.org/10.1016/0022-1910(93)90001-8

Musa, F., Bericha, D., Kelmendi, B., Rusinovci, I., Zhitia, O., Bekqeli, R., & Lushi, I. (2009). Tuber yield and other relevant parameters of some Netherlands potato varieties in agro-climatic conditions of Kosova. Bulletin of 44th Croatian and 4th International Symposium in Agriculture: 444-448.

Osman, M., & Port, G. R. (1990). Systematic action of neem seed substances against *Pieris brassicae*. *Entomologia Experimentalis Applicata, 54,* 297-300. https://doi.org/10.1111/j.1570-7458.1990.tb01341.x

Otto, D, (1996). Persistence of systemic effects of azadirachtin preparation NeemAzal W on larvae of Colorado potato beetle *Leptinotarsa decemlineata* Say. In: H. Kleeberg, V. Micheletti (eds.): Proceedings of the 4th Workshop on Practice Oriented Results on Use and Production of Neem Ingredients and Pheromones, 28 November – 1 December 1994, Bordighera, Italy:71-80.

Ozturk, G., & Yildrim, Z. (2012). Field performance of in vitro sweet potato (Ipomea batatas L. [Lam]) plantlets derived fromm seedstocks. *Turkish Journal of Field Crops, 17*, 1-4.

Ozturk G., & Yildrim, Z. (2013). Effect of bio-activators on the tuber yield and tuber size of potatoes. T Turkish *Journal of Field Crops, 18*, 82-86.

Schmutterer, H. (1990). Properties and potential of natural pesticides from the neem tree, *Azadirachta indica*. *Annual Review of Entomology, 35*, 271-297. https://doi.org/10.1146/annurev.en.35.010190.001415

Scott, I. M., Jensen, H., Scott, J. G., Isman, M. B., Arnason, J. T., & Philogene, B. J. R. (2003). Botanical insecticides for controlling agricultural pests: piperamides and the Colorado potato beetle *Leptinotarsa decemlineata* (Say) (Coleoptera: Chrysomelidae). *Archives of Insect Biochemistry and Physiology, 54*, 212-225. https://doi.org/10.1002/arch.10118

Stanković, S., Zabel, A., Kostić, M., Manojlović, B., & Rajković, S. (2004). Colorado potato beetle (*Leptinotarsa decemlineata* [Say]) resistance to organophosphates and carbamates in Serbia. *Journal of Pest Science, 77*, 11-15. https://doi.org/10.1007/s10340-003-0020-7

Thoeming, G., Borgemeister, C., Setamou, M., & Poehling, H.-M. (2003). Systemic effects of neem on western flower thrips, *Frankliniella occidentalis* (Thysanoptera: Thripidae). *Journal of Economic Entomology, 96*, 817-825. https://doi.org/10.1093/jee/96.3.817

Trdan, S., Žnidarčič, D., & Valič, N. (2006). Field efficacy of three insecticides against cabbage stink bugs (Heteroptera: Pentatomidae) on two cultivars of white cabbage. *International Journal of Pest Management, 52*, 79-87. https://doi.org/10.1080/09670870600568212

Trdan, S., Cirar, A., Bergant, K., Andjus, L., Kač, M., Vidrih, M., & Rozman, L. (2007). Effect of temperature on efficacy of three natural substances to Colorado potato beetle, *Leptinotarsa decemlineata* (Coleoptera: Chrysomelidae). *Acta Agriculturae Scandinavica B, 57*, 293-296. https://doi.org/10.1080/09064710600925984

Yang X.-B., Zhang, Y.-M., Hua, L., Peng, L.-N., Munyaneza, J. E., Trumble, J. T., & Liu, T.-X. (2010). Repellency of selected biorational insecticides to potato psyllid, *Bactericera cockerelli* (Hemiptera: Psyllidae). *Crop Protection, 29*, 1320-1324. https://doi.org/10.1016/j.cropro.2010.06.013

PERMISSIONS

All chapters in this book were first published in SAR, by Canadian Center of Science and Education; hereby published with permission under the Creative Commons Attribution License or equivalent. Every chapter published in this book has been scrutinized by our experts. Their significance has been extensively debated. The topics covered herein carry significant findings which will fuel the growth of the discipline. They may even be implemented as practical applications or may be referred to as a beginning point for another development.

The contributors of this book come from diverse backgrounds, making this book a truly international effort. This book will bring forth new frontiers with its revolutionizing research information and detailed analysis of the nascent developments around the world.

We would like to thank all the contributing authors for lending their expertise to make the book truly unique. They have played a crucial role in the development of this book. Without their invaluable contributions this book wouldn't have been possible. They have made vital efforts to compile up to date information on the varied aspects of this subject to make this book a valuable addition to the collection of many professionals and students.

This book was conceptualized with the vision of imparting up-to-date information and advanced data in this field. To ensure the same, a matchless editorial board was set up. Every individual on the board went through rigorous rounds of assessment to prove their worth. After which they invested a large part of their time researching and compiling the most relevant data for our readers.

The editorial board has been involved in producing this book since its inception. They have spent rigorous hours researching and exploring the diverse topics which have resulted in the successful publishing of this book. They have passed on their knowledge of decades through this book. To expedite this challenging task, the publisher supported the team at every step. A small team of assistant editors was also appointed to further simplify the editing procedure and attain best results for the readers.

Apart from the editorial board, the designing team has also invested a significant amount of their time in understanding the subject and creating the most relevant covers. They scrutinized every image to scout for the most suitable representation of the subject and create an appropriate cover for the book.

The publishing team has been an ardent support to the editorial, designing and production team. Their endless efforts to recruit the best for this project, has resulted in the accomplishment of this book. They are a veteran in the field of academics and their pool of knowledge is as vast as their experience in printing. Their expertise and guidance has proved useful at every step. Their uncompromising quality standards have made this book an exceptional effort. Their encouragement from time to time has been an inspiration for everyone.

The publisher and the editorial board hope that this book will prove to be a valuable piece of knowledge for researchers, students, practitioners and scholars across the globe.

LIST OF CONTRIBUTORS

Daud Kassam and Marcus Sangazi
Aquaculture and Fisheries Science Department, Bunda Campus, Lilongwe University of Agriculture and Natural Resources, P.O. Box 219, Lilongwe, Malawi

Monika Ghimire, Art Stoecker, Tracy A. Boyer and Jeffrey Vitale
Department of Agricultural Economics, Oklahoma State University, Stillwater, OK 74078, USA

Hiren Bhavsar
Hiren Bhavsar is an assistant professor at Tennessee State University, Nashville, TN 37209, USA

Daniel L. Mutisya and Clement K. Kamau
Kalro-Katumani. P. O. Box 340. 90100. Machakos

Canute P. M. Khamala
University of Nairobi. P. O. Box 30197. 00100. Nairobi

Jacob J. J. O. Konyango and Lawrence K. Matolo
Machakos University College. P. O. Box 136. 90100. Machakos

Jonas Osei-Adu, Offei Bonsu and Seth Obosu Ekyem
CSIR-Crops Research Institute, Ghana

Victor Afari-Sefa and Micheal Kwabena Osei
World Vegetable Center Eastern and Southern Africa, P. O. Box 10 Duluti Arusha, Tanzania

Munywoki James Ngelenzi, Saidi Mwanarusi and Ogweno Joshua Otieno
Department of Crops, Horticulture and Soils, Egerton University, P.O Box 536-20115 Egerton, Kenya

Simone Ubertino, Patrick Mundler and Lota D. Tamini
Department of Agricultural Economics and Consumer Science, Laval University, Canada

F. Aneani and F. Padi
Cocoa Research Institute of Ghana, P. O. Box 8, New Tafo-Akim, Ghana

Ethel Mudenda, Elijah Phiri and Lydia M. Chabala
Department of Soil Science, School of Agricultural Sciences, University of Zambia, P.O. Box 32379, Lusaka, Zambia

Henry M. Sichingabula
Department of Geography and Environmental Studies, School of Natural Sciences, University of Zambia, P.O. Box 32379, Lusaka, Zambia

Susanne Kummer, Friedrich Leitgeb and Christian R. Vogl
Division of Organic Farming, Department of Sustainable Agricultural Systems, University of Natural Resources and Life Sciences Vienna, Austria

Basu Bhandari
Dept. of Agricultural Economics and Agribusiness, Louisiana State University Agricultural Center, Baton Rouge, LA, USA

Jeffrey Gillespie
Animal Products and Cost of Production, Market and Trade Economics Division, USDA-Economic Research Service, Washington, DC, USA

Guillermo Scaglia
Iberia Research Station, Louisiana State University Agricultural Center, Jeanerette, LA, USA

Joel L. Meliya
Selian Agricultural Research Institute, P.O Box 60248, Arusha, Tanzania

Sophia Kashenge-Killenga
Chollima Agro-scientific Research Institute, P.O Box 1892, Dakawa, Morogoro, Tanzania

Kongo M. Victor
University of Dar-es-salaam P.O Box 35131, Dar es Salaam, Tanzania

Benjamin Mfupe
KATRIN Agriculture research Institute, Private bag, Ifakara, Morogoro, Tanzania

Samwel Hiza
Mlingano Agricultural Research Institute, P.O Box 5088, Muheza, Tanga, Tanzania

Ashura L.Kihupi
Sokoine University of Agriculture. (SUA), Morogoro, Tanzania

Brian J. Boman
University of Florida, 2199 South Rock Road Fort Pierce, FL 34945-3138, USA

Warren Dick
The Ohio State University, School of Environment and Natural Resources, 1680 Madison Avenue; Wooster

Masaru Sakamoto, Dong-An Kim, Keiko Imoto, Yusuke Kitai and Takahiro Suzuki
Faculty of Biology-Oriented Science and Technology, Kindai University, Wakayama 649-6493, Japan

Peter T. Wolter
Dept. of Natural Resource Ecology and Management, Iowa State University, USA

Gaurav Arora
Department of Economics, Iowa State University, USA
Center for Agricultural and Rural Development, Iowa State University, USA

Hongli Feng
Department of Economics, Iowa State University, USA
Dept. of Agriculture, Food & Resource Economics, Michigan State University, USA

David A. Hennessy
Department of Economics, Iowa State University, USA
Center for Agricultural and Rural Development, Iowa State University, USA
Dept. of Agriculture, Food & Resource Economics, Michigan State University, USA

Yonaisy Mujica Pérez
Instituto Nacional de Ciencias Agrícolas, San José de las Lajas, Cuba

Christiane Charest
Department of Biology, Faculty of Science, University of Ottawa, Ottawa, ON, Canada

Yolande Dalpé, Sylvie Séguin, Xuelian Wang and Shahrokh Khanizadeh
Ottawa Research and Development Centre, Agriculture and Agri-Food Canada, Ottawa, ON, Canada

Kheiry Hassan M. Ishag
Dhofar Cattle Feed Company, P.O. Box 1220, PC 211, Sultanate of Oman

Mona Ahmadiani, Chun (Cathy) Li and Yaqin Liu
Department of Agriculture and Applied Economics, University of Georgia, Corner Hall, Athens, GA 30206, USA

Esendugue Greg Fonsah
Department of Agriculture and Applied Economics, University of Georgia, Tifton, GA, 31793, USA

Christine M. Bliss, Brent V. Brodbeck and Peter C. Andersen
University of Florida, Institute of Food & Agricultural North Research and Education Center 155 Research Road, Quincy, FL 32315, USA

Samia Akroush, Omamah Al-Hadidi and Malek Abo-Roman
Socioeconomic Studies Directorate, National Center for Agricultural Research and Extension (NCARE), Baq'a (19381). Amman, Jordan

Boubaker Dehehibi
Sustainable Intensification and Resilient Production Systems Program (SIRPSP), International Center for Agricultural Research in the Dry Areas (ICARDA), Amman, Jordan

Bezaiet Dessalegn
Integrated Water and Land Management Program (IWLMP), International Center for Agricultural Research in the Dry Areas (ICARDA), Amman, Jordan

Priscilla F. Ribeiro, Manfred B. Ewool, Charles Afriyie-Debrah and Benedicta N. Frimpong
CSIR-Crops Research Institute, Ghana

Baffour Badu-Apraku
International Institute Tropical Agriculture, Ibadan, Nigeria

Vernon E. Gracen
West Africa Centre for Crop Improvement, University of Ghana, Legon, Ghana

Leah Pearce
Western Beef Development Centre, P.O. 1150, Humboldt, SK S0K 2A0 Saskatchewan, Canada

Herbert Andrew Lardner and Daalkhaijav Damiran
Department of Animal and Poultry Science, University of Saskatchewan, 51 Campus Drive, Saskatoon, SK S7N 5A8, Canada
Western Beef Development Centre, P.O. 1150, Humboldt, SK S0K 2A0 Saskatchewan, Canada

Žiga Laznik and Stanislav Trdan
Department of Agronomy, Biotechnical Faculty, University of Ljubljana, Jamnikarjeva 101, SI-1000 Ljubljana, Slovenia

Index